"This excellent book offers an introduction into the world of sound and interaction. Its chapters cover a wide range of subjects from auditory displays, toys and image sonification. It is an invaluable text for students, researchers and artists who want to know more about the emerging field of audio interaction, and one that should definitely sit on the shelf of anyone who works in the world of design and sound."

—Dr. Alan Chamberlain, Mixed Reality Lab,
School of Computer Science, University of Nottingham

"*Foundations in Sound Design for Embedded Media* reverberates with valuable insight for today's practitioners. This volume brings together technological and methodological know-how, aesthetic sensitivity and an understanding of how we listen and shows how these apply to the eclectic array of application areas embedded media now find themselves supporting; I am confident this book will spend more time on my desk than in my bookcase for quite some time to come."

—Paul Vickers, Associate Professor of Computer Science
and Computational Perceptualisation, Northumbria University

# Foundations in Sound Design for Embedded Media

This volume provides a comprehensive introduction to foundational topics in sound design for embedded media, such as physical computing; interaction design; auditory displays and data sonification; speech synthesis; wearables; smart objects and instruments; user experience; toys and playful tangible objects; and the new sensibilities entailed in expanding the concept of sound design to encompass the totality of our surroundings.

The reader will gain a broad understanding of the key concepts and practices that define sound design for its use in computational products and design. The chapters are written by international authors from diverse backgrounds who provide multidisciplinary perspectives on sound in its many embedded forms.

The volume is designed as a textbook for students and teachers, as a handbook for researchers in sound, programming and design, and as a survey of key trends and ideas for practitioners interested in exploring the boundaries of their profession.

**Michael Filimowicz, PhD**, is Senior Lecturer in the School of Interactive Arts and Technology (SIAT) at Simon Fraser University and coeditor of *The Soundtrack* journal. He develops new forms of general-purpose multimodal and audiovisual display technology, exploring novel product lines across different application contexts including gaming, immersive exhibitions, control rooms, telepresence and simulation-based training. He has published across disciplines in journals such as *Organised Sound, Arts and Humanities in Higher Education, Leonardo, Sound Effects, Parsons Journal for Information Mapping* and *Semiotica*. His art has been exhibited internationally at venues such as SIGGRAPH, Re-New, Design Shanghai, ARTECH, Les Instants Vidéo, IDEAS, Kinsey Institute and Art Currents, and published in monographs such as *Spotlight: 20 Years of the Biel/Bienne Festival of Photography, Reframing Photography* and *Infinite Instances*. His personal website is http://filimowi.cz.

# Sound Design

The Sound Design series takes a comprehensive and multidisciplinary view of the field of sound design across linear, interactive and embedded media and design contexts. Today's sound designers might work in film and video, installation and performance, auditory displays and interface design, electroacoustic composition and software applications, and beyond. These forms and practices continuously cross-pollinate and produce an ever-changing array of technologies and techniques for audiences and users, which the series aims to represent and foster.

## Series Editor
**Michael Filimowicz**

## Titles in the Series

**Foundations in Sound Design for Linear Media**
A Multidisciplinary Approach

**Foundations in Sound Design for Interactive Media**
A Multidisciplinary Approach

**Foundations in Sound Design for Embedded Media**
A Multidisciplinary Approach

For more information about this series, please visit:
www.routledge.com/Sound-Design/book-series/SDS

# Foundations in Sound Design for Embedded Media

## A Multidisciplinary Approach

Edited by Michael Filimowicz

Routledge
Taylor & Francis Group
NEW YORK AND LONDON

First published 2020
by Routledge
52 Vanderbilt Avenue, New York, NY 10017

and by Routledge
2 Park Square, Milton Park, Abingdon, Oxon, OX14 4RN

*Routledge is an imprint of the Taylor & Francis Group, an informa business*



*Library of Congress Cataloging-in-Publication Data*
Names: Filimowicz, Michael, editor.
Title: Foundations in sound design for embedded media : a multidisciplinary approach / edited by Michael Filimowicz.
Description: New York : Routledge, [2019] | Series: Sound design ; volume 3 | Includes bibliographical references and index.
Identifiers: LCCN 2019005862 | ISBN 9781138093874 (hardback : alk. paper) | ISBN 9781138093898 (paperback : alk. paper) | ISBN 9781315106359 (Master e-book) | ISBN 9781351603898 (pdf) | ISBN 9781351603874 (mobi) | ISBN 9781351603881 (epub3)
Subjects: LCSH: Sound—Recording and reproducing—Digital techniques. | Computer sound processing.
Classification: LCC TK7881.4 .F68 2019 | DDC 006.5—dc23
LC record available at https://lccn.loc.gov/2019005862

ISBN: 978-1-138-09387-4 (hbk)
ISBN: 978-1-138-09389-8 (pbk)
ISBN: 978-1-315-10635-9 (ebk)

Typeset in Times New Roman
by Apex CoVantage, LLC

# Contents

# Contributors

**Michael L. Austin** is Associate Professor of Music and Founding Director of the School of Music at Louisiana Tech University. His research focuses on sound and music in emerging and interactive media. He holds a Ph.D. in Humanities-Aesthetic Studies (now offered as a Ph.D. in Arts and Technology) from the University of Texas at Dallas. In addition to teaching courses in music technology, audio production, and popular music, he is also a practicing composer and sound artist.

**Mathieu Barthet** is a Lecturer in Digital Media at the Centre for Digital Music at Queen Mary University of London and Technical Director of the qMedia Studios. He was awarded a PhD in Acoustics, Signal Processing and Computer Science applied to Music from Aix-Marseille University and CNRS-LMA in 2008. His research lies at the confluence of several disciplines including Music Information Retrieval, Interaction and Perception. He is Principal Investigator of the EU-funded project "Towards the Internet of Musical Things" and Co Investigator of the EU-funded project "Audio Commons." He was chair of international computer music and audio conferences (CMMR 2012, AM'17) and served as guest editor for the *Journal of the Audio Engineering Society*.

**Johannes Birringer** is a choreographer and media artist; he codirects the Design-and-Performance Lab (DAP-Lab) with Michèle Danjoux and is a Professor of Performance Technologies in the Arts at Brunel University London. He has created numerous dance-theatre works, video installations and digital projects in collaboration with artists in Europe, the Americas, China and Japan. The DAP-Lab's crossmedia research explores convergences between physical movement choreography, visual expression in dance/film/fashion, wearable design and real-time interactive data flow environments (www.brunel.ac.uk/dap). DAP-Lab's interactive dance-work *Suna no Onna* was featured at festivals in London (2007–2008); the mixed-reality installation *UKIYO* went on European tour in 2010. The dance opera *for the time being* (2014) premiered at Sadler's Wells in London. A new series of immersive dance installations, *metakimospheres*, began touring in Europe in 2015–2018. His books include *Theatre Theory Postmodernism* (1989), *Media and Performance* (1998), *Performance on the Edge* (2000) and *Performance, Technology, and Science* (2009). He has spearheaded new transdisciplinary dance research projects, and coedited (with Josephine Fenger) the GTF books *Tanz im Kopf/Dance and Cognition* (2005), *Tanz & Wahnsinn/Dance & Choreomania* (2011), and *Tanz der Dinge/Things that dance* (2019).

**Ivica Ico Bukvic's** work encompasses aural, visual, acoustic, electronic, performances, installations, technologies, research publications, presentations, grants, patent disclosures and awards. Bukvic spent most of his career as a scholar-practitioner developing new transdisciplinary trajectories. His most recent work focuses on audio spatialization and sonification, exploring connections among the arts and human health, and recontextualizing Science-Technology-Engineering-Math for Kindergarten–12th grade education through innovative technology-mediated approaches to creativity. Having recently completed his temporary role as the interim associate dean of graduate studies and research at Virginia Tech's College of Liberal Arts and Human Sciences, since January 2019 Dr. Bukvic has been serving as the Director of Creativity + Innovation transdisciplinary community, a part of Virginia Tech's Institute for Creativity, Arts, and Technology. He is also the founder and director of Virginia Tech's Digital Interactive Sound and Intermedia Studio (DISIS) and the Linux Laptop Orchestra (L2Ork) and the co-founder of the new Creative Technologies in Music degree option, one of the first nationally accredited options of its kind in the United States. He also serves as the Institute for Creativity, Arts and Technology's Senior Fellow, Executive Committee member of the individualized interdisciplinary PhD (iPhD) program in the Human Centered Design and a member of the Center for Human-Computer Interaction with a courtesy appointment in Computer Science.

**Kristin Carlson** is an Assistant Professor in the Arts Technology Program at Illinois State University, exploring the role that computation plays in embodied creative processes. She has a history of working in choreography, computational creativity, media performance, interactive art and design tools.

**Greg Corness** is an Associate Professor in the Interactive Art and Media Department at Columbia College Chicago, working with embodied interaction in media environments. His research focuses on generative music, interdisciplinary improvisation, distributed cognition in performance and methodologies for researching experience in performance. He has developed several generative sound systems as well as computer vision and tangible interfaces for use in interactive performance and installation works.

**Michèle Danjoux** is a fashion designer, experienced educator and Research Coordinator for the School of Media & Communication at London College of Fashion, University of the Arts London. She is also co-director of DAP-Lab (http://people.brunel.ac.uk/dap/arch.html). Her artistic and research interests center on wearable design through and as performance, and involve collaboration with dancers, choreographers, musicians and media artists. She completed her doctoral study in 2017 with a practice-based thesis titled *Design-in-Motion: Choreosonic Wearables in Performance*. The project interrogates the choreographic space of real-time interactive dance performance through experimentation with specially designed audiophonic garments/accessories. Her costumes have been performed at Kibla Multimedijski Centre, Slovenia; MediaLab Prado, Madrid, Spain; Le Cube, Centre Création Numérique, Paris; Watermans Art Centre and Sadler's Wells, London.

**René van Egmond** (PhD) is an Associate Professor of Product Sound Design and Perception at the Delft University of Technology, specializing in human interaction with information environments that comprise multiple sound sources. He teaches classes such as Interactive Audio Design, Embodied Audio Design and Cognitive Ergonomics for Complex systems. He conducted research projects in cooperation with external partners, including Philips, Toyota, the Dutch National Police and the European Space Agency. Besides his academic career, he is a double bass player and a singer.

**Carole Favart** is General Manager of KANSEI Design Division and is leading the Advanced Technology Group, at Toyota Motor Europe R&D Center. She recently initiated and set up a new cross-functional Kansei-UX research department at Toyota Europe, with multidisciplinary expertise interacting with AI, Robotics, Advanced Materials, Long-term Strategy and Design. Carole's research interests lie in collaborative approaches involving scientists and design professionals bridging human needs with technical seeds in order to create emotional and pleasurable user experiences tailored to future customer expectations. Since 1990, she has been involved in design education, either as teacher, jury or member of boards and committees. Carole's design expertise ranges from interior architecture to strategy and trends analysis.

**Alexandre Gentner** (PhD) is a senior specialist for user experience strategies at Toyota Motor Europe—Advanced Technology. His activities all have in common a focus on users' affective responses to perception and cover user research, investigations related to future technology usage and experience-focused concept creation.

**Daniel Hug** has been investigating sound and interaction design related questions through artistic installations, design works and applied as well as basic research since the late nineties. In 2005, he introduced sound studies and sound design as fields of competence in the Interaction and Game Design education areas at Zurich University of the Arts, Switzerland. Hug is Lecturer and Researcher at the Chair for Music Education at the University of Applied Sciences of Northwestern Switzerland, and co-head of the Sound Design master program at Zurich University of the Arts. Furthermore, he is Visiting Lecturer for interaction and game sound at various universities in Switzerland, Austria and Finland. Hug holds a doctorate from the University of the Arts and Industrial Design Linz, is founder of the sound design and consulting company "Hear Me Interact!" and is a member of the steering committee of the conference "Audio Mostly—Conference on Interaction with Sound."

**Michael Iber** studied piano at the Royal Academy of Music in London and initially pursued a classical concert career. Since the end of the 1990s, he has been developing projects at the intersection of music, sound and media art, as well as in the field of music informatics. Since 2014 Iber has been responsible for the master class audio design at the Department of Media Technology at St. Pölten University of Applied Sciences. Besides project management of contract research projects for museum installations and music education, his main research interest lies in the design of acoustic information and data sonification. As part of his doctorate in International Logistics at Jacobs University in Bremen, Iber developed a method for the auditory analysis of production data. In addition to his teaching and research activities, Iber works as an author, director, sound designer and producer of radio features and radio plays.

**Antti Ikonen's** work covers music and sound design for contemporary dance, theater, short films, radio plays, art installations and new media. Since 2001, Ikonen has been working as a lecturer of sound design and music in Media Lab Helsinki (Aalto University), where Ikonen is in charge of Sound in New Media Master's studies and a member of SOPI (Sound and Physical Interaction) research group.

**Esther Klabbers** received her undergraduate degree in language and computer science at the University of Nijmegen in 1996. She received her PhD in speech synthesis from the Technical University Eindhoven in 2000. She performed her research there at the Institute for Perception Research working in an interdisciplinary group to develop a spoken dialogue system for providing train information (OVIS). After a short postdoctoral project in Eindhoven, she moved to Portland, Oregon, in 2001 where she first worked as a senior research associate and later as an assistant professor at the Center for Spoken Language Understanding (CSLU), which is part of Oregon Health and Sciences University (OHSU). Her research there focused on TTS as well as prosodic and acoustic analysis of speech to study the effects of Parkinson's or autism spectrum disorder. She also worked for a small startup company called Biospeech founded by CSLU director Jan van Santen from 2007 onwards. At Biospeech she was able to work on research projects that were more practical in nature, such as an app to help children with cochlear implants train their phoneme perception. In 2014, she moved back to the Netherlands to start working for ReadSpeaker. There, she is using her knowledge of linguistics, prosody and programming in R&D to develop high quality ANN synthesis and improve the quality of the unit selection synthesis voices that are currently in ReadSpeaker's portfolio. In 2018, she has returned to Portland, OR, where she continues her work for ReadSpeaker. When I first mention ReadSpeaker, I would also like to add the following sentence: ReadSpeaker is a leading company in the text-to-speech field, producing high-quality text-to-speech solutions for many different industries in many languages.

**Vincent Meelberg** is senior lecturer and researcher at Radboud University Nijmegen, the Netherlands, Department of Modern Languages and Cultures, and at the Academy for Creative and Performing Arts in Leiden and The Hague. He studied double bass at the Conservatoire of Rotterdam and received his MA both in musicology and in philosophy at Utrecht University, and an MSc in sound design at Napier University Edinburgh. He wrote his dissertation on the relation between narrativity and contemporary music at Leiden University, Department of Literary Studies. Vincent Meelberg has published books and articles about musical narrativity, musical affect, improvisation and auditory culture, and he is founding editor of the online *Journal of Sonic Studies*. His current research focuses on the relation between sound, interaction and storytelling. Besides his academic activities, he is active as a double bassist in several jazz and improvisation ensembles, as well as a sound designer.

**Stefano Delle Monache** holds a degree in Law (2000), and in Electronic Music and Technologies of Sound (2008). He received a PhD in the Science of Design in 2012 from Iuav University of Venice, Italy, where he works as assistant professor in the Department of Architecture and Arts. He has been involved in the EU projects CLOSED (Closing the Loop of Sound Evaluation and Design), SkAT-VG (Sketching Audio Technologies using Vocalizations and Gestures) and the COST Action on Sonic Interaction Design. His main area of research is on the application and evaluation of design methods and practices, including sketching, to a context of interaction design where auditory feedback is of central importance.

**Jim Murphy** is a designer and user of mechatronic musical instruments. His work focuses on the building of mechatronic instruments that are endowed with high degrees of freedom and many parameters of control; after building them, he explores the compositional and artistic possibilities of these parametrically rich instruments. Jim is a lecturer in Sonic Arts at Victoria University of Wellington's New Zealand School of Music; he received his PhD from the New Zealand School of Music in 2015 after finishing undergraduate studies at California Institute of the Arts. Jim has studied and collaborated with notable musical roboticists Ajay Kapur, Trimpin, Mo Zareei and Bridget Johnson. With Zareei, he is part of the Technical Earth kinetic sound sculpture collective (www.technical-earth.com/); his personal website may be viewed at www.jim-murphy.com.

**Mark Nazemi** is a multidisciplinary artist and educator from Vancouver, Canada. He received his PhD from the School of Interactive Arts and Technology and has been involved with teaching Sound Design, Interaction Design, Generative Art and Music Production for the past ten years. Mark has participated as a performing artist at festivals and outdoor public art installation events, and he has exhibited his work at various galleries both locally and internationally.

**Elif Özcan** (PhD) is the Director of Critical Alarms Lab at Delft University of Technology and leads the Silent ICU Project at the Adult Intensive Care Unit of the Erasmus University Medical Centre in Rotterdam (NL). Sound is the theme of Elif's professional career both in the design industry and in academia. Elif worked as a DJ as well as a self-trained sound designer at a major radio station, completed a PhD on the experience of product sounds, and has led sound design projects for Toyota Motor Europe (dashboard sounds), European Space Agency (notification system for Mission Control Rooms), and the International Standards Office (future medical alarms). Elif teaches the theory and practice of product experience and her recent research activities focus on sound-induced well-being. Elif is a mother of three young daughters and an advocate for women in science (board member of DEWIS at Delft University of Technology).

**Simon Pfaff** has a BA in music and an MS in design at the University of Arts in Zurich and specializes in Interaction Design, Physical Computing and Sound Design. Currently he is developing Birdly, the next generation of virtual reality experiences at SOMNIACS SA. Additionally, Simon does research regarding the implications of VR on CSCW and CVR at the Institute for Computer Music and Sound Technology of the Zurich University of Arts.

**Davide Rocchesso** received a PhD from the University of Padua, Italy, in 1996. He is a professor of computer science at the University of Palermo, Italy. He was the coordinator of EU projects SOb (the Sounding Object) and SkAT-VG (Sketching Audio Technologies using Vocalizations and Gestures). He has also chaired the COST Action on Sonic Interaction Design. His main research interests are sound modeling and synthesis, interaction design and evaluation of interactions.

**Jesse Seay** is an associate professor of audio arts at Columbia College Chicago, where she teaches hands-on courses in audio electronics and sound art. Her recent work has been a dialogue between her classroom teaching, her electronics workshops for makerspaces and her exploration of the intersection of textiles and analog electronics through knitting machines. Her goal is to make learning electronics approachable and inclusive. Seay has an MFA from the School of the Art Institute of Chicago and an MA in media and cultural studies from UNC-Chapel Hill.

**David Stout** (MFA CalArts) is a visual artist, composer and performer exploring hybrid approaches bridging the arts, design and technology. He is a recipient of international awards and recognitions for works that include live-cinema performance, interactive video installation, electroacoustic music and immersive theatrical events that integrate emerging technologies and multiscreen projection as an extension of performer, audience and architecture. Stout previously founded the MOV-iN Gallery and the Installation, Performance & Interactivity project (IPI) at the College of Santa Fe in New Mexico. He is currently founder and director of the Hybrid Arts Laboratory at the University of North Texas, where he coordinates the Initiative for Advanced Research in Technology and the Arts (iARTA) and holds joint positions in the Division of Composition Studies and Studio Art/New Media.

**Prophecy Sun** is a PhD candidate at the School of Interactive Arts and Technology at Simon Fraser University. Her interdisciplinary performance practice threads together both conscious and unconscious choreographies, sound and environment to create exploratory works that invoke deep body memory. Her research is supported by the Joseph-Armand Bombardier Canada Graduate Scholarship, Social Sciences and Humanities Research Council of Canada, and the British Columbia Arts Council.

**Luca Turchet** is an assistant professor at the Department of Information Engineering and Computer Science of the University of Trento. He holds master degrees in computer science from the University of Verona, in classical guitar and composition from the Music Conservatory of Verona, and in electronic music from the Royal College of Music of Stockholm. He received a PhD in media technology from Aalborg University Copenhagen. He is co-founder and Head of Sound and Interaction Design of MIND Music Labs. His scientific, artistic, and entrepreneurial research has been funded by several research grants including a Marie-Curie individual fellowship, a grant of the Danish Council for Independent research, a Leadership Fellowship of the Arts and Humanities Research Council of United Kingdom, three grants of the Swedish Innovation Agency, as well as scholarships from the Italian Minister of Foreign Affairs, EU SID-COST action, and the Lerici Foundation.

# Series Preface

## Foundations in Sound Design: A Multidisciplinary Approach

### Edited by Michael Filimowicz

This series organizes foundational topics in sound design that combine multidisciplinary perspectives across linear, interactive and embedded technologies. Such an approach is needed as today the practices of sound design are diversifying beyond what could adequately be captured by any single author or discipline. Today's sound designers need to be prepared just as much for games as for films, for programming in coding languages as for mastery of proprietary industry software, and for prototyping web applications or new industrial designs as for traditional occupations in film, television, music and audio.

The volumes are designed to be more future proof than most volumes on media technologies by focusing on high-level concepts that can be easily put into practice. The first three volumes in the series are sequenced as follows:

*Volume 1: Linear Media* covers traditional topics such as audiovisual preproduction, production and postproduction but adds other important aspects of linear media as well, such as electronica music production and basic music theory for sound designers, as well as artistic compositional practices such as soundscape design and electroacoustic music.

*Volume 2: Interactive Media* expands the cinematic soundtrack developed in Volume 1 by developing interaction approaches through consideration of gaming technologies, music programming, installations, spatial audio, real-time synthesis, performances and web-based interfaces and databases including mobile and locative media.

*Volume 3: Embedded Media* brings much needed coverage to emerging areas such as auditory display, data sonification, the role of sound in the internet of things, wearables and multimodal interaction by integrating physical computing technologies and product development in contexts ranging from toys to automobiles.

This approach to the foundations of sound design does justice to the ever growing uses, content variations, audiences and professional roles that sound design is brought to bear on in the contemporary context. Each volume in itself constitutes an introductory text to its respective area of sound design: linear, interactive and embedded media. Taken together, they comprise a comprehensive introduction to the many forms, technologies and practices of sound design.

These first three volumes set up the possibility for other books that can expand upon the foundational topics for deeper explorations of specialized topics. Prospective authors are encouraged to send ideas for other volumes that can be added to the initial three volume set, either single- and co-authored monographs or edited volumes, by submitting a proposal to the series editor, Michael Filimowicz (michael@filimowi.cz). Finally, feedback from readers on the content is always welcome.

Michael Filimowicz

# Volume Introduction

*Foundations in Sound Design for Embedded Media* introduces key concepts for the production of sounds associated with devices, environments, information displays and data. As our objects and spaces become increasingly informatic and connected to networked communication flows, new perspectives on sound design have emerged to ground inquiry and practice. The chapters in this volume cover topics related to physical computing; interaction design; auditory displays and data sonification; speech synthesis; wearables; smart objects and instruments; user experience; toys and playful tangible objects; and the new sensibilities entailed in expanding the concept of sound design to encompass the totality of our surroundings.

Chapter 1—"Sonic Interaction With Physical Computing" by Mark Nazemi— begins the volume by providing an introduction to physical computing, which entails programming microcontrollers connected to sensors, actuators, code and memory. The Arduino platform is covered as an accessible means for prototyping computational design artifacts that interact with users and the local environment.

Chapter 2—"The Electronics of Microphones and Loudspeakers" by Jesse Seay—presents key concepts of electronics associated with microphones and speakers. Different types of microphones and speakers are described, along with practical, hands-on tips for improving the performance of each. The chapter discusses many important concepts related to DIY exploration with transducers.

Chapter 3—"New Interfaces and Musical Robots: New Musical Input and Output Devices" by Jim Murphy—introduces the practice of developing new musical interfaces, and expands its treatment into the domain of musical robotics. The typical subsystems that come together to create custom controllers are presented as a means of creative expression for the sound designer.

Chapter 4—"Sketching Sonic Interactions" by Stefano Delle Monache and Davide Rocchesso—provides an overview of methods and practices for sketching sound in interactive contexts. The role and importance of provisional sound representations in conceptual sound design is explored. The cognitive benefits of embodied sound sketching are outlined by means of practical examples and exercises.

Chapter 5—"Bringing Sound to Interaction Design—Challenges, Opportunities, Inspirations" by Daniel Hug and Simon Pfaff—bridges that practices of interaction design and sound design through discussion of a series of prototypes that serve to bring the two fields together. The authors describe ways the two design disciplines can mutually benefit and argue for the necessity of a sustained exchange and effective integration.

Chapter 6—"Auditory Display in Workplace Environments" by Michael Iber—gives an overview of aspects to be considered for the design of auditory and multimodal displays in workplace environments including auditory perception and cognitive attention, the workplace design of control rooms, and sonification design.

Chapter 7—"Incorporating Brand Identity in the Design of Auditory Displays: The Case of Toyota Motor Europe" by Elif Özcan, René van Egmond, Alexandre Gentner and Carole Favart—details a sound design study for a commercial brand. The chapter focuses on the theoretical knowledge underlying auditory display design, auditory icon perception, usability of interactive systems and tools and methods for conceptual design.

Chapter 8—"Introduction to Sonification" by Ivica Ico Bukvic—offers an introduction into the world of sonification, or the translation of data and information into structured sounds: why it exists, and how we may benefit from it. The material is presented in a way that requires minimal prior knowledge and defines key terms necessary for content comprehension.

Chapter 9—"Image Sonification—a Practitioner's Account" by David Stout—serves as an introduction to the art and science of image sonification, providing a concise overview of the pioneers and evolving technical innovations that underpin contemporary approaches to visual music and photo-based sound generation.

Chapter 10—"Sound and Wearables" by Johannes Birringer and Michèle Danjoux—establishes crossovers between fashion, performance, music and sound art, investigating analog and digital techniques used for the construction of "sounding" costumes which play a significant part in the overall scenographic and choreographic organization of real-time interactive art and performance.

Chapter 11—"Smart Musical Instruments: Key Concepts and Do-It-Yourself Tutorial" by Luca Turchet and Mathieu Barthet—develops the theoretical principles and practical skills needed to understand and produce new musical instruments featuring intelligent sensing and internet connectivity, the so-called smart musical instruments (SMIs).

Chapter 12—"Text-to-Speech Synthesis" by Esther Klabbers—describes the state of the art in text-to-speech (TTS) synthesis. It gives an overview of the steps involved in going from input text to generated speech, and covers some open-source tools available to create your own TTS applications. A TTS system can be seen as consisting of two parts: (1) a linguistic front-end, and (2) an acoustic back-end.

Chapter 13—"Sound and UX Design" by Michael L. Austin—considers various ways in which user-experience (UX) design is defined among scholars and practitioners in the industry, and it briefly explores the disciplinary roots of UX and its connection to other design fields. The chapter also highlights the ways we as humans process various types of designed experiences, and it provides corresponding examples of sound design.

Chapter 14—"Toys and Playful Devices" by Kristin Carlson, Greg Corness and Prophecy Sun—explores the subject of sound design in toys and surveys the existing literature on how children learn to hear, perceive and make sense of sound in their environment. The authors provide more clarity about how and why children playfully interact with sound through various toys and objects in their world.

Chapter 15—"The Sonic Internet of Things: Sound Design for Smart Objects" by Vincent Meelberg—analyzes how sound can improve the interaction between people and smart objects by taking a material and contextual narrative approach. Smart objects generally function within an environment, and a contextual narrative approach to designing sounds for these objects takes into account this environment, as well as temporal considerations.

Finally, Chapter 16—"On the Importance of Listening" by Antti Ikonen—addresses conceptual and practical tools for embedded sound design, especially in the context of "interior design." Site-specific sound art is somewhat similar to ambient sound design in terms of the spatial and temporal features of sonic experience. Contemplative listening is the most crucial skill and fundamental method when designing sounds for specific environments.

**Other Volumes in the Series**
*Foundations in Sound Design for Linear Media*
*Foundations in Sound Design for Interactive Media*

## Acknowledgment

The chapter summaries here have in places drawn from the authors' chapter abstracts, the full versions of which can be found in Routledge's online reference for the volume.

1

# Sonic Interaction With Physical Computing

Mark Nazemi

## 1.1 Introduction

Physical computing involves the communication between our physical world and the digital world of computers (O'Sullivan and Igoe 2010). The way this is made possible is through the process of *transduction*, which is the conversion of one form of energy to another. Another way we can think of physical computing is that it provides an opportunity for learning by understanding how humans can express themselves physically and communicate through computers.

Microcontrollers are at the heart of physical computing and are miniaturized computers that feature input and output connectivity for allowing the use of sensors and actuators. The first experimenters of physical computing were musicians, specifically electronic musicians who hacked hardware devices to include additional controls for gesture-based performances and real-time manipulation of audio signals (Collins 2014). The use of microcontrollers can provide vast opportunities to develop reactive and interactive sound projects allowing the musician or sound designer to move beyond the boundaries of using a computer keyboard, mouse, screen and speakers.

To design an appropriate interactive project, we must first have a good grasp of how microcontrollers work, the types of inputs and outputs that are available to use to make transduction possible and how to process the data.

## 1.2 Input

Producing meaningful interaction begins with choosing the right input. Specifically, we want to go beyond the normal keyboard and mouse input and learn about how we can take advantage of sensors to detect changes in our environment and even our own body. Therefore, inputs can be sensors, switches, levers or sliders.

## 1.3 Sensors

There are two broad categories of sensors, digital or analog as shown in Figure 1.1. Digital refers to the number of states, usually two. For example, a digital sensor may tell us if the house cat is sitting on its mat or not. Therefore, a digital sensor's output, like the Infra-Red (IR) sensor shown in Figure 1.2, can only be in one of two possible states, it is either

**Figure 1.1** There Are Many Types of Analog and Digital Sensors Available to Accommodate Interactive Projects.

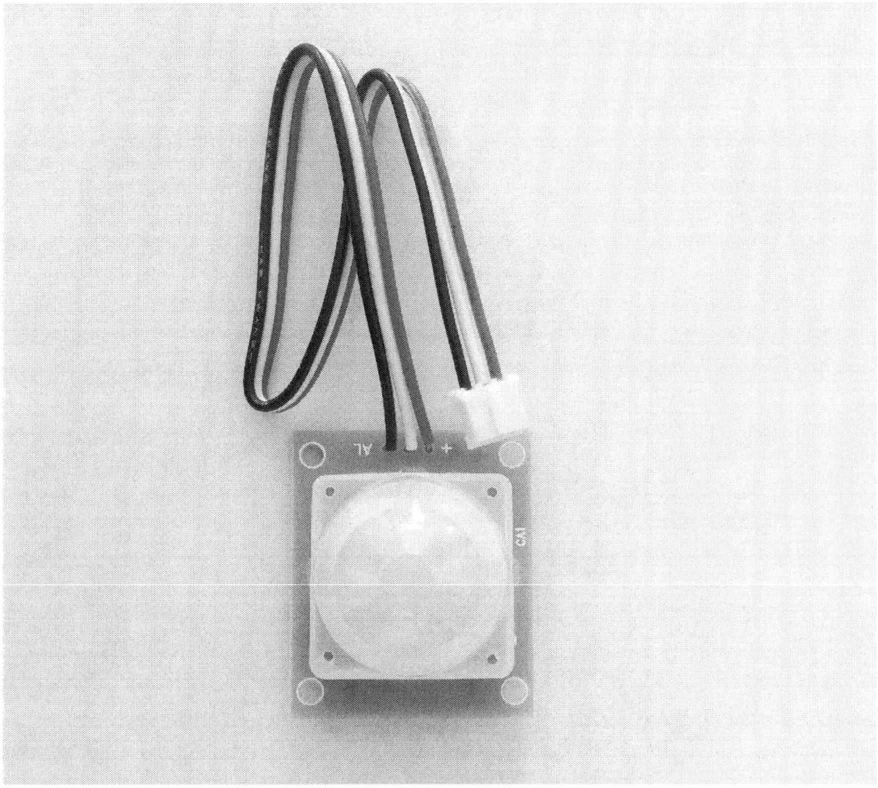

**Figure 1.2** An Infrared Motion Sensor Used to Detect the Presence of Moving Objects.

ON (1), often +5 volts or OFF (0), zero volts. Most digital sensors work with a threshold. Therefore, if the incoming measurement is below the threshold, the sensor will output one state, and if it is above the threshold, the sensor will output the other state. An example of a digital sensor is a touch sensor. A touch sensor can sense only two states, touched or not touched.

Whereas analog has a continuous range of states, therefore, an analog sensor can tell us how heavy the house cat is. Often, the output of an analog sensor is a variable resistance that can be used to control a voltage. Rather than only being able to toggle between two states, the sensor can output an almost infinite range of values.

There are many types of sensors that can work with a microcontroller, and that is why we have to decide first what parameters of the external environment, such as reading the temperature, pressure, and light, or the internal environment, like heart rate, breath, or brain activity, we want to measure. Second, we need to determine what kind of sensor is optimal for measuring that parameter.

Aside from analog or digital, sensor categories can be based on the underlying physics of their operation, the particular phenomenon they measure or a specific application. Sensors can also be passive, which means they need a separate power source to obtain the output. They can be active, self-generating, producing an electrical signal when subjected to a sensed quantity such as piezoelectric sensors, thermoelectric materials or temperature.

We can categorize commonly used sensors with microcontrollers based on their application as follows:

*Sensing pressure:* force resistive sensor (FSR), bend sensor and load sensor. Pressure sensors in general vary their resistance depending on how much pressure is applied to the sensing area. When we apply pressure the resistance becomes lower and when no pressure is applied, the resistance will be much higher.

*Sensing the environment:* temperature sensor, photoresistors (to sense light), humidity sensor, color sensor, alcohol/gas sensor, water sensor, wind sensor, sound sensor, flame sensor, methane gas sensor, barometric pressure and water flow sensor.

*Sensing movement/distance:* Accelerometer (three or more degrees of freedom), gyroscope, ultrasonic sensor, Laser Break Beam sensor, fast vibration sensor, infrared motion sensor, line tracking sensor, comparator speed sensor and proximity sensor.

*Sensing body:* pulse sensor, Galvanic Skin Response, Muscle Sensor (EMG), Brain waves (EEG), skin temperature sensor and eye tracker.

***Producing motion:***

Two basic types:

—Linear motion

—Rotary motion—DC or Servo Motors

Convert the rotary motion into a linear motion.

—Use basic mechanics and simple machines

- Lever
- Pulley
- Gear
- Cam

- Ratchets
- Linkages

This is by no means an exhaustive list, but it is meant for you to think about potential ways interactivity can be added to a project. New and exciting sensors are constantly being developed and the best resources to find the latest sensors and microcontrollers are Sparkfun Electronics[1] and Adafruit.[2]

### 1.3.1 Sensor Quality and Parameters

We now know that a sensor generates a variable signal that can be manipulated, processed, transmitted or displayed. It is also important to know that sensors have limitations or constraints which can play a significant role when we are trying to develop a stable interactive sound-based project. A sensor or instrument is calibrated by applying some known physical inputs and recording the response of the system. For expensive sensors, which are individually calibrated, this might take the form of the certified calibration curve. One of the parameters we would need to be aware of is a sensor's sensitivity, which is defined in terms of the relationship between changes in input physical signal and output electrical signal. The sensitivity is the ratio between a small change in electrical signal to a small change in physical signal. For example, a thermometer would have "high sensitivity" if a small temperature change resulted in a large voltage change.

A sensor's accuracy is essential when we are trying to create repeatable interactions. Accuracy is a measure of a difference between the measured value and actual value. Accuracy is the capacity of a measuring instrument to give results close to the actual value of the measured quantity. Accuracy is different from repeatability, which is the ability to reproduce the output signal exactly when the same measured quantity is applied repeatedly under the same environmental conditions.

To make a sensor's reading reliable, we will need to calibrate the sensor, which requires an understanding of its range and span, namely, the limits between which inputs can vary. Span is maximum value minus the minimum value of the input. Using code, we can also modify the span of a sensor to meet the requirements of a project (check out the "map()" function used in Arduino).

Lastly, all sensors produce some output noise in addition to the output signal. In some cases, the noise of the sensor is less than the noise of the next element in the circuit, or less than the fluctuations in the physical signal, in which case it is not important. Many other cases exist in which the noise of the sensor limits the performance of the system based on the sensor. Therefore, it is important to review the datasheet of a sensor to find out its limitations and strengths.

## 1.4 Output

The output is the mechanism that provides feedback to the user using movements like actuators and motors, light, video or sound. Depending on the hardware, we may have more than one output, and this can enrich the user experience or confuse the participant if the interaction model is incorrectly configured. It is also important to keep track of how much current an output device uses in a project. Some output devices such as motors or light emitting diodes (LEDs) can draw significant power, and therefore a separate power source is needed in order to operate these particular outputs.

## 1.5 Processing

So far, we have covered the physical components of computing, which include the input and output. Processing is the third part, which requires a computer to take the information from the input, make some decisions based on changes that may occur in real time and output feedback to either the user or another computer. We need to use software to make sense of the incoming data from the input and convert that data into meaningful information to be sent out as an output for the user or as a message to another computer. We learn about coding techniques later in this chapter.

## 1.6 Serial or Parallel

How the input and output of data flow over time needs to be clarified, and that is when we have to start thinking about events in serial or parallel. In serial, information or events take place one after another. For example, cars restricted to moving in a single lane one after another can be thought of as movement in serial.

Parallel is when we have events happen simultaneously. In parallel, many events or information transfers take place at once. To extend the analogy with automobile transportation, this would be like having multiple lanes for cars to move and pass each other. This idea of serial and parallel is used not only in terms of how information is relayed in a computer but also with how electrical energy flows in a circuit.

## 1.7 Basic Electronics

To build a circuit, we have to understand basic concepts of electricity and items like a 'breadboard' since in most cases additional electronic components such as resistors, capacitors and diodes are used to build our interactive or reactive projects.

Every component of a circuit has specific electrical characteristics, such as a battery providing a certain amount of electrical energy to power parts and a light bulb resisting a certain amount of electrical energy. If not enough energy is provided to the light bulb, the wires inside the light bulb will not heat up and produce light.

Voltage is the relative level of electrical energy between any two points in a circuit and is measured in volts (V), and the current is the amount of electrical energy passing through any point in the circuit, which is measured in amperes (or amps (A)). To understand how current works, we can use a water pump analogy by considering water pressure as the voltage and the how much water (electricity) is flowing past a certain point as the current.

high water pressure = high voltage
low water pressure = low voltage

The higher the amperage, the more water (electricity) flows. Resistance is the amount that any component in the circuit resists the flow of current, which is measured in ohms. A valve in the pipe acts as a resistor, limiting the current (and the voltage) flowing through the pipe. We can think of resistance as a property of a material that controls the ease with which current flows. Some materials, like isolators, which include rubber, paper, porcelain and air, have very high resistance. Since air has a high resistance, it will be difficult for a current to flow through it. *It is important to note that electricity always favors the path of least resistance to the ground (-) and that all electrical energy in a circuit must be used.* A circuit with no load is called a "short circuit." In a short circuit, the power source feeds all of its power through the wires and back to itself, and either the wires melt, or the battery may explode. When we have a short circuit using a component that is connected to our computer, you may see a warning appear that a device is drawing too much power. In this case, immediately disconnect all peripheral devices to avoid damage.

The ground is the place in a circuit where the potential energy of the electrons is zero, and sometimes this point is connected to the actual ground, either through a grounded electrical circuit or water pipe, or by some other method.

Some of the most common electronic parts are listed below:

*Diodes* permit the flow of electricity in one direction, and block it in the other direction. Because of this, they can only be placed in a circuit in one direction.

*LEDs* are diodes that emit light when given the correct voltage. Like all diodes, they are polarized, meaning they only operate when oriented correctly in the circuit. The anode of the LED connects to voltage, and the cathode connects to ground.

*Switches* control the flow of current through a junction in a circuit.

*Resistors* limit the current flow in a circuit. Resistors are not polarized and therefore can be placed in a circuit in any direction.

*Potentiometers are* resistors that can change their resistance like the volume knob on your stereo system. A potentiometer (or pot) has three connections. The outer leads are the ends of a fixed value resistor. The center lead connects to a wiper that slides along the fixed resistor. The resistance between the center lead and either of the outside leads changes as the pot's knob is adjusted.

*Capacitors* store electricity while current is flowing into them, then release the energy when the incoming current is removed. Sometimes they are polarized, meaning current can only flow through them in a specific direction. A marker on the capacitor indicates if it is polarized.

*Voltage regulators* take a range of direct current (DC) voltage and convert it to a constant voltage. For example, a regulator taking a range of 8–15 volts DC input may convert it to a constant 5-volt output. Most microcontrollers have built-in voltage regulators.

Solderless *breadboards* are the quickest tools for prototyping a new circuit or for testing parts. Many decades ago, experimenters began using kitchen breadboards to secure electronic parts and to build circuits, hence the term. Today, the components we use are miniaturized, and better methods of connecting circuits have been developed. Breadboards provide the flexibility to build and test projects without having to connect any physical parts permanently. On most breadboards, the sides called the power rails are used for connecting power and ground wires to power our circuit. The ground (-) rail is blue while the red rail is used for connecting our voltage (+). The middle area of the breadboard contains terminal strips that we use to place and connect all our parts. Once we insert our components, they will be electrically connected to anything else set in that specific row.

*WARNING:* avoid adding, removing or changing parts on a breadboard when the board is powered. Otherwise you risk shocking yourself or damaging your components.

## 1.8 Microcontroller

We have all had some experience using a controller even though we may not have been aware of it. We monitor an aspect of an environment or a process using a controller. For example, using a microwave to warm up our food uses an embedded controller to allow the user to have functionality over how long and what power to use to zap the food. The security system at our home uses an embedded controller and sensors to monitor movement in the house and trigger an alarm or notify the police if an intruder is detected. These are just two scenarios in which controllers are embedded in a particular application. They can also be found in cars, home appliances, instrumentation in medical devices, aerospace and many more.

Over time, controllers have become smaller to the point that all the components needed to function are built onto a single chip, hence the term *microcontroller*. Microcontrollers today are highly integrated chips that include the following fundamental components:

- A central processing unit (CPU or "the brain") on a single integrated circuit (IC)
- Memory (in the form of RAM, Random Access Memory, and EPROM, Erasable Programmable Read Only Memory)
- Input/output (I/O; to allow for the connection of sensors and output devices like motors in both serial and parallel)
- Interrupt Controller
- Timers
- Power (an option on some microcontrollers)

There are several advantages to using microcontrollers. Aside from being small, they are cheap, efficient, compatible with a broad range of sensors and actuators, can be open source and typically have a large community of supporters that share knowledge about projects, code and custom-made microcontrollers. There are many types of microcontrollers on the market, and the choice of which to use depends on the application, how much memory you need, number of I/Os, wireless communication, power consumption and size. For our purpose, we are going to look at the world's most popular microcontroller, the Arduino, as shown in Figure 1.3, and briefly talk about other microcontrollers available on the market.

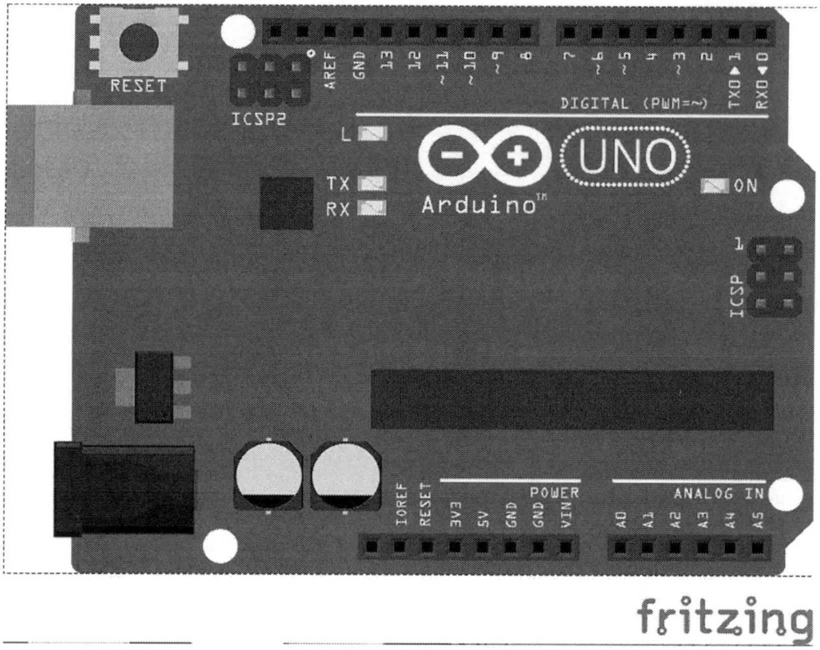


**Figure 1.3** The Arduino Uno.

## 1.9 The Arduino

The Arduino is not just a microcontroller but also an open-source platform that includes easy-to-use hardware, a software programming language called "Wiring" and the Arduino Software IDE (integrated development environment), which is based on the Java language. The Arduino was initially built at the Ivrea Interaction Design Institute to provide students with a way to quickly prototype with electronics without the need for an advanced background in engineering or computer programming. Since it is open source, many have built powerful versions of the Arduino for all types of applications including wearables, 3D printing, embedded environments and the Internet of Things (IOT) applications.

For the purpose of this chapter, we will take a closer look at the most common Arduino board, the Uno.

The Arduino Uno has the following components:

- Digital inputs/outputs: pins 2 to 13
- Analog inputs: pins 0 to 5

- The Atmel ATMega microcontroller
- A USB port to communicate with a computer
- Reset button
- Transmit and receive LEDs
- Connection for external power supply (9–12V DC)

The Arduino IDE is compatible across operating systems.[3]

The Arduino language called "Wiring" reads a program from the top down. The programs written for Arduino are called sketches, shown in Figure 1.4, and are broken into three sections: initialization, setup and loop.

All Arduino programs run in two parts:

- Setup()—preparation
- Loop()—execution

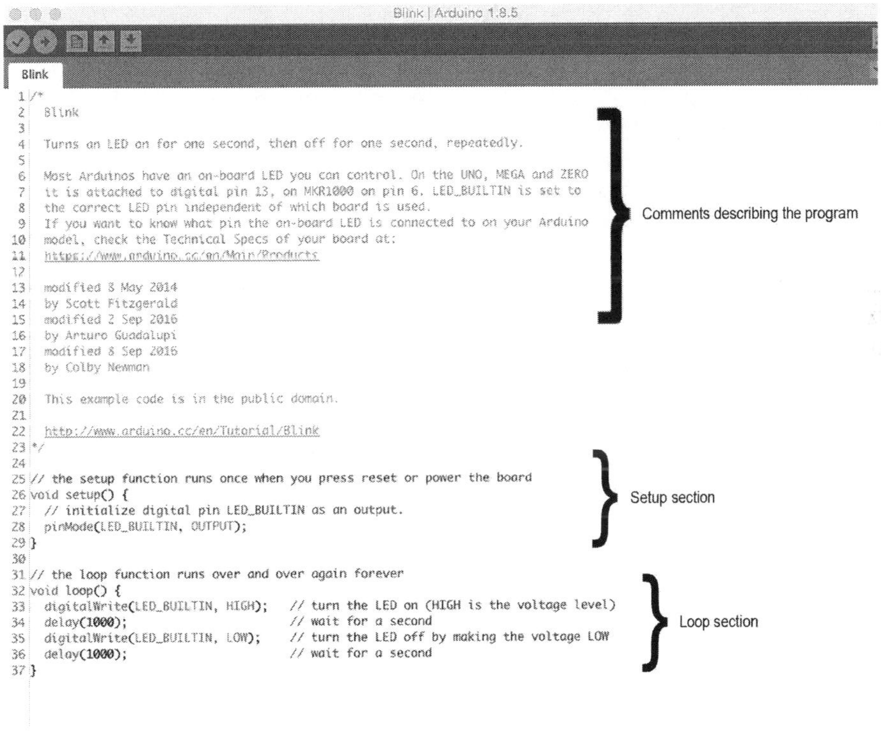


**Figure 1.4** The Arduino IDE.

The setup section, always placed above your loop(), is for setting pin modes, initializing serial communication and so forth. The setup() is the first thing called in an Arduino application. All applications must have a setup() method, even if nothing is done in it. Some devices also need to be initialized when the microcontroller starts up. The compiler will check for this method, and if it is not defined, an error will occur.

The loop section is the code to be executed—reading inputs, triggering outputs and so forth. The loop() contains anything that needs to happen repeatedly in the application, for example:

- Checking for a new value from an input—sending a signal to a pin
- Sending debug information

Any instructions in this method will repeatedly run until the application is terminated.

The third part (initialization), at the beginning of the code, is where we would declare our variables, import specific libraries for add-on hardware like Shields, and place bulk comments.

Wiring comes with a series of examples to try out without the need to code. To test that you have your Arduino Uno set up properly, run the most basic code called Blink. First, connect your Arduino to your computer via USB board. Then select the right board and serial port under the "Tools" menu in the Arduino software. Go to Examples > Basics > Blink. Press the Compile button (check mark symbol). If you have correctly chosen your board and serial port, Arduino should compile without errors. Next, press the arrow button to upload to your board. You should see the transmit (TX) and receive (RX) LEDs blink rapidly on your Uno. That is a good sign! It means the code is appropriately being uploaded onto your Uno. Next, you should see an LED next to pin 13 blink every second. Congratulations, you have successfully executed your first program. To add more complexity to your program, you will need to add some inputs and outputs to get started. There are many Arduino starter kits available for purchase that come with electronic parts and how-to guides to get you up and running in no time.

### 1.9.1 Arduino Input/Output

The most important part of the Arduino hardware that will be used for connecting input and output devices are the onboard pins. All the pins can be individually addressed and assigned through the Arduino IDE. The analog pins A0 to A5 are the pins commonly used to connect sensors. The digital pins can be configured as either inputs or outputs. Digital pins (3, 5, 6, 9, 10 and 11 on the Uno) with the ~ symbol act similar to an analog pin by measuring voltage changes between 0 and 5 volts. These are called PWM (pulse width modulation) pins that operate at a frequency of approximately 490Hz. PWM is a type of digital signal used to change the proportion of the signal in a high position (5V or ON) to when it is low (0V or OFF) over a particular time interval. This proportion is referred to as the "duty cycle," which controls the percentage the component remains in the ON position. A duty cycle is like a square waveform whereby one can switch rapidly between the on and off positions. For example, we use the analogWrite() command to vary the brightness of an LED or the rotation angle of a Servo Motor. The analogWrite() is on a scale of 0–255. Therefore if we code analogWrite(127) it would be at 50% duty cycle, and so our LED brightness would be at half brightness, whereas if we code analogWrite(255), it would be at full brightness. The analogWrite() function is only meant for the PWM pins and will not affect the analog pins on the Arduino. The PWM pins play an integral role when we want to add complexity to our interaction or when we just want a change that is not binary (on or off).

You may be wondering why we are limited to a range of only 0–255. That is because the digital pins on the Arduino use an 8-bit ADC (analog-to-digital converter), while the analog pins have a more significant range 0–1024 because they use a 10-bit ADC. A converter is always used to change an analog signal into a digital signal.

### 1.9.2 Communication

For many projects, it is common to have computers communicating with other devices. One of the most common configurations for physical computing systems is to have a microcontroller read a sensor and then send the value of the sensor to a multimedia computer. The computer processes the input from the microcontroller and performs an action. For example, the multimedia computer changes the playback of a video or the pitch of a sound or activates some other multimedia response. The reverse of this configuration is also common, where a computer sends the coordinates of the mouse to the microcontroller to position a motor.

The most common form of communication between electronic devices is serial communication. Communicating serially involves sending a series of digital pulses back and forth between devices at a mutually agreed-upon rate. The sender transmits pulses representing the data at the agreed-upon data rate, and the receiver listens for pulses at that same rate. As explained in the car analogy in section 1.5, the word *serial* means "one after the other." Serial data transfer is when we transfer data one bit at a time, one right after the other.

We use serial communication between the Arduino board and a computer or other devices. This communication happens via the Arduino board's serial or USB connection and on digital pins 0 (RX) and 1 (TX). Thus, if you use these functions, you cannot also use pins 0 and 1 for digital I/O. There is sometimes confusion because an Arduino has a USB connector, but Arduino is all about serial and not USB. Therefore, interfacing to things like USB flash drives, USB hard disks, USB webcams and so forth is not possible. The USB, however, can be used to power your project. USB power packs are a comfortable and convenient way to power up your project without relying on heavy batteries.

So how does serial communication work? We use a protocol—the set of parameters according to which the two devices agree to send information. There are many different protocols for serial communication, each suited to a separate application. We also need a physical connection (serial port)—in the case of an Arduino, we use the USB cable to connect the hardware to the computer.

A timing speed (bps) has to be set regardless of what serial protocol you are using. In order to count the pulses, there has to be agreement about how fast they are coming. The timing of the pulses is called the data rate or the baud rate. The most common rate for Arduino is 9600 pulses per second. Typically eight pulses are grouped together. This means that one group of eight pulses (also called a byte) is sent per millisecond, which is faster than we can possibly perceive or detect. Some sensors, such as an accelerometer or a pulse sensor, may require a faster baud rate, so it is essential to refer to the datasheet to make sure the correct timing is set.

An electrical connection is also required, and this refers to the voltage level of the electrical pulses that will be sent. With Arduino, we use 0 and 5 volts.

Lastly, the package size has to be determined because there has to be some agreement as to how the sequence of pulses is interpreted. By interpreting them in groups of eight (a byte), you can send numbers between 0 and 255. Serial data is passed byte by byte from one device to another. Therefore, it is up to the programmer to decide how each device (computer or microcontroller) should interpret those bytes: the beginning and end of a message is and what to do with the bytes in between.

## 1.10 Sound Interaction Design

A sound interaction that is well planned and executed can significantly enhance communication and augment meaning or emotional qualities in interactive contexts. The potential of sound is now being realized in particular fields such as virtual reality, multi-linear game sound engines, augmented reality, mixed reality and IOT. New Wave recording artist Thomas Dolby envisioned the idea of sonifying the web and in collaboration with Beatnik released a series of toolkits for developers to use sound for the internet. That project came to a close in 2009, and while the dream of sonifying the web may not have come to fruition, sound interaction is far from dead! The role of sound in product design, IOT, AI-powered devices like Alexa and immersive art-based projects is growing. As a sound designer, we have to carefully consider the relationship between interaction and communication to avoid over-designing sounds that lose their significance and become an annoyance to the user. Parameters such as tonal quality, loudness, repeatability and organic versus digital timbre can quickly shift the user or listener's perception of sound from profoundly meaningful to noise. Therefore, sonic experience design must be well thought out when combined with physical computing practices. The level of interaction produced is dependent on how much control we allow the user to have over the sound (Noble 2009). If the user has no control over the sound, then the level of interaction can be considered to be very low, and the user in such an instance is more of a spectator. Furthermore, there are various ways we can provide user feedback that may not necessarily be part of an interactive system. For example, the clicking sound heard when pressing buttons at an ATM and the many sonic notifications designed for medical instruments are all part of providing users feedback.

### *1.10.1 Sound Interaction as Input*

We can allow the user to create a sound input using a tool such as musical controllers like the Monome or OSC-based applications like TouchOSC. In this case, data such as MIDI can be mapped to buttons, sliders, keys or joysticks to have precise control over the sound. We can also create a customized tool for providing sound feedback via MIDI using sensors and Arduino (Simeonidis 2014). To do so, we will need a few items to get started:

- Arduino Uno
- A breadboard (solderless)

- A MIDI out connector
- Resistors (10k- and 220-ohm)
- An FSR sensor
- A switch
- Jumper wires

First, connect the 5V and ground connection using wires to your breadboard as shown in Figure 1.5. Set up your FSR sensor and switch on the breadboard as shown in the illustration. Make sure to use a jumper wire to connect the analog pin A0 of your Arduino to the FSR and a jumper to digital pin 10 for the switch.

Let us add the MIDI out jack and a 220 ohm resistor as shown in Figure 1.7.

Now that we have completed building our circuit, we can experiment with the code. The analog FSR sensor and the digital switch will allow for physical interaction for adding some performative quality to our setup. Connect a musical device that produces sound to the MIDI jack and copy the following code to your Arduino sketch.

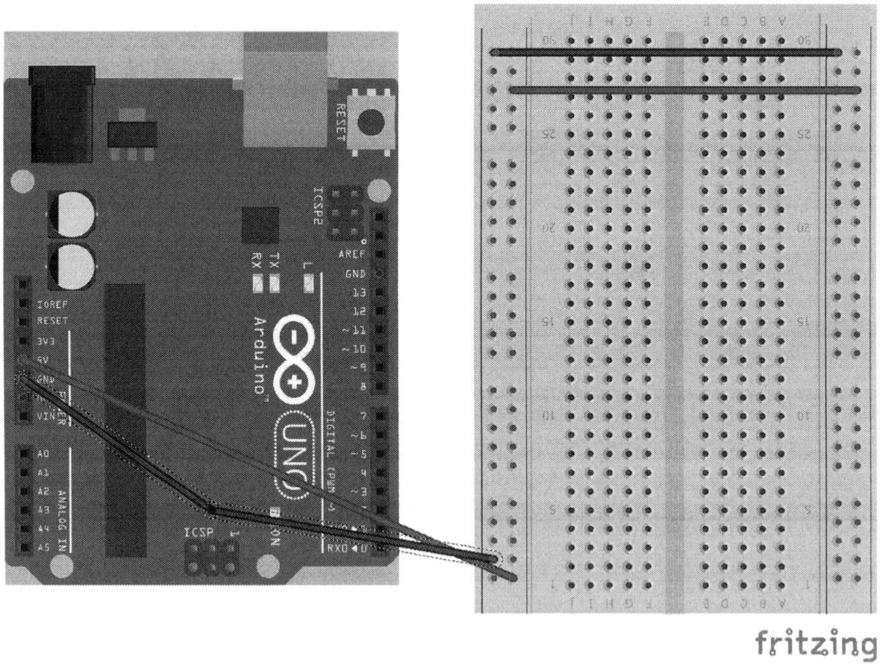


**Figure 1.5** Connecting the Voltage and Ground from the Arduino to the Breadboard.

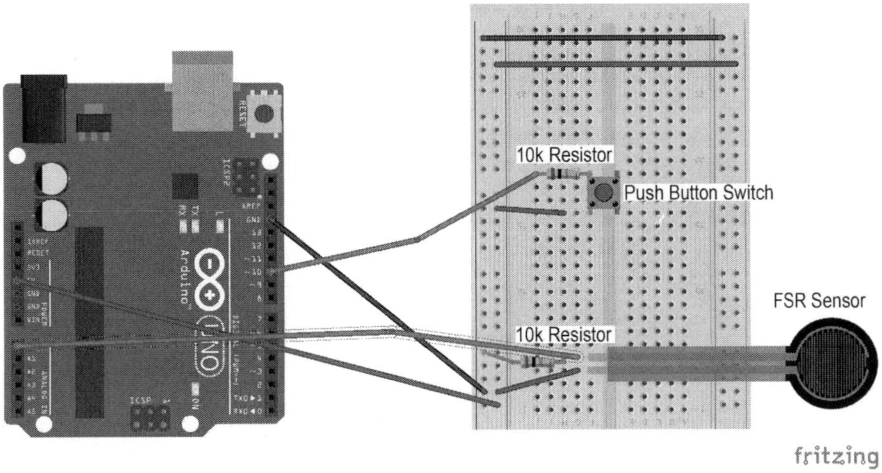


**Figure 1.6** Connecting the FSR Sensor, Button and Resistor.

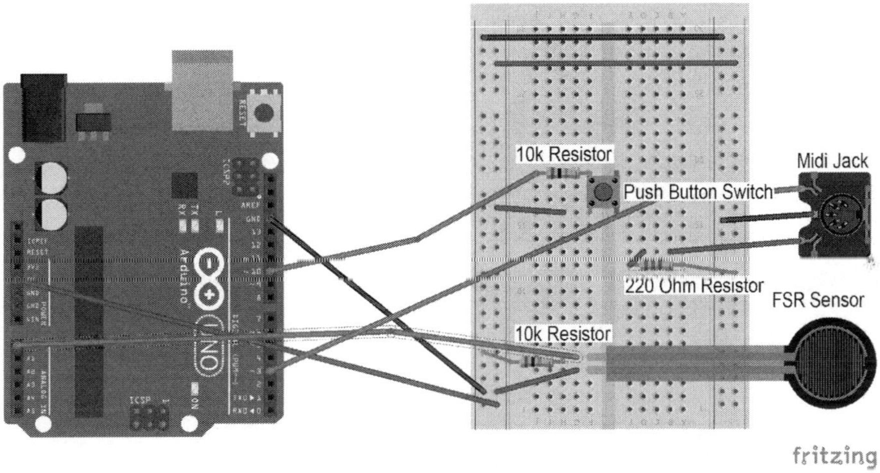


**Figure 1.7** Connect the MIDI Out Jack, and Make Sure to Double Check the Ground and Voltage Connections.

## *Example 1 Code:*

```
#include <SoftwareSerial.h>
 // Variables:
  byte note = 0;            // The MIDI note
value to be played
```

```
//software serial
SoftwareSerial midiSerial(2, 3);
// digital pins that we'll use for soft
serial RX & TX
  void setup() {
    // Set MIDI baud rate:
    Serial.begin(9600);
      midiSerial.begin(31250);
  }
  void loop() {
    // play notes from F#-0 (30) to F#-5
(90):
    for (note = 30; note < 90; note ++) {
      //Note on channel 1 (0x90), some note
value (note), middle velocity (0x45):
      noteOn(0x90, note, 0x45);
      delay(100);
      //Note on channel 1 (0x90), some note
value (note), silent velocity (0x00):
      noteOn(0x90, note, 0x00);
      delay(100);
    }
  }
  // plays a MIDI note. Doesn't check to
see that
  // cmd is greater than 127, or that data
values are less than 127:
  void noteOn(byte cmd, byte data1, byte
data2) {
    midiSerial.write(cmd);
    midiSerial.write(data1);
    midiSerial.write(data2);
```

```
    //prints the values in the serial mon-
itor so we can see what note we're playing
    Serial.print("cmd: ");
    Serial.print(cmd);
    Serial.print(", data1: ");
    Serial.print(data1);
    Serial.print(", data2: ");
    Serial.println(data2);
}
```

The first example will merely play notes without interaction. It is always good to check and make sure we are sending MIDI data correctly to our device, and that sound is produced. Now, let's modify the code and add some basic interactivity.

*Example 2 Code:*

```
#include <SoftwareSerial.h>
const int switchPin = 10; // The switch is
on Arduino pin 10
const int LEDpin = 13;    // Indicator LED
 // Variables:
 byte note = 0;              // The MIDI
note value to be played
 int AnalogValue = 0;        // value from
the analog input
 int lastNotePlayed = 0;     // note turned
on when you press the switch
 int lastSwitchState = 0;    // state of the
switch during previous time through the
main loop
 int currentSwitchState = 0;
//software serial
SoftwareSerial midiSerial(2, 3); // digital
pins that we'll use for soft serial RX & TX
```

```
void setup() {
  // set the states of the I/O pins:
  pinMode(switchPin, INPUT);
  pinMode(LEDpin, OUTPUT);
  // Set MIDI baud rate:
 Serial.begin(9600);
  blink(3);
   midiSerial.begin(31250);
}
void loop() {
  // My potentiometer gave a range from 0
to 1023:
  AnalogValue = analogRead(0);
  // convert to a range from 0 to 127:
  note = AnalogValue/8;
  currentSwitchState = digitalRead(switch
Pin);
  // Check to see that the switch is
pressed:
  if (currentSwitchState == 1) {
    // check to see that the switch wasn't
pressed last time
    // through the main loop:
    if (lastSwitchState == 0) {
      // set the note value based on the
analog value, plus a couple octaves:
      // note = note + 60;
      // start a note playing:
      noteOn(0x90, note, 0x40);
      // save the note we played, so we
can turn it off:
```

```
        lastNotePlayed = note;
        digitalWrite(LEDpin, HIGH);
      }
    }
    else {  // if the switch is not pressed:
      // but the switch was pressed last
time through the main loop:
      if (lastSwitchState == 1) {
        // stop the last note played:
        noteOn(0x90, lastNotePlayed, 0x00);
        digitalWrite(LEDpin, LOW);
      }
    }
    // save the state of the switch for next
time
    // through the main loop:
    lastSwitchState = currentSwitchState;
  }
  // plays a MIDI note. Doesn't check to see
that
  // cmd is greater than 127, or that data
values are less than 127:
  void noteOn(byte cmd, byte data1, byte
data2) {
    midiSerial.write(cmd);
   midiSerial.write(data1);
    midiSerial.write(data2);
   //prints the values in the serial monitor
so we can see what note we're playing
    Serial.print("cmd: ");
     Serial.print(cmd);
```

```
      Serial.print(", data1: ");
  Serial.print(data1);
     Serial.print(", data2: ");
   Serial.println(data2);
}
// Blinks an LED 3 times
void blink(int howManyTimes) {
  int i;
  for (i=0; i< howManyTimes; i++) {
    digitalWrite(LEDpin, HIGH);
    delay(100);
    digitalWrite(LEDpin, LOW);
    delay(100);
  }
}
```

In example 2, we have assigned the FSR sensor to control the pitch and the digital switch to start and stop the note. Now that you have a basic grasp of how MIDI can be implemented with Arduino try building on this example by making your musical instrument and add different sensors for creating a more complex interaction. What if we wanted to create a stand-alone musical instrument that didn't require an external MIDI-based sound-generating device? We can do so by building a small amplification circuit and an Arduino.

You will need the following items:

- Arduino Uno
- Jumper wires
- 100 ohm resistors
- Two photocell variable resistors (or one Lux sensor with built-in resistor)
- An 8 ohm speaker

Prepare your Arduino board like the first example by connecting the 5V and ground to the breadboard. Next, let's connect the necessary parts to our breadboard as shown in the illustration.

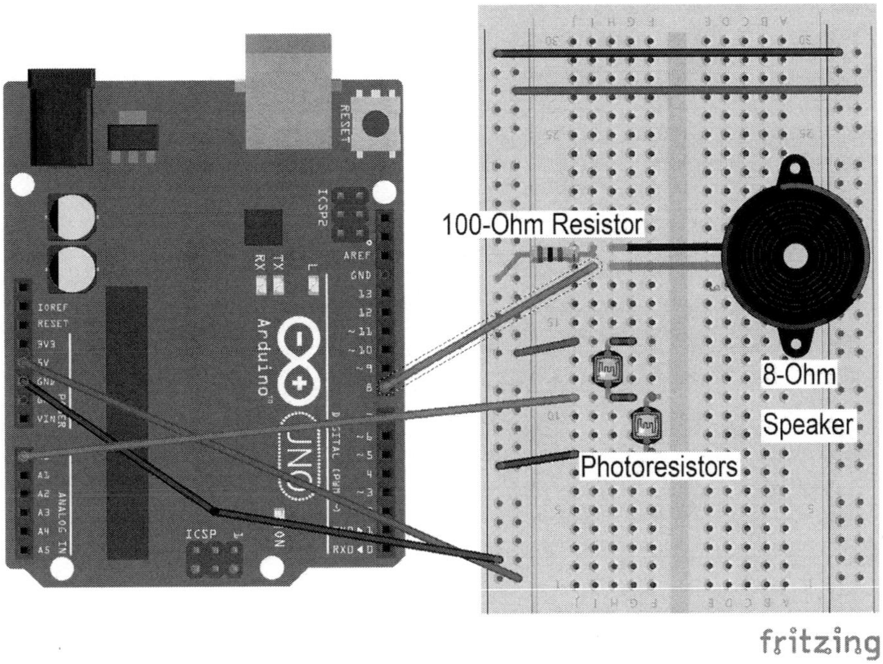


**Figure 1.8** Illustration Showing How Two Photoresistors and an 8 ohm Speaker Are Connected to Arduino.

The two photoresistors are connected to A0 in a voltage divider circuit. You can replace the two photoresistors and use a single Lux sensor instead to produce the same result. Our 8 ohm speaker is connected to digital pin 8 or any digital pin, but make sure to keep track of the pin number so you can modify the code. Once you have your circuit built, we need to calibrate the range of our photoresistors so we can obtain the minimum and maximum values. Those values will then be used to map to a range of frequencies we want to control. Run the following code and observe the minimum and maximum range by looking at the incoming values in the serial monitor on your Arduino IDE. Try covering the sensor and then exposing it to light to take note of the values produced.

```
void setup() {
  Serial.begin(9600);      // initialize
serial communications
}
```

```
void loop()
{
  int analogValue = analogRead(A0); // read
the analog input
  Serial.println(analogValue);      // print
it
}
```

Once you have determined your input range, let use the map function to sweep a frequency range from 100 to 1000 Hz. For this example, our minimum and maximum sensor values are 100, 800. Yours may be different so adjust the code accordingly to match your setup.

```
void setup() {
    // nothing to do here
 }
 void loop() {
   // get a sensor reading:
   int sensorReading = analogRead(A0);
   // map the results from the sensor read-
ing's range
   // to the desired pitch range:
   float frequency = map(sensorReading, 100,
800, 100, 1000);
   // change the pitch, play for 10 ms:
   tone(8, frequency, 10);
 }
```

Once you have uploaded the code, try moving your hand over the light sensor. You will notice the frequency change in value from 100 Hz to 1000 Hz since we used our map() function to set this range. We also used the tone() command to change the pitch for a specific duration. You can build a simple theremin-inspired instrument based on this example.

To add more variety, we can replace our light sensor and add three FSR sensors that will act as our keys as shown in Figure 1.10. To add more variety, we can replace our light sensor and add three FSR sensors as shown in the illustration.

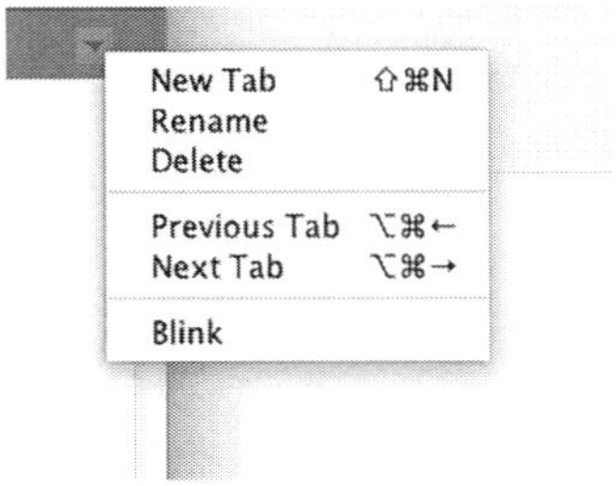


**Figure 1.9** Clicking on the Arrow Allows for Addition of a New Tab.

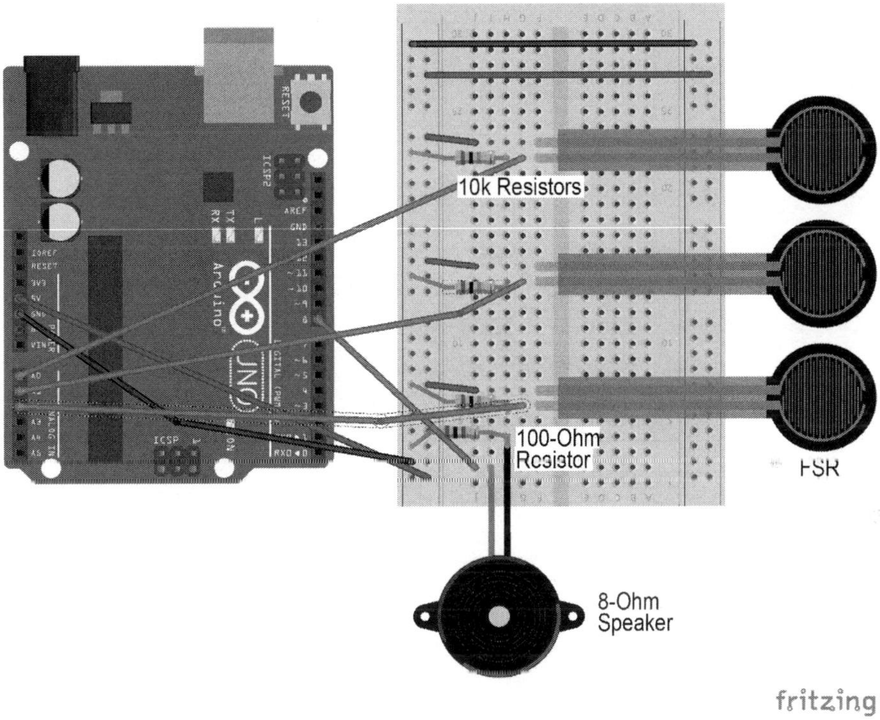


**Figure 1.10** Configuration showing how to connect three FSR sensors to the analog inputs of an Arduino.

For this example, we will use some predefined pitches to create a standard Western diatonic scale. To add the predefined pitches in your sketch, click on the arrow as shown in the illustration and add a new tab. Then open the accompanying file called pitches and copy paste the code in the newly created tab. In your main sketch, copy the code below.

```
#include "pitches.h"4

const int threshold = 10;      // minimum
reading of the sensors that generates a
note

const int speakerPin = 8;      // pin number
for the speaker

const int noteDuration = 20;   // play notes
for 20 ms

// notes to play, corresponding to the 3
sensors:

int notes[] = {

  NOTE_A4, NOTE_B4,NOTE_C3};

void setup() {

}

void loop() {

  for (int thisSensor = 0; thisSensor < 3;
thisSensor++) {

// get a sensor reading:

int sensorReading = analogRead(thisSensor);

// if the sensor is pressed hard enough:

if (sensorReading > threshold) {

// play the note corresponding to this sen-
sor:

      tone(speakerPin, notes[thisSensor],
20);

    }

  }

}
```

Advanced users can send the FSR sensor values directly to Processing or Max/MSP to generate much more complex sounds. We can also make any conductive object into an input so you can create more natural user interfaces as input. There are several ways you can accomplish this.

### 1.10.2 *Natural User Interface*

Interfaces that provide the ability to interact using modalities like touch, gesture or voice are often referred to as natural user interfaces (NUI). Using NUIs allows for fun and easy ways for users to interact with technology. Specifically for music, NUIs open the door for performers to control musical parameters in new ways and conform to behaviors we are already used to, therefore reducing the cognitive load of learning something new (Mortensen 2018). Imagine modifying several controls of a synthesizer simply by waving your hand. Using NUIs we can create interfaces that feel less technological and more experiential. For example, you are given the task of creating an interactive meditative experience for a client. What if by merely touching water we can produce sound? Wouldn't that impress them! And it is entirely possible using the appropriate technological tools. One caveat is that people tend to make interfaces overly complicated and unintuitive, such as by having multiple gestures control something. After a while, it may no longer feel natural. Therefore, it is essential to create a clear user experience by mapping the points of interaction and what they control. Here are some suggestions for creating a meaningful NUI:

- Take advantage of a user's existing skill set.
- Reuse common human skills to expedite the learning process. For example, tapping is an easy way to control the rhythm or BPM of a song.
- If you are designing a musical instrument, take into account your user. Is it for a novice or skilled musician?
- Make the behavior easy. Simplifying the behavior by breaking it down into tiny steps can reduce the learning curve if your user is unfamiliar with the interface.
- Imitating interaction can make the feedback easy to understand for the user. For example, raising your hand automatically raises the volume or pitch of a sound.

Let's learn about some of the existing devices that allow for the creation of NUIs.

### 1.10.3 *Makey Makey*

This handy device has a series of connectors that act as touch sensors. It is a plug and play unit that does not require an Arduino. It has a USB connection that connects directly to your computer and controls any software that uses a mouse or keyboard for input. The capacitive pins on the board are set up to work with alligator clips. You can attach one end of an alligator to an input on the Makey Makey and another to a conductive object like a fruit. Now the fruit becomes the controller. So now you can have a bowl of fruit that can be mapped to samples triggered in a music software like Ableton Live. Check out a variety of projects created using Makey Makey.[5]

### 1.10.4 *Capacitive Shields*

There are capacitive shields like the Bare Conductive Touch Board for the Arduino that provide the same functionality as the Makey Makey with more features and functionality. We can use conductive ink or pens to make paper-based controllers or build soft-circuit buttons using conductive thread and fabric. The Bare Conductive Touch Board has a built-in audio interface that allows for triggering of sound effects and music directly from the board without additional circuitry.

**Figure 1.11** Adafruit Circuit Playground Has Built-In Capacitive Pads and Multiple Sensors for Creating Wearable and Interactive Projects.

Manufacturers of electronic parts Sparkfun and Adafruit offer their own capacitive shields that allow for converting objects into natural user interfaces.

### 1.10.5 Open Sound Control

So far we have discussed ways of working with the Arduino platform to capture data from sensors and convert that into musical data like MIDI. One of the ways we can improve the usability of such devices is by having the ability to wirelessly communicate the information to our devices.

The research team at UC Berkeley Center for New Music and Audio Technology (CNMAT) created a protocol called Open Sound Control (OSC). The use of OSC enables wireless communication using network technology among various devices such as computers, synthesizers and multimedia devices like Wii Remote from Nintendo. Data is transmitted in real time and one of the advantages of OSC is that we are not limited to the range of MIDI (0–127). We can decide how to map the data so we can control a wide range of parameters. Here are some of the features of OSC according to the CNMAT organization:

- Open-ended, dynamic, URL-style symbolic naming scheme
- Symbolic and high-resolution numeric argument data
- Pattern matching language to specify multiple recipients of a single message
- High-resolution time tags
- "Bundles" of messages whose effects must occur simultaneously
- Query system to dynamically find out the capabilities of an OSC server and get documentation

The basic settings that we need to know to get communication between devices working are the following:

- Host and local IP addresses: All the devices that are meant to send OSC messages must be connected to a wi-fi network. Each device has to be configured by inputting the appropriate IP addresses for the host and local devices.
- Port information: Data is sent and received through ports. Therefore, each device must use the same port for incoming and outgoing data. The standard ports used are 8000 for outgoing and 9000 for incoming. However, the user has the ability to change these values as long as they are changed on all devices used for communicating OSC messages.

Software manufacturers have recognized the power of OSC, and many, like Logic Pro, Native Instruments, Ableton and Max Msp, now natively support it.

Now we can build our own wireless interfaces on our mobile devices that control our musical equipment.

### 1.10.6 TouchOsc

One of the most popular tools for creating digital musical control interfaces is TouchOsc by Hexler. The TouchOsc editor, which is installed on a computer (Mac, Windows or Linux), is used to build the interface for an iPad, iPhone or Android devices. The editor provides the ability to add and organize buttons, sliders, dials and menus that can be assigned to control specific musical parameters. Once the interface is designed, it is wirelessly uploaded to the mobile device. To do so, make sure the host and local IP addresses are entered correctly. Once the editor is successfully uploaded to the mobile device, the user can control musical parameters of another device that recognizes OSC messages.

### 1.10.7 Osculator

Osculator is a handy tool for "bridging" multiple devices together. For example, Osculator provides the ability to use the Wii remote to control synthesizer parameters via gestures. In fact, any Bluetooth device with built-in sensors can be used to convert data to either MIDI or OSC messages using Osculator. The software comes with examples to quickly get started connecting multiple devices to control each other. Currently, Osculator works on Mac computers only.

### 1.10.8 OSC for Arduino

Using OSC in Arduino allows for building custom hardware controllers. In order to do so, we would need to use wi-fi or Bluetooth enabled devices to send the data wirelessly. There is an OSC library available for Arduino, which we can use to convert sensor data into OSC messages. Now we can send the messages to any software that can receive OSC data.

There are also wi-fi shields available for the Arduino Uno, but recently there has been a surge in powerful wi-fi microcontrollers that can be used to build IOT devices. The Particle Photon is an example of a wi-fi enabled microcontroller that works like a regular Arduino. One significant advantage of such devices is the ability to code in the cloud. That means we can wirelessly update, track and grab data from our devices from anywhere!

**Figure 1.12** (Left) Wi-Fi Enabled Chip Called the Photon That Can Allow for Cloud-Based Arduino Programming; (Right) The Adafruit Bluefruit That Allows for an Easy Way to Have Wireless Communication via Bluetooth.

We can create musical performances without having to be present in the same space. Using the cloud, we can modify sequences of music or use sensor data from another part of the world to control our musical devices. The possibilities are endless with IOT-enabled microcontrollers.

One of the most versatile wi-fi enabled microcontrollers is the ESP8266. This tiny module packs a lot of power because the chip can be directly programmed using its software development kit (SDK). Therefore, you can have an Arduino-compatible version of the ESP8266 at a fraction of the cost. There is a large online community that provides support for the ESP8266.[6]

## 1.11 Conclusion

In this chapter, we introduced the concept of physical computing: the ability to augment the musical and sound design experience by using sensors, microcontrollers and actuators to provide an easy way to modify our sounds. The Arduino open-source platform was used to explain how we can program microcontrollers to take data from sensors and manipulate the information. As we move towards wireless controls, the possibilities become endless by using sensors that can communicate wirelessly to multiple software platforms to control both sound and visual information. Any technology comes with constraints that need to be accounted for when we are developing an interactive project. Therefore, it is essential to precisely map out the user interaction and the type of experience we want to create. Doing so allows us to determine the most appropriate types of sensors to use and the kinds of algorithms we need to work with to make sure our feedback is clear and optimized for the user. There is a large online community that supports the Arduino platform, making it easy for a novice user to learn and develop projects quickly. Furthermore, Arduino itself provides powerful examples to help expedite the learning process.

## Notes

1 www.sparkfun.com
2 www.adafruit.com
3 www.arduino.cc/en/Main/Software
4 This header file can be downloaded from http://bit.ly/pitchesheaderfile
5 https://labz.makeymakey.com/remixes
6 www.esp8266.com

## References

Collins, N., 2014. *Handmade Electronic Music: The Art of Hardware Hacking*. Hoboken, NJ: Taylor and Francis.

Mortensen, D., 2018, February. *Natural User Interfaces—What Are They and How Do You Design User Interfaces that Feel Natural?* Viewed April 5, 2018 <www.interaction-design.org/literature/article/natural-user-interfaces-what-are-they-and-how-do-you-design-user-interfaces-that-feel-natural>.

Noble, J., 2012. *Programming Interactivity: A Designers Guide to Processing, Arduino, and OpenFrameworks*. Beijing: O'Reilly.

O'Sullivan, D., and Igoe, T., 2010. *Physical Computing: Sensing and Controlling the Physical World with Computers*. Mason, OH: Course Technology.

Simeonidis, B. P., 2014, August 23. *MIDI Output Using an Arduino*. Viewed February 5, 2018 <https://itp.nyu.edu/physcomp/labs/labs-serial-communication/lab-midi-output-using-an-arduino/>.

2

# The Electronics of Microphones and Loudspeakers

Jesse Seay

## 2.1 Introduction

This chapter is intended as a practical introduction to audio concepts for embedded audio. It assumes the reader has some experience with and understanding of microcontrollers, beginner-level DC electronic circuits and circuit diagrams.[1] We will informally review the fundamentals of electricity and acoustics to give the reader a basis for understanding microphones, loudspeakers and the AC signals between them. We will then consider various types of microphones, loudspeakers and amplifiers that are suited for embedded applications.

This chapter is not a survey of specific models. Instead it aims to deepen the reader's understanding of how these components work and introduce terminology one is likely to encounter while perusing and pursuing the ever-changing array of available options. The chapter also includes practical tips on improving the performance of each component, based on the author's experience. Users will find that following these tips can significantly improve the loudness and fidelity of their audio. Understanding these tips can also mean the difference between breaking equipment and getting years of reliable use from it.

## 2.2 Transducers

Microphones and loudspeakers are both transducers—devices that convert energy from one form to another. Your toaster is a transducer, converting electric energy to heat energy. Your car converts gas to mechanical energy. Microphones and loudspeakers convert energy between acoustic and electric forms. In order to understand them, it's helpful to understand how energy behaves in these forms.

## 2.3 Acoustic Energy

Imagine striking the head of a drum. The drum surface sinks in, then bounces back out, vibrating back and forth until it comes to rest in its original position. This vibration moves the air molecules both in front and behind the drumhead. When molecules are being pushed by the drumhead, we call it compression, as the molecules are forced closer together. When the drumhead bounces back, the molecules move away from each other and it is rarefaction, the opposite of compression. A single cycle of a sound wave consists of a compression phase followed by a rarefaction phase. The number of these cycles occurring per second is a wave's frequency, which we hear as "pitch." How much the molecules are displaced is the wave's amplitude; we hear it as volume or loudness.

Note that the individual molecules don't go very far. Like the drumhead, which comes to rest in its original position, the air molecules eventually return to their original positions. The energy that moves through them, however, can go long distances, passing through air molecules until it reaches the ears of the listener. We call energy in this form "acoustic energy."

## 2.4 Electricity

We often discuss where electricity is (in a battery, in a wall outlet), or what it does (powers your hairdryer). We rarely discuss what it *is*. Electricity is a stream of moving electrons.

Electrons, protons and neutrons are the subatomic particles that make up atoms. We describe protons as being positively charged, and electrons as being negatively charged. Electrons and protons attract each other. One positive proton plus one negative electron equals a neutral charge. Thus, a neutral atom contains an equal number of protons and electrons.

It's possible to excite electrons and bump them out of an atom, or for extra electrons to get dropped into an atom. When that happens, the extra electrons attempt to flow from where there are too many electrons to where there are not enough electrons. Once there is an equal number of electrons to protons, the current of electrons ceases to flow. When we generate electricity, we are exciting electrons to move between atoms, en masse.

Electricity is defined by its current and voltage. You can think of current as *how many* electrons are flowing and voltage as how excited the electrons are. Current is measured in amps. Voltage is measured in volts. When we multiply the volts and amps of an electric circuit together, the result, measured in watts, is power, the ability to do work. (We'll discuss watts later in the section on power amplifiers.)

We conventionally think of electricity as flowing from a point of high voltage to a point of low voltage.[2] You have seen these points on the terminals of a battery: high voltage is marked "+" and low voltage is marked "-". These points are also referred to as positive and negative, or as power and ground.

When current flows continuously from positive to negative, it's called direct current (DC). All batteries provide DC. Electricity can also take the form of alternating current (AC), in which the electrons reverse direction periodically. The difference between DC and AC is akin to the difference between brushing one's hair and brushing one's teeth. A hair brush typically travels down one's hair in only one direction. A toothbrush typically moves back and forth. The toothbrush bristles don't travel far but still get the job done because the brush alternates directions, moving forward and backward. Likewise, the electrons in AC alternate between positive voltage (moving forward) and negative voltage (moving backward). This approach works well for carrying electricity great distances, which is why power plants use AC to deliver electricity to the outlets of your house. It's this type of current that is delivered by microphones.

## 2.5 Microphones

Microphones convert acoustic energy (air molecules in motion) into electric energy (electrons in motion). The acoustic fluctuations in pressure become changes in electrical voltage. The air molecules' compression creates a positive change in voltage while rarefaction creates a negative change in voltage. The louder the sound, the more electricity generated. In this way, information about a sound wave is captured and converted to an audio "signal," which can be sent to the input pin of a microcontroller. Different types of microphones accomplish this in different ways. We'll consider two types here: piezo mics and electret mics. Both are well suited for embedding due to their small size, low cost and ease of use.

## 2.6 Piezo Discs

A piezoelectric (or "piezo") disc is an easy, inexpensive way to add a sound sensor to a microcontroller. It consists of a brass disc, with a thin layer of crystal attached to one side. The crystal produces voltage in response to mechanical stress. This makes it a great method for detecting knocks or taps. It will also detect sound traveling through other objects. This works best when the brass face is mounted flush on the surface of the sound

**Figure 2.1** Electret Mic and Piezo Disc.

source, using glue or double-sided tape. For this reason, flat surfaces are ideal, to maximize surface area contact.

Light, stiff materials are most effective since heavy, flexible material absorbs vibration. Piezo discs work particularly well mounted on thin wood (hence their popularity as acoustic guitar pickups). They also do fine with thin plastic or metal. Attach the disc to the bottom of a thin plastic cup or empty tin can and it becomes an electric tin-can telephone. Piezo discs can also be worn on the body. Slide the disc under a wide elastic band fitted as a snug choker and it detects the human voice, as well as coughs, gasps, swallows and other epiglottal disturbances. (Turn the crystal side out to avoid skin contact with the solder points.)

The amount of current that piezo discs produce is very small. For this reason, keep the wires connecting the disc to the microcontroller short as longer wires may pick up stray electrical noise that interferes with the piezo disc's weak signal. Twisting the wires together can also help. Even with short wires, noise can still interfere with the piezo disc's signal. To minimize this interference, and to protect the microcontroller from the piezo disc's high voltage spikes, connect a 1 Mohm resistor between the input pin and ground, attaching the resistor as close to the microcontroller as possible (Lindblom 2016). The resistor removes noise by providing a path to ground for it (this also lessens the voltage of the signal). If more noise reduction is needed, decrease the value of the resistor.

Piezo discs are difficult to solder. Save yourself time and frustration by using discs with pre-attached wires. If you do need to solder to a wireless disc, dab a blob of solder flux onto the disc before tinning it. Remove the sticky flux afterward with rubbing alcohol or a flux remover. Don't be discouraged if you melt a hole in the crystal with the soldering iron. Just pick another spot and try again.[3]

## 2.7 Electret Microphones

An electret condenser mic uses a small metal capsule containing two parallel capacitive plates to convert sound waves into electric current. Electret mics capture sound with great fidelity and are popular as a means for nice, clean audio recordings. As with all mics, this requires care in placing and mounting the microphone in space, just the right distance from an acoustic source. *Embedding* an electret mic can alter the signal in multiple ways. If the mic is placed inside a solid enclosure, the sound may be muffled, particularly high frequencies. Microphones worn on the body can pick up noise from the wearer's movements, especially if the mic or its cable is handled. If the sound source moves away from the mic, the signal strength decreases exponentially.

Nonetheless, an embedded electret mic can be a useful sensor. Like the piezo disc, it produces a signal voltage that the microcontroller's analog-to-digital converter (ADC) converts to an integer based on the sound's amplitude (loudness). Unlike the piezo disc, the electret mic's signal voltage is so low, it needs a preamplifier to connect to the board.

A preamplifier, or preamp, increases the voltage of a weak signal so that it can be usable in other parts of a circuit. (A preamp is different from a power audio amplifier, which is designed to drive a loudspeaker. We'll return to power amps later in this chapter). There are a wide range of preamp designs, for use in a variety of applications, some unrelated to audio. Our purposes require a preamp fast enough to process the high frequencies of an audio signal, with an output between 0 and 5 V, the safe operating range for the microcontroller. For convenience, it should run off the 3.3 V/5 V provided by the board. Vendors such as Adafruit and Sparkfun sell electret mics hardwired with an appropriate preamp, mounted on a breakout board complete with power supply. These mics have the additional advantage of adding DC bias to the electret mic's signal.

## 2.8 DC Bias

A microcontroller like the Arduino expects to read signal voltages between 0 and +5 V but a microphone creates voltage that swings between positive and negative voltage. The microcontroller will only see the positive half of this signal. This is what occurs with the piezo disc—the microcontroller ignores the negative voltage from the piezo. The ADC maps voltage between 0 V and +5 V to an integer between 0 and 1023. In the Arduino programming language, this integer is returned with the AnalogRead function. Let's consider how AnalogRead maps an audio signal that swings between +1 V and -1 V.

In Figure 2.2, AnalogRead samples the signal eight times over one cycle. While the voltage level changes continuously, AnalogRead returns the same value, 0, for all negative voltage levels.

What if we shifted the voltage of the entire waveform up by one volt? This is illustrated in Figure 2.3. Now the signal consists entirely of positive voltage and AnalogRead returns analogous values for the entire waveform. Shifting an entire AC signal this way is called DC bias (or DC

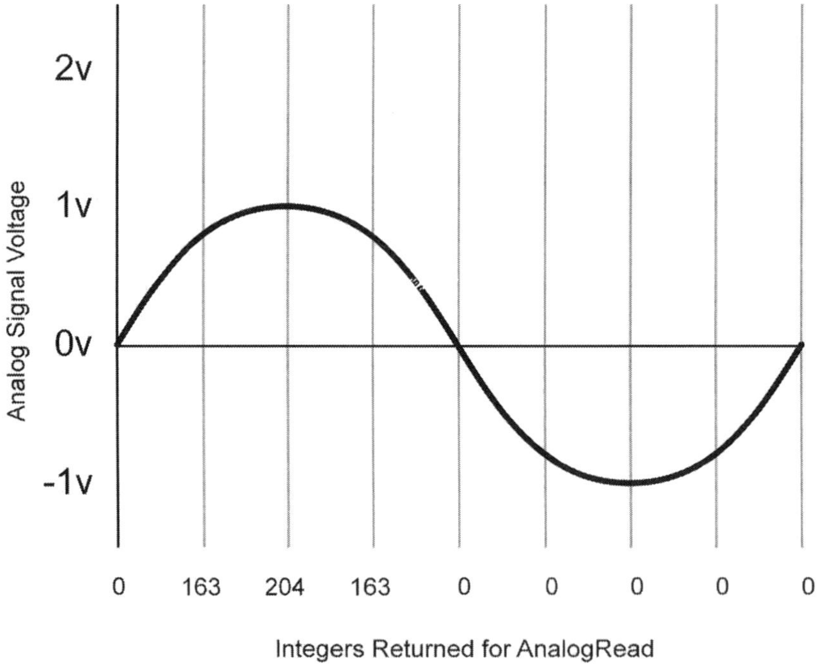


**Figure 2.2** Integers Returned by AnalogRead for a Fluctuating Signal Voltage.

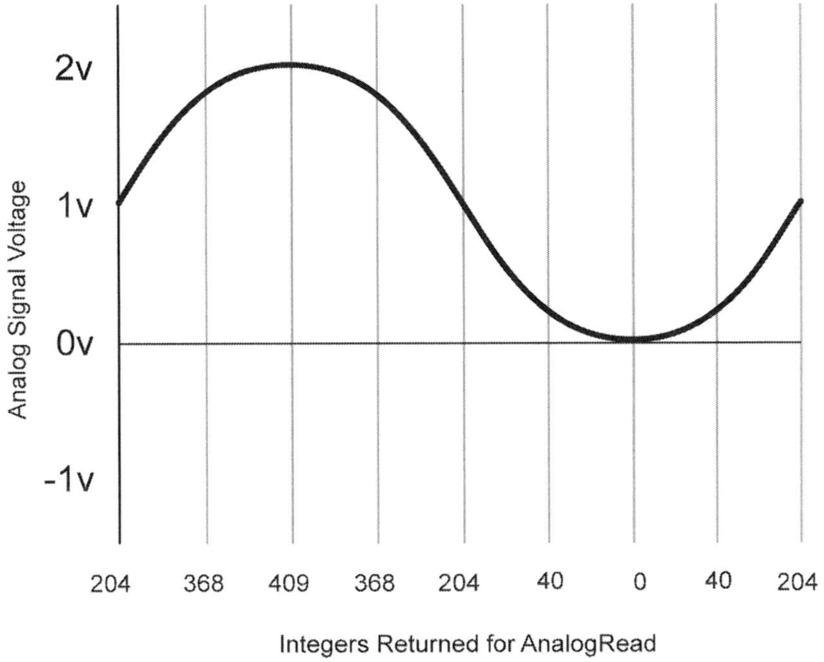


**Figure 2.3** Integers Returned by AnalogRead for a Fluctuating Signal Voltage With DC Bias.

offset). The electret mic's preamp adds a DC bias for us. While the signal in our example is biased 1 V, preamps will typically bias the signal to half of the voltage of the preamp's power supply. For boards running off the Arduino's 3.3 V or 5 V power, this would result in a bias of 1.65 V or 2.5 V, respectively.

Without DC bias, AnalogRead returns 0 for silence. Any number above 0 indicates the presence of sound (or noise). The higher the number, the higher the amplitude.

With a 1 V bias, AnalogRead returns 204 for silence. A high amplitude can be indicated by either a very higher integer or a very low integer. This necessitates a change in code.

We could calculate amplitude by subtracting the offset value (204 with a 1 V bias) but this value would change if the voltage of the power supply changes. Adafruit Industries demonstrates a useful approach in the Arduino code it provides for its Electret Microphone Amplifier (pictured in Figure 2.1). Here the amplitude is calculated by subtracting the minimum value from the maximum, a measurement referred to as peak-to-peak (Earl and Ada 2013).

## 2.9 Loudspeakers

Like an electric fan, a loudspeaker's job is to move air molecules. The bigger and more expensive a fan is, the more air it moves.

A tiny clip-on fan is perfect for one person.
A twelve-inch table fan works for a living room.
A heavy three-foot floor fan cools a large industrial shop.

This principle applies to loudspeakers as well. Bigger, more expensive loudspeakers are generally louder and have more bass. (They also require more power and bigger amplifiers.) To choose the right loudspeaker, know your needs beforehand and keep your expectations realistic. Here are factors to consider when choosing your loudspeaker:

- Frequency response: what is the frequency range of your sound? How important are low frequencies?
- Volume: how far away is the listener? How loud is the environment around your loudspeaker?
- Design limits (size, weight, budget): how big and heavy can the loudspeaker be? How much can you spend?

## 2.10 Frequency Response

Frequency refers to the speed of vibration. We hear frequency as *pitch*. High frequency vibrations sound high-pitched, low frequencies sound low-pitched. Frequency is measured in hertz, defined as the number of vibrations (cycles) per second.

Different mics and loudspeakers work better or worse for different frequency ranges. To illustrate this, use a tone generator app to sweep through the human hearing range (20 Hz to 20 kHz) on your computer or phone. Listen to it with the built-in loudspeaker. Now connect it to external loudspeakers: plug-in computer loudspeakers, home stereo, car stereo, whatever you can find. Note the differences. You'll find that a bigger and more expensive sound system produces louder sound with more bass.

It's important to note that "loud enough" is subjective and depends on context. If you are unsure how much treble or bass you require, try playing your project's audio through a stereo system with bass and treble control knobs.

- What happens when you turn down the bass?
- What happens when you turn down the treble?

If you don't need much bass, small loudspeakers should work fine.

## 2.11 Volume

Trying to compete with other loud sounds is usually futile. One man's noise is another man's music. If you get louder, they get louder, until finally there is so much noise, no one can stand it and your audience leaves.

Before resorting to volume-control violence, try this:

- Move the loudspeaker closer to the listener's ear (or vice versa).
- Remove or mute other sounds. This may be easier said than done, but if you can do it, do it.
- Deaden the acoustics of the space. You can't compete with your own echo. An empty room with hard floors, bare walls and big glass windows sounds terrible. Sound bounces off all those hard, flat surfaces. The addition of a wool rug, a plush couch, thick curtains and a few bookshelves filled with books improves the listening experience considerably.

## 2.12 Design Limits

Design limits such as size, weight and budget vary by project. The following annotated list of types of loudspeaker drivers includes details for choosing a solution appropriate to your specific needs. Note that loudspeaker drivers are sometimes referred to as loudspeakers. When we use the term "driver," we mean the transducer that is producing the sound, separate from its enclosure.

These driver options are available from vendors that sell electronic components. The number of available models can be overwhelming. Some vendor websites offer detailed search filters that will narrow the search, based on your design criteria.[4]

## 2.13 Piezo Discs

These vibration sensors also work surprisingly well as loudspeakers. Inexpensive and lightweight, piezo discs are a great choice for high frequency sounds, buzzes and beeps (like those produced by the Arduino "tone" function). For this reason, vendors sometimes list them as buzzers and package them inside of plastic housings that improve the sound and provide mounting tabs. You can improve the frequency range and volume of a bare disc by gluing it to a stiff, flat surface. In Figure 2.4, it is attached to corrugated plastic. Unlike the other loudspeaker drivers presented here, piezo discs can be connected directly to a microcontroller's output pins

**Figure 2.4** Piezo Disc Attached to Corrugated Plastic.

**Figure 2.5** A Variety of 8 Ohm Dynamic Loudspeaker Drivers.

without a power amplifier. Piezo discs are also the least expensive option on this list, especially when purchased in bulk.

## 2.14 Dynamic Loudspeakers

Dynamic loudspeakers come in a range of sizes. Small models, a few inches in diameter, are a popular solution for tabletop projects, where the listener is within a few feet. The most conventional driver on this list, it is also the easiest option for high-fidelity playback of audio files. Some models include a metal frame with screw holes for easy mounting. Others can be held in place with hot glue. We'll discuss mounting later in the chapter, in the section on enclosures.

## 2.15 Surface Transducers

Also called audio exciters, these "hide" your loudspeaker in plain sight. Surface transducers vibrate any object to which they are attached,

**Figure 2.6** Surface Transducer.

Figure 2.7 Bone Transducer.

turning the object itself into a speaker cone. The choice of object affects the sound. See this chapter's section on diaphragms for a guide to choosing materials.

## 2.16 Bone Transducers

The bone transducer is a curious variation on the surface transducer. This type of exciter is designed to vibrate the listener's skull. The sound bypasses the outer ear entirely and the audio is only heard by the affected listener. The listening experience is not high fidelity, but it is certainly unique, and a little uncanny. Take care when mounting it. The wire connections are particularly fragile at the solder points and the use of strain relief is recommended.

## 2.17 DIY Loudspeakers

There are a number of functional DIY loudspeaker designs documented online. Some involve a drinking cup, copper wire and magnets.[5] Others use traditional craft methods to produce e-textile loudspeakers from cloth or paper.[6] Building your own DIY loudspeaker is not too difficult, particularly if you understand how a conventional dynamic loudspeaker works.

## 2.18 How a Dynamic Loudspeaker Works

Loudspeakers are based on a simple principle of electromagnetism. Electric current running through a copper wire produces an electromagnetic field. If this electrified wire is placed within an existing magnetic field, the

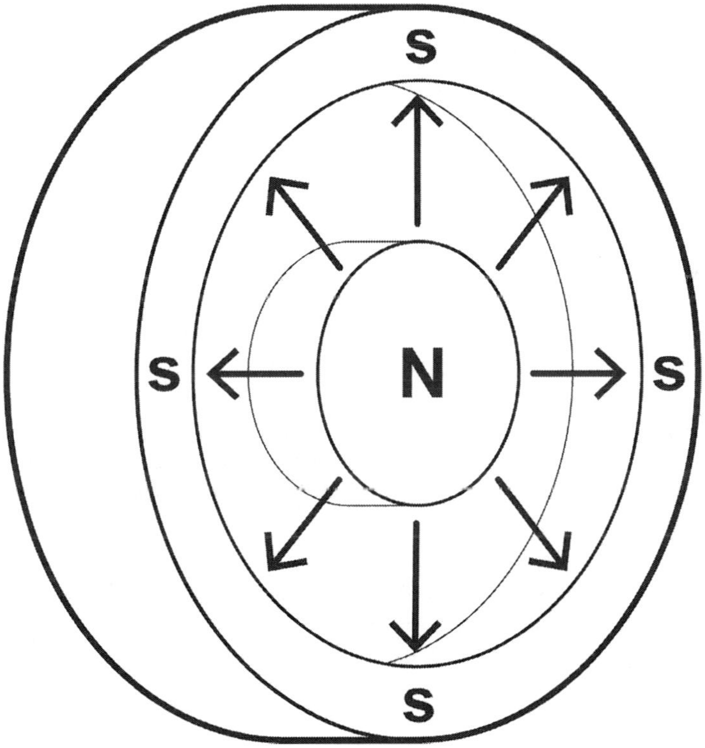


**Figure 2.8** Loudspeaker Magnet.

wire experiences physical force, and that force is strong enough to produce sound waves.

A dynamic loudspeaker uses a specially mounted ring magnet to supply the magnetic field.

The center "pole piece" is the magnet's north pole (labeled "N") while the exterior circumference forms the south pole (labeled "S").

The magnetic fields radiate out from the pole piece, across an o-shaped slot. The audio signal provides the current. It runs through our loudspeaker's *voice coil*, a single, coiled length of wire, which fits into the o-shaped slot. The magnetic fields cause the coil to push outward, vibrating the loudspeaker's cone-shaped diaphragm. Like a drumstick landing on the head of a drum, the force of the coil moves the cone, which moves the air molecules, producing sound.

Dynamic loudspeakers are found in everything from your home stereo to your earbuds. When creating your own embedded loudspeakers, a few elements deserve special attention for improving sound quality.

## 2.19 The Diaphragm

In both microphones and loudspeakers, the *diaphragm* is the thin piece of material that serves as the physical connection between the air molecules and the electronics. In microphones, sound waves set the diaphragm in motion to make electricity. In loudspeakers, electricity sets the diaphragm in motion to make sound waves.

As illustrated in Figure 2.9, off-the-shelf loudspeakers use a paper or plastic cone as the diaphragm. DIY loudspeakers and audio exciters require the user to attach an appropriate diaphragm. The choice of diaphragm has a noticeable impact on a loudspeaker's sound. The most efficient diaphragms are stiff and lightweight, with large, flat surface areas.

To find an appropriate diaphragm, test the material as a "fan." Any material that makes an effective impromptu "hand fan" will move air molecules for audio. Given a Styrofoam plate or a loose piece of leather, the Styrofoam plate is the clear winner. In fact, Styrofoam works particularly well. For a more rugged option, experiment with corrugated plastic (the stuff of realtor yard signs, available at art supply shops).

Some lightweight materials that lack the required stiffness can be effective if they are stretched taut across a frame, like a drumhead. Try Mylar plastic or cooking parchment. Soft fabrics and glass vessels are typically less effective as loudspeakers, or as hand fans for that matter (which is not to say that you shouldn't experiment).

The attachment of the voice coil to the diaphragm is also important. Care should be taken to maximize surface area contact between the two.

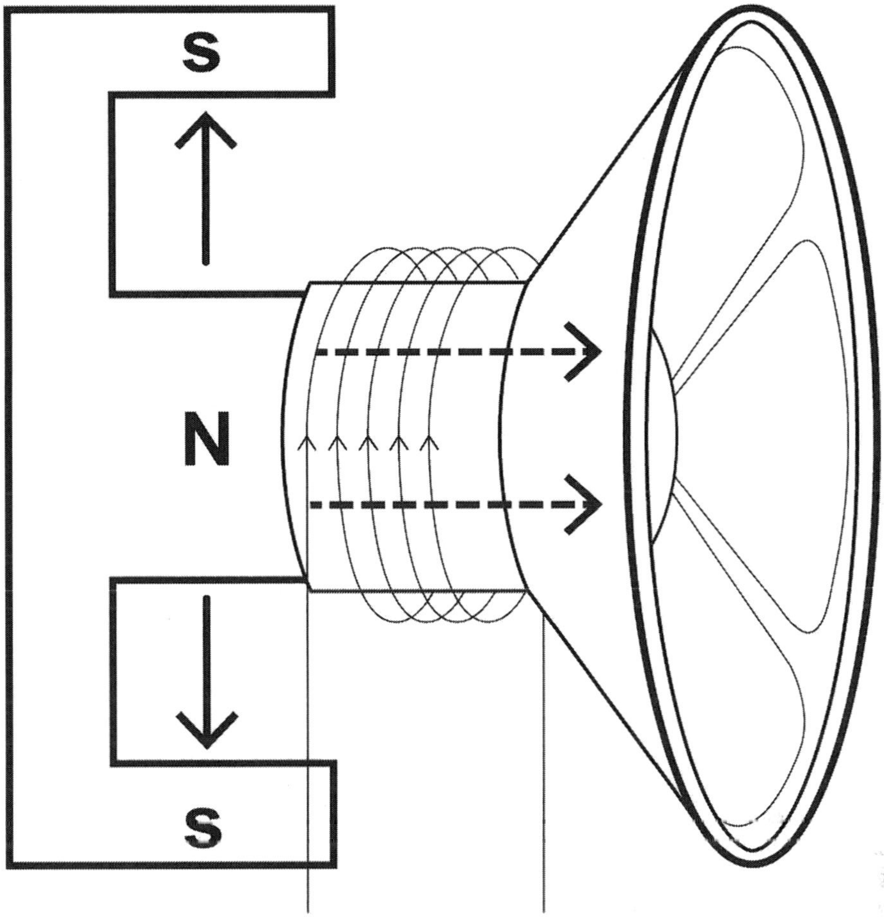


**Figure 2.9** Dynamic Loudspeaker Diagram With Arrows Illustrating the Magnetic Fields (thick arrows), Current Flow (thin arrows) and Physical Force (dashed arrows).

In audio exciters (which contain the coil and magnet of a traditional loudspeaker, but no cone) the coil is topped with a flat plastic ring, designed to attach to a flat surface. Some exciters have a pre-applied adhesive on the surface of the ring for convenience. If yours does not, use a glue suited to your materials, poster tack or double-sided tape (the latter have the advantage of being removable). For best results, cover the entire surface of the ring. A piezo's brass disc serves as its diaphragm. Improve the sound by attaching the brass side to a larger diaphragm (as seen in Figure 2.9). For both piezo discs and voice coils, the attachment should be firm, as anything wobbly will rattle.

These guidelines are useful for e-textile loudspeakers as well, where the textile is the diaphragm. Wire (or conductive thread) is stitched, knitted, woven or glued to the textile, and placed over magnets.

Here the flatness of the textile presents a design challenge, as the vectors of current, magnetic field and movement run perpendicular to each other, creating an X, Y and Z axis. You can visualize these vectors using your left hand, a technique known as Fleming's Left Hand Rule.

- The first finger is the magnetic field, flowing from north (knuckles) to south (fingertip).
- The middle finger is the electric current traveling from positive (knuckles) to negative (fingertip).
- The thumb is the physical force, the direction the wire moves.

A textile's lack of depth limits the third axis. Therefore, the arrangement of the magnets is key. To be audible, the wire must be located between both a north and a south pole. This can be accomplished by gluing two magnets, side by side, to a stiff backing, with opposite poles facing out. For a ready-made solution, use a scavenged hard drive magnet or the ends of a horseshoe magnet.[7]

E-textile loudspeakers make for loud tapestry, but there are limitations to wearing them. The volume of the loudspeaker depends on the strength of the magnet. However, wearing a powerful magnet on the body can result in injury and is not recommended.

**Figure 2.10** Wires Knitted Into a Textile.

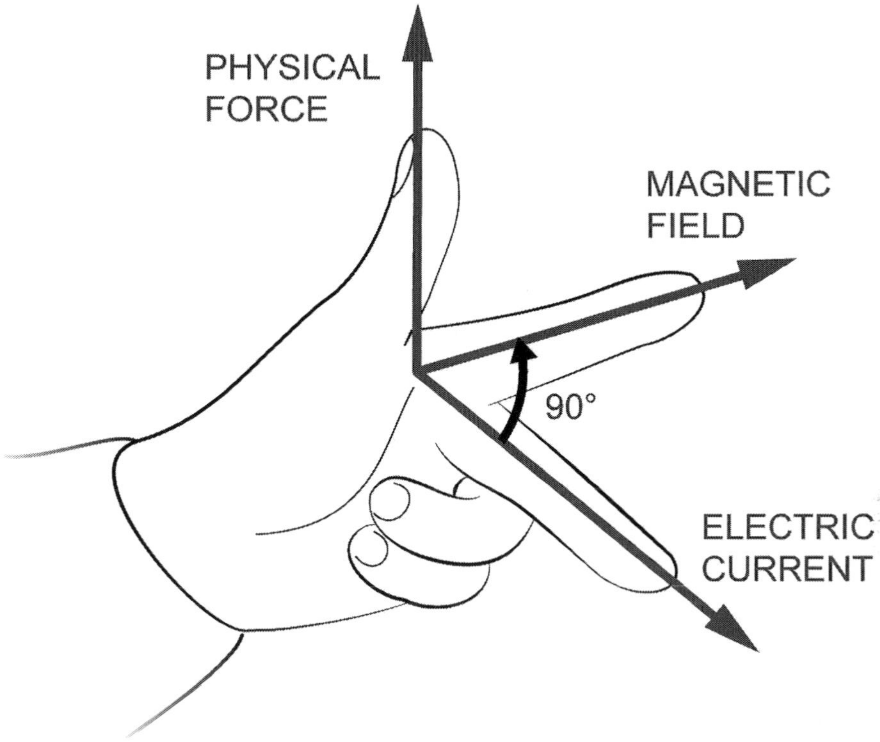


**Figure 2.11** Fleming's Left Hand Rule.

## 2.20 Loudspeaker Enclosures

While a raw loudspeaker driver has a certain aesthetic appeal, mounting the driver inside an enclosure improves the sound. Why? In the moment that compression of air molecules is occurring on the front of the diaphragm, rarefaction is occurring on the back.

When compressed molecules meet rarefied molecules, the effect is canceled out, and the energy of the sound wave is reduced. For this reason, loudspeaker drivers are usually mounted in an enclosure sealing off the back and sides, while the front of the driver is left open to the room. The goal is to isolate the air molecules on the sides and back from the front of the diaphragm.

Any material covering the front should be acoustically transparent (meaning air molecules can easily pass through) to avoid muffling the

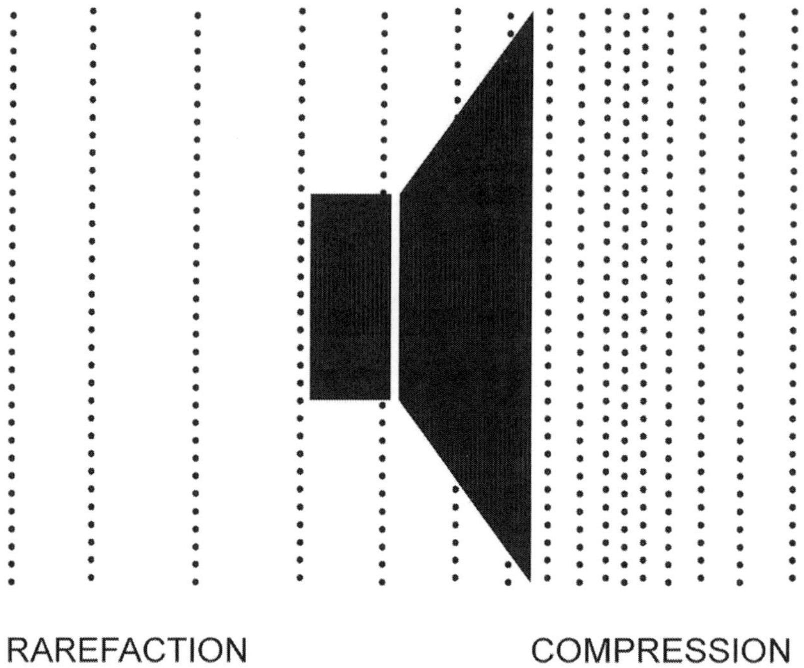


**Figure 2.12** Compression Occurs in Front of the Loudspeaker Driver as Rarefaction Is Occurring Behind It.

**Figure 2.13** Loudspeaker Driver, Mounted Flush.

sound. A solid surface with plenty of holes works, so does a breezy cloth (like burlap) stretched across the opening. Simply setting the loudspeaker inside a box with holes cut in the front is insufficient, and may even sound worse. The barrier between the loudspeaker driver front and its sides needs to be sealed off, with minimal air gap. A box with holes cut in the front can work, provided the loudspeaker face is mounted flush against the surface with the holes.

## 2.21 Power Amplifiers

Earlier, we discussed preamplifiers as a way to increase a microphone signal's voltage so a microcontroller can read it. If we want to send the signal to a loudspeaker driver, we'll need to increase its current as well. This is because we need to increase the *power* of the audio signal. Power is the product of voltage and current. It is measured in watts. Moving air molecules requires a lot of power. For a signal to provide that kind of power to the loudspeaker, we use an audio power amplifier.

Power amplifiers are rated for power in watts, and for impedance in ohms. Loudspeakers have similar ratings. We'll discuss what these numbers mean in a moment, but first here is how to use these ratings.

*Rule #1: the impedance of the loudspeaker should be equal to, or higher than, the load impedance of the amplifier*. A loudspeaker with a lower impedance could damage the amplifier. A loudspeaker with a significantly higher impedance may not produce sound.

*Rule #2: the power rating of the loudspeaker should be equal to, or higher than, the power rating of the amplifier*. Connecting a loudspeaker with a lower power rating may damage the loudspeaker.

It isn't hard to match a loudspeaker to an amplifier. Some online vendors make it easy by suggesting appropriate loudspeaker/amplifier pairings. Small dynamic loudspeaker drivers, 1–3" in diameter, typically have an impedance of 8 or 4 ohm, and are generally rated for amplifiers between 0.25 watt (250 mW) and 3 watts. (The manufacturer's datasheet provides exact values.) If you'd like to use anything more complicated than a single 8 ohm loudspeaker connected to a small amplifier, it's helpful to understand what the power and impedance ratings mean, to avoid irreparable damage to your hardware. We'll cover these terms next.

## 2.22 Power

The higher the power rating of the amplifier, the more power it can deliver to the loudspeaker. More power enables the loudspeaker to move more air molecules, which results in a louder sound. However, there is a limit to how much electric energy a loudspeaker can convert to acoustic energy. The loudspeaker's power rating indicates the maximum power it can handle. If the amplifier delivers more power than this, the extra energy is converted to heat that will damage the loudspeaker.

Keep in mind that there is some "wiggle room." The power rating indicates the power an amplifier is *capable* of delivering. If the amplifier is turned down or playing a quiet signal, it will deliver less.

## 2.23 Impedance

*Impedance* is the term used to describe a circuit's opposition to AC current (which includes all audio signals). You've likely heard the term *resistance*, which is a circuit's opposition to DC current. Impedance includes resistance as well as *reactance*, a form of opposition that is frequency dependent.

Impedance is a tricky concept, but we can attain a functional understanding of it by comparing it with a simple LED circuit. To do that, we must understand the terms *series* and *parallel*.

*Series* describes components that are wired one after another, like beads on a string, where the current must flow through each one to pass from points A to B.

The total resistance is the sum of the resistors in series. In Figure 2.14, the total resistance between A and B is 200 ohm.

*Parallel* describes components that are wired next to each other, like lanes on a highway. The current splits between the different paths. Calculating the math of multiple resistors in parallel can get tricky, but it's easy

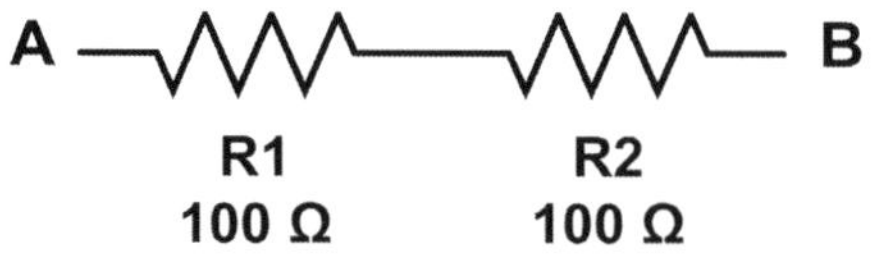


**Figure 2.14** Two Resistors in Series.

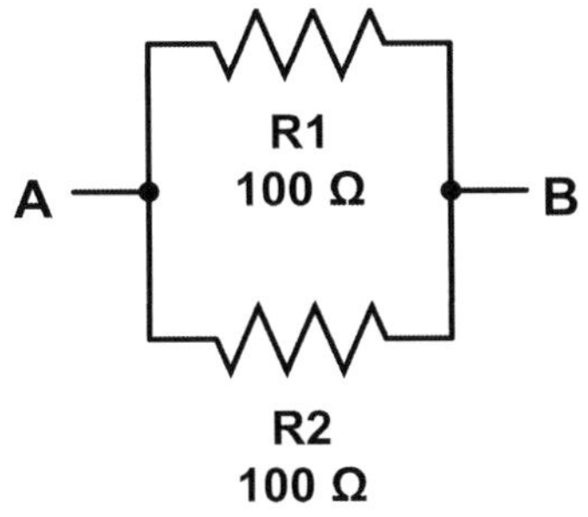


**Figure 2.15** Resistors in Parallel.

if all of the resistors are the same value. Simply divide the total resistance by the number of resistors. In Figure 2.15, the total resistance between A and B is 50 ohm.

Now on to our demo circuit.

This diagram shows the standard wiring for an LED. A resistor is placed in series with the LED to restrict the amount of current that can flow through the circuit—too much current would burn out the LED. But with the correct resistor in place, the LED operates at the current level the manufacturer intended.

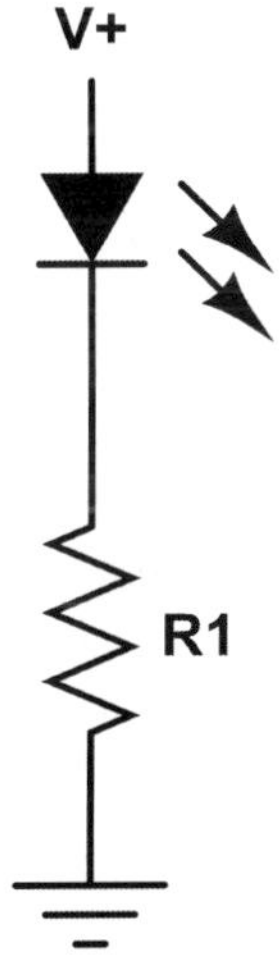


**Figure 2.16** A Basic LED Circuit.

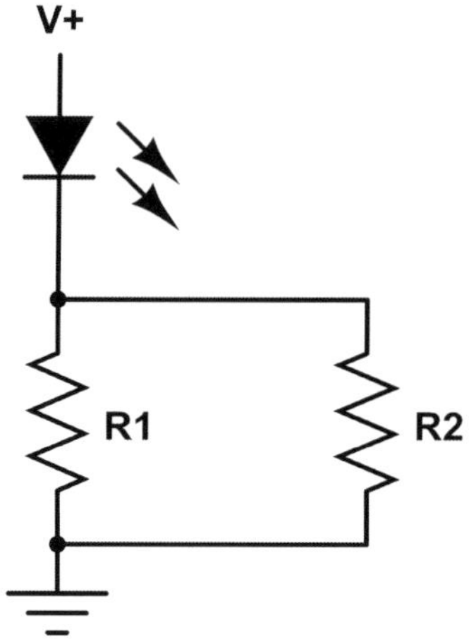


**Figure 2.17** Adding a Second Resistor to our LED Circuit Lowers the Resistance.

Let's play around with this circuit just a bit. What would happen if we added an identical resistor in parallel with the first? We've made a second, equal path for the current to flow through, and the resistance is cut in half as a result. More current can flow. So much current can flow, it might even burn out the LED.

What if that second resistor were of a much higher resistance than the first? We've opened another path for the current, but it's a small one, and the extra current that flows isn't enough to destroy the LED.

Let's see that same circuit, in a different context.

Instead of an LED, we'll use a triangle to represent the circuitry inside a power amplifier. R1 now represents the "output impedance" of the amplifier, and is located between the output terminals.

When we connect a loudspeaker to the output, it's as if we're adding a second resistor in parallel with R1. This is fine provided that the impedance of the loudspeaker is high enough that it doesn't overload the circuitry. Small power amplifiers meant to drive a loudspeaker will specify the "load impedance" to indicate a minimum impedance for the loudspeaker.

What if you added a second loudspeaker, in parallel with the first?

The impedance drops even further. More current can flow, and the circuit may heat up. The amplifier may not be designed to provide this much current and could be damaged as a result.

A safer solution is to wire the second loudspeaker in series.

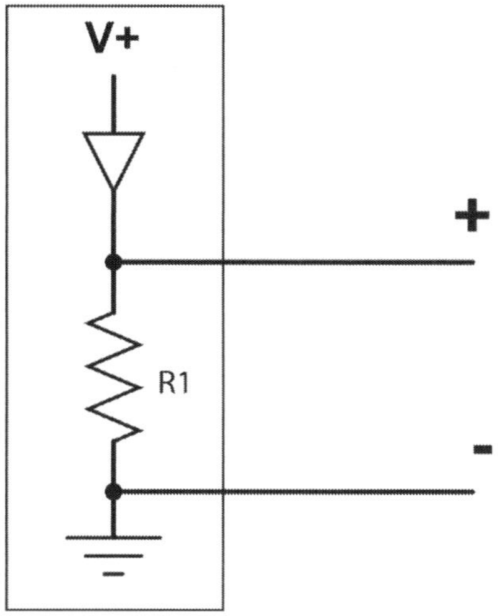


**Figure 2.18** Simplified Diagram of a Power Amplifier.

This increases the total impedance. The higher the impedance of the loudspeakers, the easier it is on the amplifier. However, the volume of each loudspeaker will be quieter. You can compensate by turning the amplifier up (at least, until the amp can't provide any additional power).

Impedance matching is a particular challenge with e-textile and other DIY design loudspeakers, which typically have unusually low impedance. Here are a few strategies to avoid damaging the amplifier:

- Add a resistor in series with the loudspeaker. This will reduce the volume but it's better than a blown amp. Select a resistance value that matches the amp's rated impedance (for instance, 8 ohm). It's important that the power rating of the resistor be at least as high as the amplifier. Educational and hobbyist workshops tend to stock ¼ W carbon film resistors, which will work for small ¼ W amplifiers.
- For amplifiers rated above ¼ W, use a non-inductive 8 ohm ceramic resistor. Much of the amplifier's power will dissipate through the resistor as heat, which can make it unsafe to touch. Protect the components from melting by attaching a heat-sink to the resistor.
- There are amplifiers designed for low-impedance loudspeakers. Specifically, a Class-T or Class-D amplifier will shut off if it is in danger of being overdriven. Adding a protection resistor to this setup extends how hard the amplifier can drive the speaker before it shuts itself off.

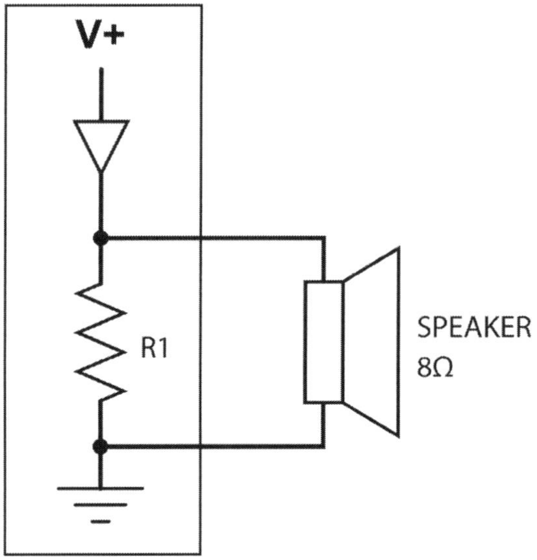


**Figure 2.19** Power Amplifier Driving One Loudspeaker.

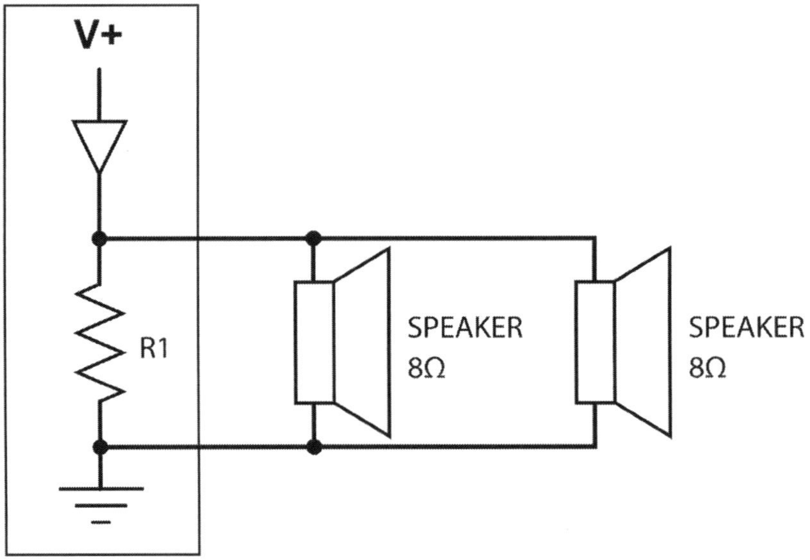


**Figure 2.20** Power Amplifier Driving Two Loudspeakers Wired in Parallel.

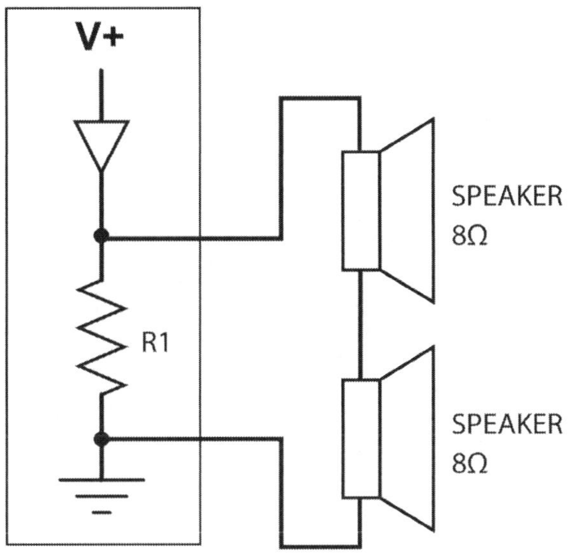


**Figure 2.21** Power Amplifier Driving Two Loudspeakers Wired in Series.

## 2.24 Stereo and Mono Recordings

When you hear an ambulance siren, you can locate the vehicle because you have two ears: if it's louder in your right ear, you look to your right. A stereo system has a separate audio channel for each ear, to create the illusion of sounds originating from different locations. Adjusting the balance knob of your home stereo increases the volume of one channel while decreasing it for the other.

For stereo playback, each channel needs its own amplifier and loudspeaker. A system with only one loudspeaker can only play mono. For some recordings it makes little or no difference. A recording of a single voice (or instrument) is usually fine in mono. But for other recordings, switching to mono can change the balance of sounds in the recording. For example, a stereo recording of music might have voice in the middle, guitar to the left and piano to the right. If you were to use only the left channel

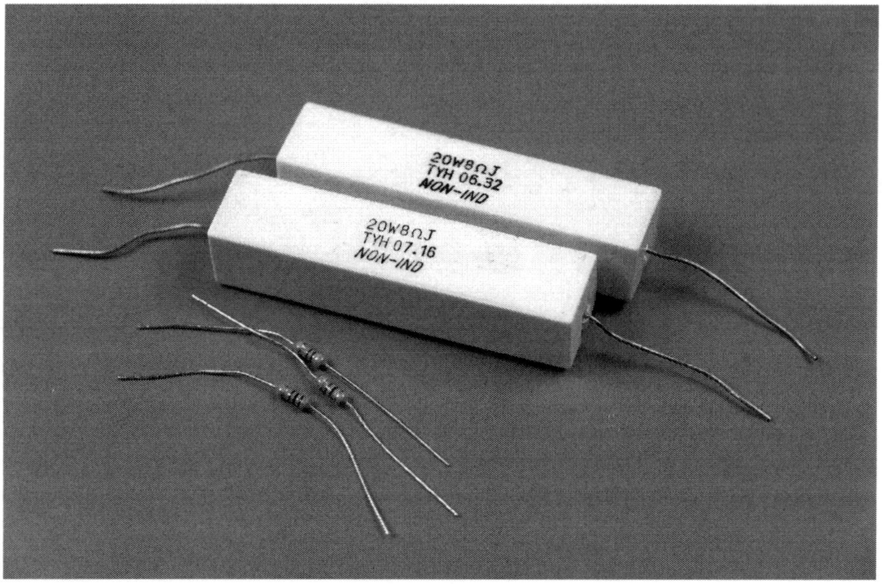


**Figure 2.22** Carbon Film Resistors and Ceramic Resistors.

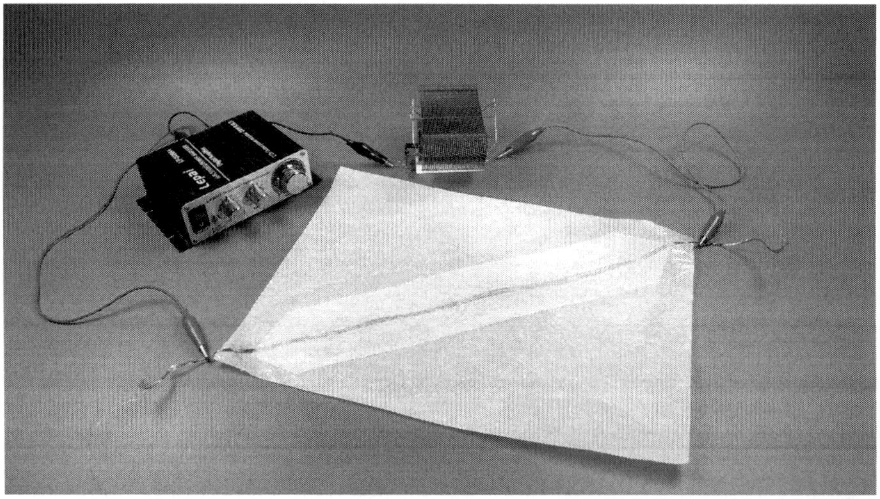

**Figure 2.23** A Simple E-Textile Paper Speaker Connected to a Class T Amplifier, in Series With Resistors Attached to a Heat Sink.

of this recording, the piano would be quieter. If you used only the right channel, the guitar would be quieter.

It is possible to combine the left and right channels into a single mono track using audio editing software. The result may or may not work for your needs, but it will sound different. Test the results before deciding.

## 2.25 Conclusion

This chapter intended to give readers an understanding of the physics of microphones and loudspeakers as transducers of acoustic and electric energy. We started with a brief description of energy in acoustic and electric forms then moved on to microphones, specifically piezo discs and electret mics, as a means for converting acoustic sounds into audio signals. We also covered preamps to connect microphones to microcontrollers and introduce DC bias.

Next, we discussed loudspeakers: the factors involved in choosing a loudspeaker, and specific loudspeaker driver options: dynamic loudspeakers, piezo discs and audio exciters. For those wishing to build a DIY loudspeaker, we explored the science of how a loudspeaker works. We provided a few practical methods for improving the sound quality of different drivers. We also touched on design issues for e-textile loudspeakers. As loudspeakers require power amplifiers, we covered power and impedance ratings and their role in matching a loudspeaker to an appropriate amplifier. We closed with a comparison of stereo and mono and the factors involved in adapting stereo audio recordings to single-channel playback.

The chapter contained both scientific theory and practical solutions. It is hoped that this combination will provide readers an easy point of entry with plenty of room to expand as their experience and knowledge grow to match their creativity.

## Notes

1 We will use the Arduino as our microcontroller reference platform. The reader should also be familiar with connecting an LED with a resistor to a DC power supply, and with the circuit symbols and functions of these components. Platt (2015) offers a thorough introduction to this material.

2 This concept is referred to as "conventional flow." Another way of conceptualizing it, "electron flow," envisions the electrons flowing from negative to positive. The

difference between the two is worth exploring further but beyond the scope of this chapter.

3 Collins (2009, p. 31–44, p. 64–70) provides a detailed guide for wiring and soldering both piezo discs and electret microphones.
4 For example, see the websites www.digikey.com/ or www.mouser.com/.
5 Krupczak et al. (2004) gives detailed instructions for building a plastic cup speaker.
6 https://www.instructables.com/id/Hard-Drive-Magnet-Speakers/.
7 https://www.instructables.com/id/Hard-Drive-Magnet-Speakers/.

## References

Collins, N., 2009. *Handmade Electronic Music: The Art of Hardware Hacking*. 2nd ed. New York: Routledge.

Earl, B., and Ada, L., 2013. *Measuring Sound Levels*. Viewed July 6, 2018 <https://learn.adafruit.com/adafruit-microphone-amplifier-breakout/measuring-sound-levels>.

Krupczak, J., et al., 2004. *A Simple Loudspeaker Which Students Can Build and Take Home*. Viewed July 6, 2018 <http://studylib.net/doc/10527357/a-simple-loudspeaker-which-students-can-build-and-take-home>.

Lindblom, J., 2016. *Piezo Vibration Sensor Hookup Guide*. Viewed July 6, 2018. <https://learn.sparkfun.com/tutorials/piezo-vibration-sensor-hookup-guide>.

Platt, C., 2015. *Make: Electronics: Learning Through Discovery*. 2nd ed. San Francisco: Maker Media, Inc.

3

# New Interfaces and Musical Robots

## New Musical Input and Output Devices

Jim Murphy

### 3.1 What Is a Musical Interface?

We can use software to design sounds, sequence audio and create music. In order to do this, though, a user needs to be able to input commands into a computer. The computer is not, in and of itself, a musical instrument. For the computer to become a usable instrument, a device that sits between the user and the computer is needed. This device, diagrammed in Figure 3.1, translates human gestures into signals usable by the computer. This interfacing has historically been accomplished with keyboards, mouses and other general-purpose human interface devices. While these everyday devices have utility in music and sound design applications, the unique gestural requirements of music performance and composition have led to a rich field of musical interface design. This section focuses on these musical interfaces (which are different from the similarly named "audio interfaces" that serve to bring audio into and out of a computer), focusing on the burgeoning field of new musical interface design and custom controllers for studio use, performances and beyond.

Early computers received user input in the form of punched paper tape or cards or via the laborious toggling of banks of switches. Clearly, for expressive musical and sonic gestures to be realized in an intuitive manner (particularly by non-programmers), a new generation of input devices was needed. Artist/engineer Perry Cook's 2001 paper "Principles for Designing Computer Music Controllers" provides an overview of significant milestones in musical interface design (Cook 2001).

Many computer music tasks can be completed with the most basic of interface devices: keyboards and mouses. These fundamental human interface devices (HIDs) are particularly well suited for "offline" musical use, where the user has time to take advantage of the high resolution and

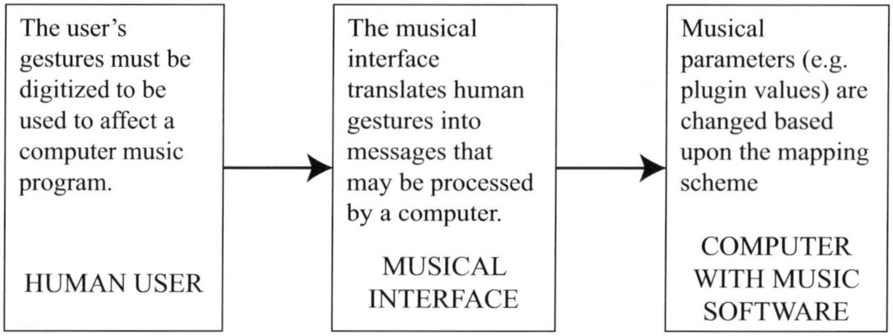


**Figure 3.1** Musical Interfaces Serve as the Connective Tissue Between the Human Musician and the Computer Instrument.

precision afforded by the keyboard and mouse. Digital Audio Workstation (DAW) software, for example, is developed with the keyboard and mouse as the primary interface devices. These everyday HIDs, though, are less well-suited to the realization of real-time musically expressive gestures or to the capturing of events in a manner that mirrors how traditional musical instruments are configured. While the computer keyboard and a piano keyboard both contain rows of keys, the computer keyboard's layout makes little sense when trying to input musical notes. Using keyboards and mouses requires the user to make potentially unintuitive leaps of logic, resulting in a non-expressive disconnect between the sonic idea and the means of realizing it.

Among the earliest digital interfaces to address this problem of non-intuitive control over musical software were digital piano-style keyboard controllers. These controllers (such as the one shown in Figure 3.2) allowed a user to input musical events into a computer in a manner that closely approximated the playing of traditional instruments or using traditional studio hardware. Computer music luminary Max Mathews took this idea further, developing controllers based on instruments such as drums and violins (Park 2009). With these controllers, musicians could use their expertise at creating expressive musical gestures to produce new computer-based music.

Interfaces based upon traditional instruments do not always provide the most intuitive control paradigm: new arrangements of buttons, knobs and sensors sometimes provide the most effective means of using computer music tools. For example, the popular digital audio workstation Ableton Live is often controlled with a mouse and keyboard. Using Ableton Live with keyboard and mouse can be challenging in a live performance

**Figure 3.2** A Typical Musical Interface: An array of sliders, buttons and knobs serves to translate a user's gestures into values that can be used to change musical parameters on a computer.

context: the performer must make precise mouse gestures that might be better realized with a customized button array. To address this challenge, a number of application-specific or customizable computer music controllers have been developed. While some of these (such as the Monome (Arar and Kapur 2013)) are hardware-based, there are also a number of mobile apps (such as TouchOSC, described in Roberts 2011) that allow smartphone and tablet users to precisely control musical software. So, in addition to the musical instrument-inspired interfaces such as keyboard and percussion controllers, a growing number of novel interfaces are available that allow for more intuitive control over a digital audio workstation.

Arguably the most exciting recent development in musical interface design is the advent of a number of relatively easy-to-use development platforms that allow for custom-designed interfaces to be created. Instead of making do with off-the-shelf interfaces, developers can rapidly create a one-of-a-kind interface that fulfills a particular goal. A rich community of builders of these new interfaces has emerged; the annual conference on New Interfaces for Musical Expression (NIME) is a gathering of designers and users of these cutting-edge interfaces, and their freely accessible archives of publications at www.nime.org/archives/ is a rich library of information on the state of the art of new musical interface design.

Since musical interfaces have become relatively easy to build from scratch, perhaps the best way to understand and use them is to build your own. The following subsections deal with the development and use of a simple exemplar musical interface. Along the way, key design considerations are addressed and emphasis is placed on the decision-making processes that can lead to compelling and musically expressive new interfaces.

### *3.1.1 An Example Interface*

Hardware-based musical interfaces come in a wide variety of shapes and forms. These range from wearable sensor-laden gloves (such as Laetitia Sonami's Lady's Glove interface, described in (Bongers 2000)) to so-called hyperinstruments (exemplified by Cléo Palacio-Quintin's Hyper-Flute (Palacio-Quintin 2008)), consisting of electronically augmented acoustic instruments. In spite of this range of looks and feels of musical interfaces, there are a number of commonalities across the board. Most musical interfaces, for example, can be illustrated at a high level by the block diagram in Figure 3.3.

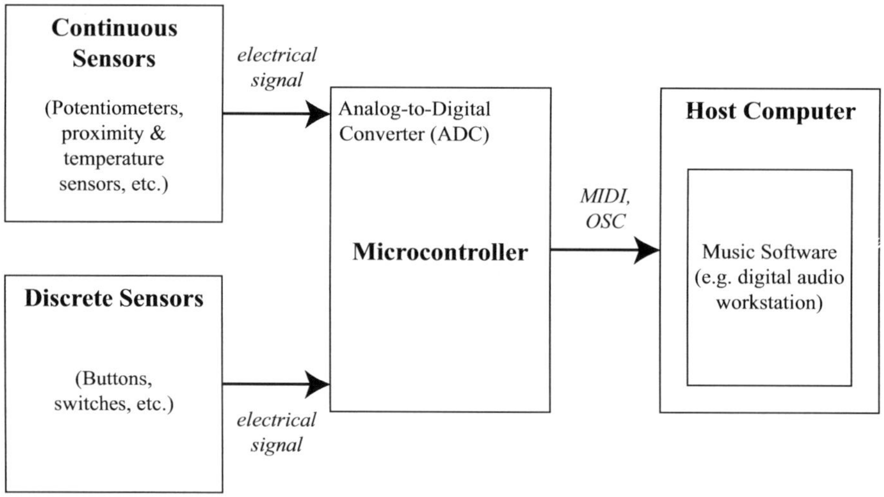


**Figure 3.3** A Block Diagram of a Simple Musical Interface: Sensors (either discrete or continuous) send electrical signals into a microcontroller. The microcontroller converts these electrical values into digital ones and then transmits them to a host computer. Software on the host computer interprets these values according to its mapping scheme.

As shown in Figure 3.3, musical interfaces consist of a number of sensors that are connected to a central microcontroller. This microcontroller is in turn connected to a host computer, communicating with it through the use of some sort of communications protocol. As can be seen in the illustration, it is clear that the microcontroller is at the heart of a musical interface. A basic understanding of microcontrollers, then, is a very useful thing when developing new musical interfaces.

Microcontrollers are small computers in and of themselves. They are designed to be inexpensive, robust and able to receive signals from the outside world; because of their low cost and robustness, they are used in many consumer applications. The same features that make them attractive for everyday uses make them good for musical interface design: a microcontroller can be programmed to read a number of sensors and send messages (based upon the sensor readings) to a computer that is equipped with musical software. In a musical interface, the microcontroller sits at the boundary layer between the user's input and the communication of messages to the host computer.

Microcontrollers have been around for decades. Historically, they were difficult to program: early examples required the programmer to be adept at difficult-to-learn assembly language. In recent years, this paradigm has shifted: a large number of easy-to-use frameworks have emerged, allowing those without an electronics engineering background to program microcontrollers to fulfill tasks. The most notable microcontroller framework is Arduino (Shiloh and Banzi 2014), which is a programming environment and set of software tools that allow a wide variety of microcontrollers to be programmed using a common set of commands. Using Arduino, a musical interface designer can connect a number of buttons, knobs and other sensors to the microcontroller and easily write code that can send the messages to a host computer.

### *3.1.2 Sensors in Musical Interfaces*

Musical interfaces are, as shown in Figure 3.3, an assortment of electrical sensors whose values are read by a microcontroller and transmitted to a host computer. But what is a sensor? How can they be used to receive a user's gesture, and how might an interface designer choose an interfaces' sensors based upon the sorts of gestures for which the interface is designed?

An electrical sensor is a type of transducer. Transducers are mechanisms that take one form of energy as input and convert it to another form of energy as output. The microphone is a transducer that converts air pressure to electrical signals; as a device that converts one sort of energy into electricity, the microphone is a good example of an electrical sensor. Digital thermometers (which transduce temperature into electrical signals) are another type of electrical sensor. Buttons and knobs (also known as potentiometers) convert physical displacement into electrical signals. Clearly, there are a wide variety of means by which some real-world value can be converted to electricity: any of these electrical sensors can be used to transduce real-world values into the electrical signals that a microcontroller can process. Once the signal is transduced from its original medium into an electrical signal, the signal can be further processed onboard the microcontroller.

Sensors can be divided into two overarching categories: analog (or continuous) sensors and digital (or discrete) sensors. Continuous sensors can distinguish between a range of values. The sliders on a mixing console are continuous, able to be moved into a huge range of positions. Discrete sensors can distinguish between only two states: on and off. A key on a computer keyboard or a button on a Monome controller is discrete. At its simplest, a discrete sensor such as a button is connected to a microcontroller by wiring one terminal of the button to a known reference voltage (such as 5 V) and the other terminal to the input of the microcontroller (shown in Figure 3.4). The microcontroller's input must be assigned some value,

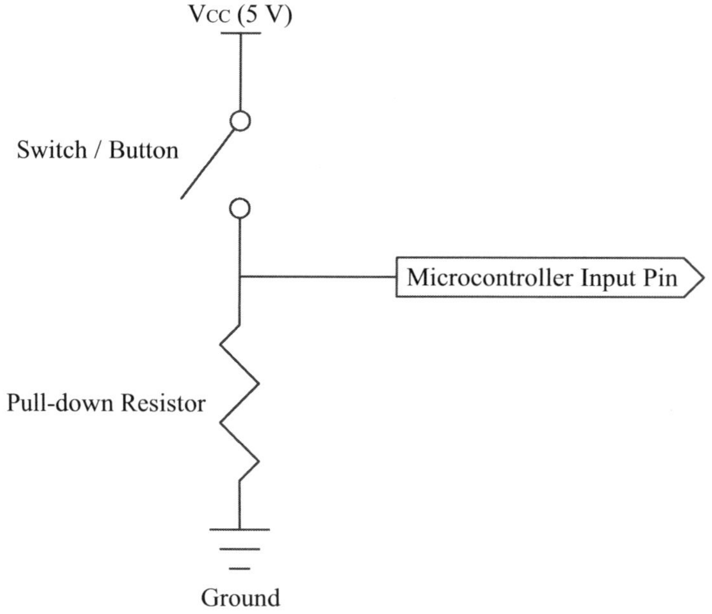


**Figure 3.4** A Button Circuit, Consisting of a Button (or Switch) Connected to a Microcontroller Input Pin: The pull-down resistor (typically with a value of 10 kΩ) pulls the input pin low when the switch is open. This circuit is typical of buttons, switches and other discrete sensors on musical interfaces.

though, so it is usually connected to the value opposite of the known reference voltage through a resistor. In the case of the button being connected to 5 V, the resistor is connected to ground. This resistor (called a pull-down resistor in the event that it pulls the microcontroller input to ground) is the default state of the pin. When the lower-resistance path to 5 V is presented upon the button being pressed, the resistor is bypassed and current flows along the path of least resistance.

Continuous sensors output an analog value: a varying voltage typically between 0 V and 5 V. While there are many types of continuous sensors, they can often be illustrated by the circuit shown in Figure 3.5. This circuit represents a variable voltage divider and is exemplified by an

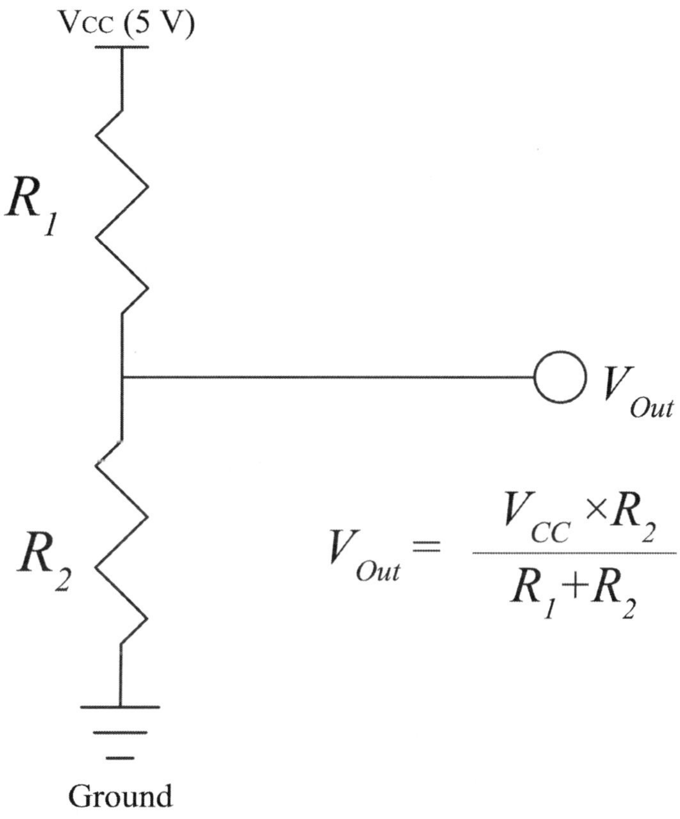


$$V_{Out} = \frac{V_{CC} \times R_2}{R_1 + R_2}$$

**Figure 3.5** A Resistive Voltage Divider Circuit: The values of the input voltage (VCC), R1, and R2 set the output voltage. By varying the ratio between R1 and R2 (as in a potentiometer), different values of VOut may be obtained. In musical interfaces, these varying voltages are read into the microcontroller's analog inputs for conversion to values that may be transmitted to a host computer.

analog potentiometer. In this voltage divider, one input is connected to the microcontroller's ground pin (representing 0 V). The opposite input leg is connected to a reference voltage (often the microcontroller's 5 V pin). The middle leg, often called the wiper, is connected to the microcontroller's Analog-to-Digital converter (ADC). The wiper's value varies depending on the transduced signal's value; on a potentiometer, the wiper is physically moved between the ground and reference pins, resulting in a wiper voltage closer to or further away from the reference voltage (as shown in Figure 3.5). This varying voltage (which might be any value within the sensor's range) gets converted by the ADC from an analog value to the digital numbers that may be understood by the interface's microcontroller. This process is very similar to the conversion from analog microphone signals to digital audio values that happens on a computer's audio interface.

Why use one type of sensor over another? One might wish to use a continuous sensor in a situation where fine-grained control is needed. A knob can be turned to a particular angle and read by the microcontroller to set a value in a digital audio workstation to a particular user-defined amount. Not all musical actions require this level of precision, though; some musical actions benefit from easy-to-control clarity. Muting and unmuting a track in a digital audio workstation, for example, doesn't warrant fine-grained control, meaning that a button or other discrete sensor is perfectly acceptable.

Historically, sensors were difficult to configure and use: some electrical engineering knowledge was needed in order to deploy all but the simplest buttons and potentiometers. Recently, though, vendors such as Adafruit (www.adafruit.com) and Sparkfun (www.sparkfun.com/) have made widespread easy-to-use breakout boards that allow sensors to be connected to a microcontroller in a plug-and-play manner. A number of novel interfaces have been developed in recent years that take advantage of such exotic sensors. The NIME archives, mentioned above, are a trove of reports of designing and using such interfaces.

### *3.1.3 Musical Interface Communications*

Microcontrollers give musical interface designers much flexibility in terms of the sensors used and the means by which the microcontroller communicates with a host computer. While any number of sensors might be used to realize different gestures, the choice of communications protocol is more rigid when developing musical interfaces. Most off-the-shelf musical interfaces (such as the popular Korg Nano Controller or Akai MPD controllers) use MIDI (outlined in MIDI 2018) as a communications protocol to allow the interface's microcontroller to send and receive messages with a host computer.

MIDI, introduced in 1983, remains popular due to its simplicity and widespread compatibility: most DAW software is directly compatible with MIDI hardware. MIDI is showing its age: it is a relatively low-resolution protocol, meaning that most of its messages are only capable of communicating 128 discrete values. Other protocols (such as the very capable Open Sound Control; Wright 2005) address some of MIDI's shortcomings but haven't yet gained the widespread acceptance of MIDI. For those interested in developing interfaces usable by the majority of popular audio creation software, MIDI remains the protocol of choice when looking to communicate between microcontroller and host computer.

MIDI messages contain information about the message type as well as the value of the message that is being sent. The microcontroller can send different messages with various values in response to different user input events; these messages are received by audio software on the host computer and can be used to affect the performance or composition.

Traditionally, MIDI messages were sent using MIDI-specific hardware. This hardware, consisting of a five-pin circular connector, required the host computer to be equipped with a similar input jack. Modern computers typically don't have this jack, and MIDI for musical interfaces is now usually sent over a USB connection in a mode known as MIDI HID. Not all Arduino-compatible microcontrollers can serve as a MIDI HID device. One popular and inexpensive Arduino-compatible microcontroller that can easily be configured to serve as a MIDI HID-based interface is the Teensy LC (as well as other Teensy 3-series microcontrollers) (Stoffregen 2018).

While the communications protocol is fixed in order to allow compatibility between an interface and the host software, the *meanings* of the messages are not so rigid. When a user presses a button, the MIDI message, or messages, that are sent are decided upon by the designer of the interface. This relationship of user input to interface output is known as mapping, and it is a core concern of those seeking to create engaging musical interfaces (Tanaka 2010). A very simple interface can be made into a compelling instrument with the use of creative mapping schemes. The Monome, mentioned above, is one example of this: a simple grid of buttons can be assigned a huge array of mapping schemes, resulting in limitless usage paradigms.

As explored in prior subsections, a musical interface consists of three main elements. The first element is the input sensor block, which serves to translate real-world values into the electrical signals read by the microcontroller. The microcontroller itself makes up the second element of a musical interface, converting the electrical signals from the sensors into mapped messages that might be transmitted to a host computer. This communications scheme makes up the third part of the interface, consisting of messages that are sent from the interface to the host and may be used to affect things like the values of a plugin or the parameters of a DAW.

## 3.2 Mechatronic Music: Actuated Instruments

Transducers are everywhere in electronic music and sound design. Microphones are one type of transducer, converting sound pressure into an electrical signal. The sensors used in musical interfaces (discussed in the prior section) are a second type, taking a user's input and converting it to a message used to control audio. Both of these types of transducers may be seen as input devices, allowing "real world" signals to be converted to a type of signal that might be read by a computer.

Transducers often serve as outputs, as well: loudspeakers are a ubiquitous example of a transducer that takes an electromagnetic signal and converts it to physical motion (and, in turn, sound waves). Loudspeakers are not the only type of output transducer present in the production of sound. A small but growing subset of new musical instruments makes use of motors and other sorts of motion-producing transducers (which are collectively called actuators) to convert computer-coded messages to physical motion. While the motives of the designers of these musical automata are myriad, some have cited the potential to use computer-aided composition techniques (such as DAW-based production and MIDI sequencing) to produce physically mediated sounds without the need for loudspeakers (Long 2018).

This section focuses on musical instruments and sound-making devices that use these actuators to make sound. After examining the rich history of automatic musical instruments, the types of output transducers used in many contemporary systems are presented. Some simple circuits that might be used to drive motors in a musical manner are shown, and the future of the field of musical robotics is presented.

### *3.2.1 Musical Automata*

Inventors have been fascinated with automatic musical instruments for hundreds of years. The Bānu Mūsā Brothers, Baghdad-based inventors active in the ninth century, developed a water-powered automatic organ capable of being programmed to play different songs. Throughout the

European Renaissance, carillons and organs were driven with clockwork mechanisms that allowed them to be automated. This innovation of automatic instruments continued beyond the Renaissance, with notable composers such as Haydn and Beethoven composing for clockwork musical automata. The Industrial Revolution saw a rapid growth in the complexity and popularity of automatic instruments: player pianos and music boxes proliferated, satiating people's desires to hear music in a time when the only other option was to wait for a chance to hear a piece played by a live ensemble.

The ascendency of automatic music was abruptly cut short by the arrival of the phonograph and, in the following decades, the loudspeaker. Large, complicated pneumatic and clockwork instruments were rendered obsolete by the compact and relatively inexpensive phonograph. Though automatic instruments fell largely out of favor with most people during much of the twentieth century, composers such as George Antheil and, later, Conlon Nancarrow continued to create otherwise impossible compositions using player pianos.

The 1970s saw many musical trends challenged: electronics and studio-oriented production techniques began to overturn decades-old music recording paradigms. In parallel with this trend was the rebirth of automatic instruments, catalyzed by artists such as Trimpin and Godfried-Willem Raes. These artists saw the potential for electrical actuators to be used much as the clockwork mechanisms of automated instruments from prior centuries. Unlike those difficult-to-program devices, though, motors and other actuators could be computer controlled and used to produce morphing, changing, generative physical soundscapes.

Since the 1970s, many inventors, artists and engineers have developed automatic musical instruments. Some refer to this new generation of electronically controlled actuated instruments as "musical robots," while others prefer the term "musical mechatronics" (which foregoes any implications of artificial intelligence that might accompany the use of the word "robot"). These musical mechatronic instruments include automatic drum-playing mechanisms (exemplified by Ajay Kapur's MahaDeviBot (Kapur 2010)), the author's own mechatronic chordophones (one of which is shown in Figure 3.6) and huge arrays of percussive devices such as those of Eric Singer (2004). Musical roboticist Jason Long's 2017 history serves as an excellent starting point for those interested in an in-depth survey of trends in musical automata (Long 2017). All of these instruments are computer-controlled, allowing them to respond to traditional MIDI notation or to input from new musical interfaces such as those discussed in the previous section.

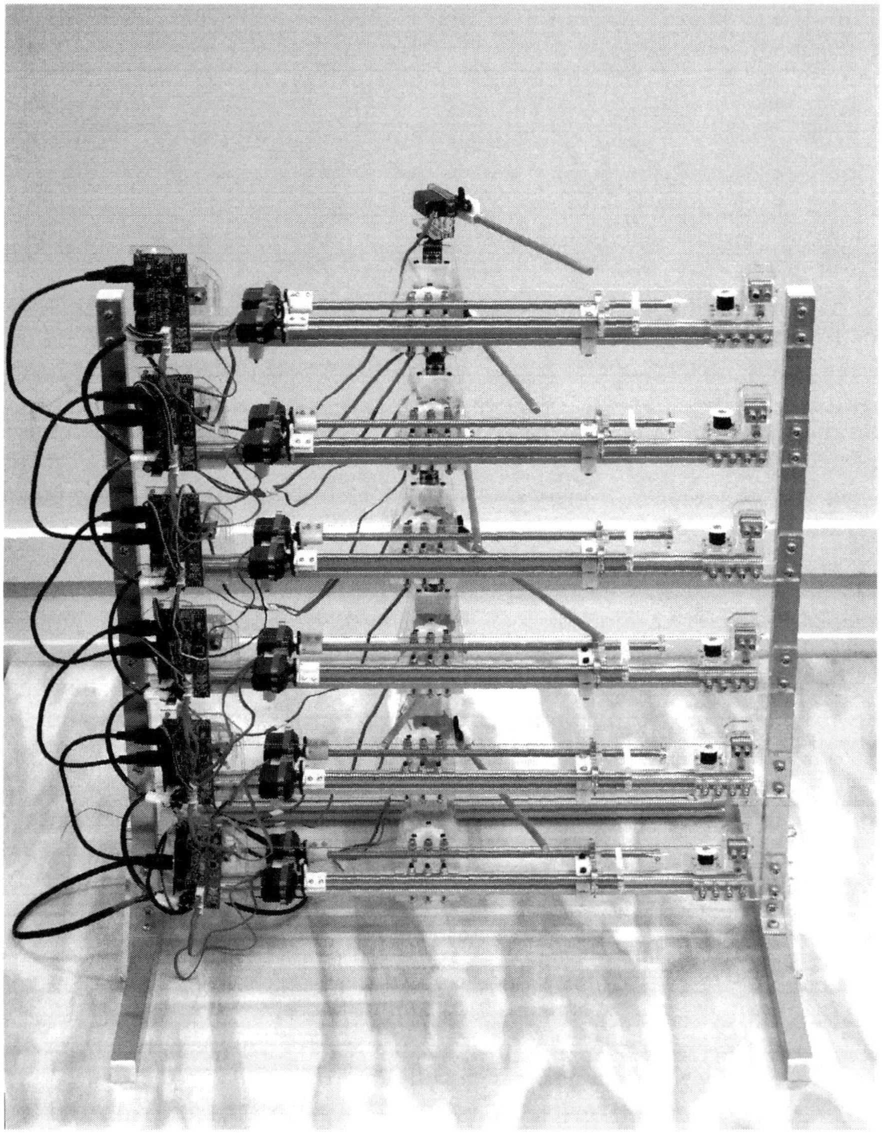

**Figure 3.6** Swivel 2, a Mechatronic Slide Guitar Designed by the Author: This instrument consists of arrays of servomotor actuators that respond to MIDI messages. Users can program Swivel 2 to produce emulations of human-played guitar music or to produce entirely new types of music.

### 3.2.2 *Types of Actuators*

There are many types of modern mechatronic musical instruments: self-playing slide guitars, tuning forks that are automatically struck, and drumsticks capable of striking myriad positions on a snare head. While diverse in function, each mechatronic instrument shares something with all others: a dependence on actuators in order to produce sound in response to a computer-generated command. In addition to various "exotic" actuators (air-powered and water-powered systems, for example), there are two main families of actuators used by mechatronic musical instruments: rotary motors and solenoid actuators. The following paragraphs introduce and provide some usage contexts for these two main families of actuators.

Electric motors are everywhere in modern life: from the large motors that power electric cars to the humble motors that turn the carousels in our microwave ovens, these actuators that convert electromagnetic signals to continuous rotary motion are versatile and commonplace. Electric motors convert an input voltage and current to output rotational motion; different types of motors accomplish this in different manners, but most use magnets that rotate in response to applied electrical power. Figure 3.7 shows a

**Figure 3.7** A Variety of Actuators: From left to right, a DC motor with gearbox, a stepper motor, a solenoid and an RC servo. Each type is extensively used in musical mechatronic applications.

motor of a type often used in mechatronic instruments: the motor's body contains coils that become energized in the presence of direct-current electrical power. These energized coils attract magnets that are positioned radially around the coils. This attraction results in rotary motion. The motion is often too fast and weak to be useful on its own, necessitating some form of gearbox to slow it down and render it stronger (higher-torque) in order to be able to bow a string, rotate a loudspeaker or move a mechanical hand up and down a mechatronic guitar's fretboard.

In addition to the DC gearmotor shown at left in Figure 3.7, there are other types of motors able to produce rotary motion in response to input power. One popular type is the stepper motor. Unlike DC gearmotors (which rotate in a continuous manner), stepper motors can be instructed to rotate in precise steps. With a stepper motor, a musical instrument's drumstick-holding arm could be instructed to rotate through precisely 45 degrees with no need for any other sensors. Like stepper motors, servo motors are also able to rotate through precisely specified angles. Servomotors (or "servos" for short) consist of a normal DC motor (often with a gearbox to increase torque and reduce output velocity) whose shaft is connected to a potentiometer or similar sensor. The potentiometer turns along with the motor shaft and sends its signal back to some motor controller circuitry capable of comparing the computer-specified angle with that reported by the motor's potentiometer. Voltage is applied to the motor until the angle on the potentiometer matches that specified by the computer.

The rotary motors mentioned above are useful whenever a mechatronic musical instrument's mechanism must be rotated. Bow-wheels, pickers and dampers often make use of rotary motors: some instruments, such as MechBass (McVay 2015) and GuitarBot, use large numbers of motors. Such motors, though, can be somewhat noisy, slow and difficult to digitally control. For these reasons, many inventors of mechatronic musical instruments opt for the simpler and quieter solenoid actuators.

Solenoid actuators are among the simplest type of actuators, consisting of a coil wrapped around a magnetic plunger. When electrical power is applied to the coil, the magnetic plunger moves in response to the electromagnetism of the coil. In effect, this is a sort of electrically switchable magnet: the plunger moves towards the magnet when the magnet is switched on. Solenoids are often quite fast and quiet, allowing for high-speed percussive events to be generated: these actuators are particularly popular for robot drumming mechanisms, allowing drumsticks to rapidly strike drumheads or piano keys. Musical roboticist Gil Weinberg's percussion instruments (Hoffman and Weinberg 2010), as well as the popular Yamaha Disklavier player pianos, are examples of instruments that make extensive use of solenoid actuators.

### 3.2.3 Controlling Mechatronic Musical Instruments

Without a means of controlling the motors, a mechatronic musical instrument is merely a collection of metal and wires. By adding circuitry that allows a composer, sound designer or performer to switch the motors on and off in a controlled manner, this assemblage of parts turns into a controllable instrument.

The majority of mechatronic musical instruments make use of similar control schemes. This scheme consists of a host computer with some form of output interface that allows the computer to communicate with the mechatronic instrument, a computer onboard the mechatronic instrument, and power electronics to convert the low-powered signals that are generated by the onboard microcontroller into high-powered ones to drive the instrument's actuators. The remainder of this section will break down each of these steps, providing an overview of the way that a human-controlled computer can interface with a mechatronic musical instrument.

The musical interfaces mentioned earlier in this chapter were input devices, allowing a user to input control gestures into a computer. When working with mechatronic musical instruments, the opposite scenario is needed: messages created inside a computer (often with the aid of user-controlled musical interfaces) need to be output from the computer into the "brain" of the musical robot. To do this, musical communications schemes are used: these schemes are understood by both the host computer and the musical robot. While a variety of communications schemes exist, the one most commonly used for mechatronic musical instruments is MIDI: easy to implement and compatible with many DAWs, MIDI remains nearly ubiquitous in spite of more modern protocols such as Open Sound Control (OSC).

MIDI messages can be used to control a wide range of mechatronic musical instrument parameters: mechatronic percussion instruments, for example, often use the MIDI NoteOn command to switch on a solenoid at a particular voltage level that corresponds to the MIDI note's velocity. A mechatronic guitar might use the MIDI NoteOn to move a motor to the correct position on its neck and, once there, to pluck the string at a particular intensity.

When using MIDI to control mechatronic musical instruments, a MIDI message is transmitted from a DAW on a PC to the musical robot's microcontroller. The MIDI signal is often converted, with the aid of a MIDI interface, from USB MIDI to the more traditional (and easy to implement in hardware form) DIN5-based hardware MIDI. Once the MIDI message is received by the musical instrument, it must be interpreted: custom software running on the onboard microcontroller (which, in recent mechatronic instruments, is often Arduino-based) parses the MIDI message and powers the actuators that correspond to those specified by the MIDI message.

After parsing the MIDI with its onboard software, the musical instrument's microcontroller sends a control signal to actuator driver electronics. The control signals sent by the microcontroller are low voltage and low current, lacking sufficient power to cause motors to spin or solenoids to move. The role of the actuator driver electronics, then, is to convert this low-power signal sent by the microcontroller to a higher-powered signal capable of moving whatever actuators are on the instrument. The simplest actuator control electronics are those that drive solenoid actuators. This circuit, shown in Figure 3.8, consists of a MOSFET (a transistor that serves as a sort of voltage-controlled switch, letting high power flow between the transistor source and drain in response to a low-power input signal at the transistor's gate). One of the solenoid's two wires is connected to the positive terminal of a DC power supply. (Caution: when building this circuit, be extremely careful and seek outside assistance when working with power sources!) The other terminal, plugged in to the MOSFET's source, is connected to ground (via the MOSFET's drain pin) once the control signal from the microcontroller arrives at the MOSFET's gate pin. With this circuit implemented, electrical signals from a microcontroller can be used to switch a motor on and off. If the microcontroller is configured to respond to musical messages, then the motor can be made to play musical phrases, realizing the core functionality of any musical robot.

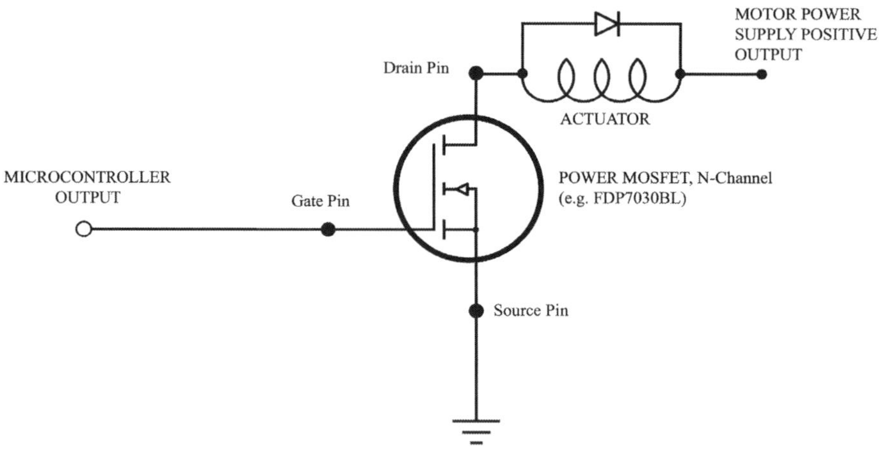


**Figure 3.8** A Basic Power MOSFET Circuit: This circuit may be used to control an actuator such as a solenoid or DC motor.

The role of the mechatronic musical instrument designer is to create instruments that produce sonically interesting, usable and creative outputs in response to user-controlled inputs. These outputs might be photo-realistic emulations of human input (as in the solenoid-powered Yamaha Disklavier pianos), or they might be noisy timbres (as in sonic artist Mo H. Zareei's noisemaking instruments (Zareei et al. 2013)). Or, of particular interest to the sound designer, they might be mechatronic loudspeakers like instrument builder Bridget Johnson's that can use motors to aim at different areas of a space in response to MIDI messages and, once aimed, "play" the room's acoustics (Johnson et al. 2016). Indeed, mechatronic musical instruments design is a field rife with innovation, and many yet-to-be-tried paradigms exist for new inventors and composers to explore.

## References


Arar, R., and Kapur, A., 2013. A history of sequencers: Interfaces for organizing pattern-based music. *Sound and Music Computing Conference*. Stockholm, Sweden: Logos Verlag Berlin.

Bongers, B., 2000. Physical interfaces in the electronic arts. *Trends in Gestural Control of Music*, pp. 41–70.

Cook, P., 2001, April. Principles for designing computer music controllers. *Proceedings of the 2001 Conference on New Interfaces for Musical Expression*. National University of Singapore, pp. 1–4.

Hoffman, G., and Weinberg, G., 2010, April. Shimon: An interactive improvisational robotic marimba player. *CHI'10 Extended Abstracts on Human Factors in Computing Systems*. ACM, pp. 3097–3102.

Johnson, B. D., Kapur, A., and Norris, M., 2016. Speaker.motion: A mechatronic loudspeaker system for live spatialisation. *Proceedings of the 2016 Conference on New Interfaces for Musical Expression*. Brisbane, Australia.

Kapur, A., et al., 2010. The machine orchestra. *Proceedings of the International Computer Music Conference*. New York.

Long, J., 2018. *Closing the Loop: Implementing Real-time Audio Feedback Systems in Musical Robots*. PhD Dissertation, Victoria University of Wellington.

Long, J., Murphy, J., Carnegie, D., and Kapur, A., 2017. Loudspeakers optional: A history of non-loudspeaker-based electroacoustic music. *Organised Sound* 22, no. 2, pp. 195–205.

McVay, J., Carnegie, D., and Murphy, J., 2015, February. An overview of MechBass: A four string robotic bass guitar. *Proceedings of 6th International Conference on Automation, Robotics and Applications (ICARA)*, 2015. IEEE, pp. 179–184.

MIDI Association, 2018. *MIDI DIN Electrical Specification*. [Online] Viewed June 20, 2018 <www.midi.org/specifications-old/item/midi-din-electrical-specification>.

Palacio-Quintin, C., 2008, June. Eight years of practice on the hyper-flute: Technological and musical perspectives. *Proceedings of the 2008 Conference on New Interfaces for Musical Expression*. Genova, Italy, pp. 293–298.

Park, T. H., 2009. An interview with Max Mathews. *Computer Music Journal* 33, no. 3, pp. 9–22.

Roberts, C., 2011, July. Control: Software for end-user interface programming and interactive performance. *Proceedings of the 2011 International Computer Music Conference (ICMC)*. Huddersfield, England.

Shiloh, M., and Banzi, M., 2014. *Make: Getting Started with Arduino*. 3rd ed. San Francisco: Maker Media, Inc.

Singer, E., Feddersen, J., Redmon, C., and Bowen, B., 2004, June. LEMUR's musical robots. *Proceedings of the 2004 Conference on New Interfaces for Musical Expression*. Shizuoka, Japan, pp. 181–184.

Stoffregen, P. J., 2018. *Teensy USB Development Board*. [Online] Viewed June 20, 2018 <www.pjrc.com/teensy/>.

Tanaka, A., 2010. Mapping out instruments, affordances, and mobiles. *Proceedings of the 2008 Conference on New Interfaces for Musical Expression*. Sydney, Australia.

Wright, M., 2005. Open sound control: An enabling technology for musical networking. *Organised Sound* 10, no. 3, pp. 193–200.

Zareei, M. H., Kapur, A., and Carnegie, D. A., 2013. Noise on the grid: Rhythmic pulse in experimental and electronic noise music. *Proceedings of the 2013 International Computer Music Conference*. Perth, Australia.

4

# Sketching Sonic Interactions

Stefano Delle Monache and Davide Rocchesso

## 4.1 Introduction: Why Sketch Sound?

The word *sketching* is rarely associated with the idea of sound. In its common usage, the concept of "sketch" is typically associated with the idea of a visual depiction, mostly rough, and yet expressing something, an object, a landscape, a situation, and so forth, normally by means of few but clearly marked lines and strokes. According to the online dictionary *WordReference*, the noun *sketch* means (1) "a simply or hastily executed drawing or painting, esp. a preliminary one, giving the essential features without details"; (2) "a rough design, plan, or draft, as of a book"; (3) "a brief or hasty outline of facts, occurrences, etc."[1] As a verb, *to sketch* not only means to actually produce a sketch of something or someone, but also "to set forth or describe in a brief or general way." The etymology can be dated back to the fifteenth century, with the widespread availability of paper, pens and inks, and dry media such as chalk and graphite. The noun originated from the Dutch *schetz*, via the Italian *schizzo* (literally splash, of imitative origin), and back to the Latin *schedium* (a form of improvised and extemporaneous speech) and to the Greek *σχέδιοζ* (schédios—temporary, off-hand, done in a moment, unprepared).

Thus, sketching, either as product or as process, suggests a peculiar form of representation inherently capable of eliciting meaning and interpretation, even unintended. Experienced designers, and creative professionals in general, make a wide use of provisional, rough representations, physical props and drawings, especially in the beginning of the design process, when ideas need to be conceived, communicated, compared, developed and selected. In this respect, and regardless the nature of the representation, the sketch is a point of stability, a temporary placeholder around and through which the designer thinks, reasons, makes decisions and advances in the process towards the solution to the particular problem at hand.

In this chapter, we approach the basic characteristics of sketching and the cognitive benefits of such activity. By leveraging the growing body of work around sketching in the visual, interaction and sonic domains, we will see why it is beneficial to consider sketching when approaching a sound design project from scratch. In this respect, the chapter structure is inspired by the workbook *Sketching User Experiences*, by Greenberg and colleagues (2011). That workbook is a valuable resource of practical examples and exercises. Sketching user experiences captures the essence of temporality, dynamics and interaction in designs that cannot be represented by means of static forms of displays only (i.e. drawings). The temporal dimension is inherently essential for the formation and unfolding of the sonic experience, and many of the sketching methods and techniques described in the workbook are valid for and can be applied to the aural domain. Indeed, sonic sketching is not only about designing sound, but rather about designing *with* sound, that is, considering the interplay between the auditory aspect and the other formal, interactive, visual and haptic features of the artifact at hand, from the very beginning of the project.

### 4.1.1 Anatomy of Sketching

The central role of sketching in contemporary design has been advocated by many educators, designers and researchers (Goldschmidt 2017; Buxton 2007; Tholander et al. 2008; Tohidi et al. 2006). Sketching is not about drawing, but about designing, that is, considering multiple ideas and solutions first, reflecting and distilling a subset worth being further elaborated and iteratively transformed and refined.

This process results in a continuous, fluent and uninterrupted loop of thinking while crafting, towards the search for the optimal solution. Whatever the tool or the medium used to produce the representation are, there are, however, certain qualities or prerequisites of the representational means that have been acknowledged to be beneficial for effective sketching (Goldschmidt 2017): a speed of production coherent with the immediacy of thought; an adequate malleability so that the attention is focused on the search, and not spared on the production process; a tolerance for inaccuracy and incompletion so that it is possible to stop the production at any time, without loss of information; the possibility to manipulate the representation, and yet to move back and forth through the various transformations. It follows that the self-generated displays must be *quick* and *evocative*, and since they provide impressions, they are *timely*, *economical* and especially

**Figure 4.1** Thumbnail Sketch of the Ferrari Dino 206 Competizione, Superimposed on the Car Model Prototype.

*Source:* Designer Paolo Martin, 1968—courtesy of the designer.

*disposable*. They can be provided at a glance, and thrown away, since the investment is in the concept and not in the execution, as it can be appreciated in the sketch in Figure 4.1. In a nutshell, these features represent the reason why paper and pencil are still the preferred tools by designers.

However, there is more to it than that. The incompleteness and the provisional nature of the sketch make it *ambiguous* , and thus prone to perceptual reinterpretation and new discoveries. Scribbling is a fundamental form of wandering around in the early stage of the search, as it provides access to interactive imagery (Purcell and Gero 1998): internal representations stored in memory can be recalled and ascribed to the external representation on paper or other sketching media. In turn, mental images inform the making of the sketch. Hence, the sketch talks back to the designer by revealing new relationships and directions, expanding and shrinking the boundaries of the design space towards the resolution of the problem at hand (Goldschmidt 2003).

### *4.1.2 (Sonic) Sketching as Embodied Practice*

Certainly, this process appears to be quite intuitive and sound if framed in the visual domain. But how does it apply to the sketching of embedded media in general, and of sonic interaction in particular? The premise is that sketch-thinking is not dependent on the medium chosen, whether a drawing, a photograph, physical props or sounds. Yet, each medium is more or less appropriate to represent and evoke a specific concept. Hence, as sketching does not equal *drawing*, so sonic sketching is not about *drawing* sound; it is rather a reflective activity occurring while being engaged in representing the *sonic concept* under scrutiny. As a consequence, further disambiguation is needed in the context of embodied sound sketching: drawing-to-sound converters and processors, such as MetaSynth[2] and alike, are creative tools that exploit graphical paradigms to generate and manipulate sound, rather than conceptual representations.

The effectiveness of sketch-thinking resides in the direct involvement of the body. In the case of drawing, it is the hand that helps encode and actively explore the information. The active engagement of the body becomes all the more essential to catch and express the dynamics of designs, behaviors and relations unfolding in time and space, as interactive sound is. In this respect, doing design is primarily an embodied practice made of talk, action, gesture and sketching. Representations of situations, processes, systems and more in general configurations emerge as multimodal explanations that bridge dynamic and static forms of expression (Tholander et al. 2008). Especially when working in a team, designers combine language and physical actions with static displays (e.g. props, whiteboards) to construct and specify complex representations.

We believe that such an attitude can be extremely beneficial to the sound design practice, as well. Effective sonic sketching requires tools and methods that foster embodied thinking, that is, that exploit the body as natural mediator between internal, mental representations and external representations of sound (Kirsh 2013; Leman 2008).

### 4.1.3 Sketching in the Sound Design Funnel

Typically, the conceptual design process can be represented as a funnel, shown in Figure 4.2. The design space expands and shrinks according to the alternation of concept generation and convergence, towards the final concept selection. It is an iterative process in which a certain number of concepts are elaborated and resolved through exploration and clarification. Depending on the stage in the funnel, the representations of the sonic concepts may require a certain degree of detailing. Diverse techniques and methods of *audiolization* (Özcan and Sonneveld 2009) may come into play, before the actual embedding of synthetic sound models and physical computing.

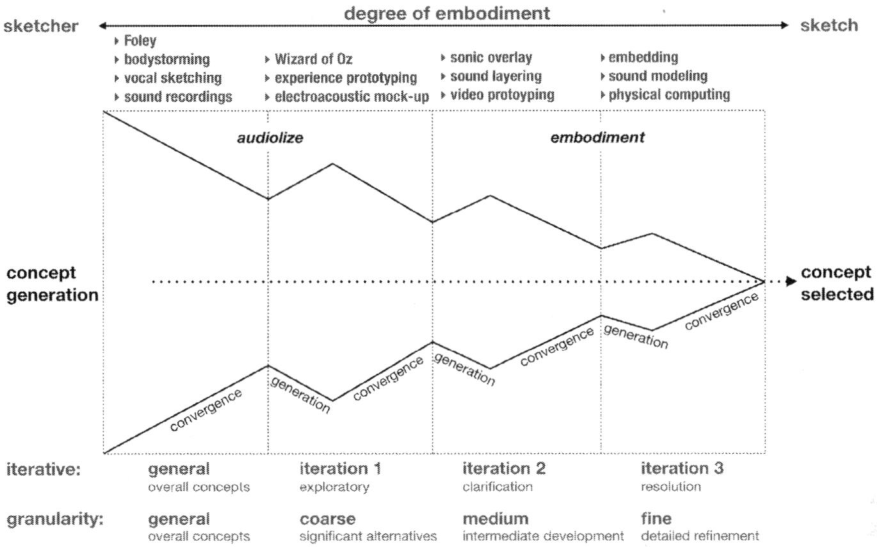


**Figure 4.2** Generation, Elaboration and Reduction of Sound Ideas Represented as a Funnel, With Diverse Methods Suitable to "visualize" the Sonic Concepts According to the Degree of Embodiment Centered on the Sketcher or the Sketch.

*Source:* Adapted from Greenberg et al. 2011, p. 11.

The conceptualization starts with the analysis of the design brief, the rationale. Constraints, requirements, abstract concepts and values, usually organized in diagrams, mind maps, annotations and storyboards, are progressively explored and embodied within physical and tangible properties of the design, that is, form, shape, materials, interactive behaviors, sound functions and configurations, and so forth.

So, where to start for audiolizing abstract conceptual associations? How can we evoke the idea of a gentle and friendly squeezing through sound? We start from the human body. Humans are equipped with extremely powerful and expressive tools to produce nonverbal sounds on the fly, namely, voice and gestures, and the innate ability to communicate sonic concepts through vocal imitations (Ekman and Rinott 2010; Lemaitre et al. 2013 ). In addition, through our body, we can act out scenarios while manipulating and performing sounding objects (Kemper and Hug 2014). Much of the next sections are dedicated to practical ways to organize and optimize the search in the conceptual stage of the sound design process.

## 4.2 Properties and Materials of a Sonic Sketchbook

The sketchbook is a powerful resource for recording, developing, showing and archiving ideas. Designers are used to collect and annotate ideas and pictures, explore variations, develop alternatives and so on.

Sketches, which can be defined as "representations of either direct percepts, or ideas and images held in the mind" (Goldschmidt 1991 ), may fulfill at least four main functions (Van der Lugt 2005). *Thinking* sketches are essentially private and individual, and act as direct support to the design-thinking process, where the designer and the representation are in the continuous loop of understanding while sketching. *Talking* sketches are meant for communication purposes among peers. The representation surface is shared and available for cooperative manipulation and discussion. *Storing* sketches allow the designers to archive their ideas and return to them later. Finally, sketches may have a *prescriptive* function, when they are meant to specify features or decisions to actors not directly involved in the design process. The last two functions somehow freeze, rather than develop, design ideas, although the first is meant for retaining information, while the second for communication purposes. The four functions can represent another way to descend the design funnel, according to the degree of embodiment from the sketcher to the sketch (see Figure 4.2).

Designers and architects are used to collect "external" images, physical and digital , as sources of inspiration for their projects. Collecting is an ongoing activity aimed at fostering wandering around in the early stage of concept generation. Generally, the collection is fashioned in order to facilitate a serendipitous encounter with others' works and casual browsing (Keller et al. 2006 ). It has been demonstrated how early and repeated exposure to examples prior to prototyping improves the creative work (Kulkarni et al. 2014).

Certainly, sound ideas can be annotated on paper, yet a proper sonic sketchbook can be organized by means of systematic collection of physical objects and use of software tools, digital repositories and archives.

### *4.2.1 Collecting Found Objects*

Sounding objects are inherently evocative. Sound carries information about processes, sources, intentions behind the sound-producing actions and their meaning. Found objects can be collected not only as resources for physical sound-making, as in Foley-oriented approaches (Kemper and Hug 2014), but also as sounding tools for reflection. In this respect, if the aim is to have a sketchbook supporting *thinking*, *talking* and *storing* functions, it becomes crucial to organize sounding objects in categories that can enhance serendipity in casual browsing and creative audiolization. Metaphors and conceptual associations are powerful tools used by designers to explore and communicate the design space. A chair not only communicates its function, but also its physical properties (visual, auditory, tactile) and its context of use, and these concrete aspects can evoke specific abstract associations, such as luxury, softness, pleasure, discomfort, and so on.

Therefore sound-producing tools can be organized in boxes or shelved according to their concept-evoking sound, becoming sort of sonic concept generators.

Özcan et al. (2014) categorized product sounds in six categories (air, alarm, cyclic, impact, liquid, mechanical) and nine basic conceptual associations capable of mentally evoking these categories (sound source, action, location, sound type, onomatopoeias, psychoacoustics, temporal description, emotion, abstract meaning).

Lemaitre et al. (2015) studied the use of vocal imitations of nonverbal sounds, that is, vocally self-generated concepts, as perceptually relevant means to categorize environmental sounds. The resulting classification is organized in three main families, namely, *machines*, *basic mechanical interactions* and *abstract sounds*, further split into classes of sounds reflecting perceptually relevant sound concepts, unambiguously discriminated in terms of interaction, as well as temporal and timbral properties. The *machine* family includes mechanical and electromechanical objects, and it is further arranged in ten classes: *alarms*, *button and switches*, *doors*, *hand tools*, *fridges*, *blenders and mixers*, *photocopiers*, *windshield wipers*, *car revs up* and *car interior*. The *basic mechanical interactions* family includes sounds that are produced mechanically as a result of the interaction between two objects or by the deformation of one object. They can be the result of interactions between *solids*, *liquids* and *gasses*. Since human gestures form the basic level of the cognitive representations of action sounds, the resulting ten classes are organized into *blowing*, *whipping*, *shooting*, *crumpling*, *rolling*, *scraping*, *hitting*, *dripping*, *filling*, *gushing*. Finally, the *abstract sounds* family contains sounds that cannot be ascribed to any mechanical source and that become relevant for their spectrotemporal morphologies. *Abstract sounds* can be categorized in six classes of sound morphologies, that is *up*, *down*, *low-high-low*, *impulse*, *repeated*, *stable*.

Other strategies for sound categorization can be used as well in the organization of collected sounding objects, as long as they facilitate conceptual association and exploration. For instance, Hug elaborated a conceptual strategy based on the abstract affective dimensions of sounds, starting from the interpretive role of sound in fictional media, like film or games. Thus, sound may evoke abstract concepts related to qualities of use, qualities of power and energy, structural states, manifestation of life, gesturality and motion, atmosphere or mood, and others relevant themes in user-artifact interaction (Hug 2010).

*Exercise 1:* Collect one sounding object a day. Write down the reasons why its sound captures your attention. The day after revise the text and try to label the reasons according to sound categories and conceptual associations. Skim the most meaningful label(s) representing the sounding object and start populating the corresponding boxes.

### 4.2.2 Repositories and Archives

Repositories, archives and private collections, on the web or locally stored on hard drives, represent a primary inspirational resource. Many commercial libraries of sound effects are available on the market. *Sound Ideas,*[3] *Hollywood Edge*[4] and *BlueBox*[5] libraries, just to name a few, provide rich collections of sound databases thematically arranged. *Crowdsource SFX*[6] is an interesting platform for the creation and distribution of crowdsourced sound effects libraries. Participating members can access the entire library, yet retain the copyright ownership on their contribution.

Despite the richness and the high quality of the sound recordings, these collection turns out to be difficult to browse. They are normally organized according to metadata, tags of various kinds, encoded in the sound files, which respond to the needs and workflow of sound composers and designers working with sound effects for linear media. The organization reflects the rationale of an efficient selection task, rather than a serendipitous query and exploration.

*Freesound.org*[7] is an online collaborative database, where sounds are uploaded and shared under Creative Commons licenses. Compared to previous databases, there are no quality constraints, and sound contributions can be browsed through text query, content-based similarity search, or via tags and geotags. Notably, it is possible to browse the history of a specific sound based on the remixes by users. In addition, the *Freesound Explorer*[8] is a visual interface that allows the exploration of the database in a two-dimensional space, according to initial text query (Font and Bandiera 2017).

### 4.2.3 Personal Recordings

An alternative to ready-made libraries is to use self-generated representations, that is, personal recordings that can be organized according to individualized criteria. The recordings may include both found and *especially performed* sounds, that is, sounds made with an *intention*, through vocalizations or manipulation of physical objects. However, the physicality of collecting tangible media (i.e. objects, magazines, catalogs and the like) renders collecting a continuous *background* activity, with reduced cognitive load. Searching is facilitated by visual and spatial memory, making the retrieval an activity close to doodling (Keller et al. 2006). On the contrary, interaction with digital collections requires careful preparation, whereas browsing mostly relies on verbal keywords.

Hence, building an audio database can be a very laborious task. Normally, sound files are arranged hierarchically in a tree-structure of folders and subfolders, according to a given rationale. The description, indexing and retrieval of sounds represent the main challenge (Font et al. 2018), although several metadata tools for media browsing are available on the market. Music-oriented software, such as Soundly,[9] Metadigger,[10] BaseHead[11] and Soundminer,[12] are conceived in order to be readily integrated in the typical workflow of digital audio workstations. The cons are that such a traditional workflow results in a disembodied interaction.

Instead, the rationale of the (sonic) sketchbook is primarily to collect food for thought, *thoughtful* sonic materials to facilitate critical and reflective approaches (Löwgren and Stolterman 2004), rather than the immediate use of sound recordings in audio editors and plug-ins. In this respect, more general-purpose media organizers and note-taking tools, such as Evernote,[13] OneNote[14] and Hackpad,[15] become handy to manage digital media, sounds, pictures, diagrams and textual annotations. Social platforms, such as SoundCloud[16] and Pinterest,[17] can be creatively misused to collect and organize self-generated sonic concepts.

As a final remark, whatever the chosen organizational strategy is, it is important to maintain the focus on sound. Making a good plan is crucial. As an example, Figure 4.3 shows a possible arrangement that merges the three sound categorizations described in Section 4.2.1. At this stage, subjectivity comes into play, as there is no unique position for a given sound, especially as far as interpretation is concerned. Hence, we may pin that particular whoosh, produced by squeezing an empty toothpaste tube, as a happy manifestation of life, just like the Skype "contact available" notification.

*Exercise 2:* Choose one of the sounding objects from your Foley box. Explore variations of the categorized sound. You can both play the object and imitate its sound with your voice. Record at least twenty variations (consider video-capturing the performance as well). Choose the three most compelling variations and label them according to the emerging conceptual associations. You are practicing the solfège of the sounding object!

| | Basic Mechanical Interactions | | | Machines | Abstract | | | | | |
|---|---|---|---|---|---|---|---|---|---|---|
| | #solids | #liquids | #gasses | | #up | #down | #l-h-l | #impulse | #repeated | #stable |
| #source | | | *toothpaste tube* | *shaver* | *skype* | | | | | |
| #action | | | *squeezing* | *switching on / off* | *contact available* | | | | | |
| #location | | | bathroom | bathroom | *mobile phone* | | | | | |
| #onomatopoeia | | | *whoosh* | *rattling* | *n.a.* | | | | | |
| #psychoacoustics | | | sharp | sharp | *pitched* | | | | | |
| #temporal description | | | short, impulsive | impulsive, repeated | short | | | | | |
| #emotion | | | happy | *fear* | happy | | | | | |
| #abstract meaning | | | manifestation of life | *mood* | manifestation of life | | | | | |

**Figure 4.3** Example of Sonic Concepts Arrangement and Labeling.

Collecting sounds is a fundamental practice of the design activity. Whether they are physical or digital media, they are intended to foster concept generation and discussion. Moreover, well-curated collections can be shared with peers and collaborators.

## 4.3 Audio Sampling the Real World

Audiolization is the aural sketching mode of quickly representing concepts conveyed through sound. In the previous sections, we showed why it is beneficial to collect and arrange sound ideas. If you, reader, have practiced the exercises on sound collecting, you should have noticed how demanding is to produce and capture variations of a given sound in terms of time and manual or vocal dexterity. Developing fluency and expertise in sonic sketching is essential to preserve cognitive resources. These abilities pertain not only to the ease of expression but also to the capability to see and hear cues and relationships while making the representation, as summarized in the diagram in Figure 4.4.

Fluency and expertise result in almost simultaneous acts of creation—*hearing that*—and understanding—*hearing as*— (Goldschmidt 1991). For example, I *hear that* switching on that specific shaver produces a repeated and impulsive sound, I *hear it as* a rattle with a mood, I *hear that* when manipulating the sounding object in a certain way, I *hear* the mood *as*

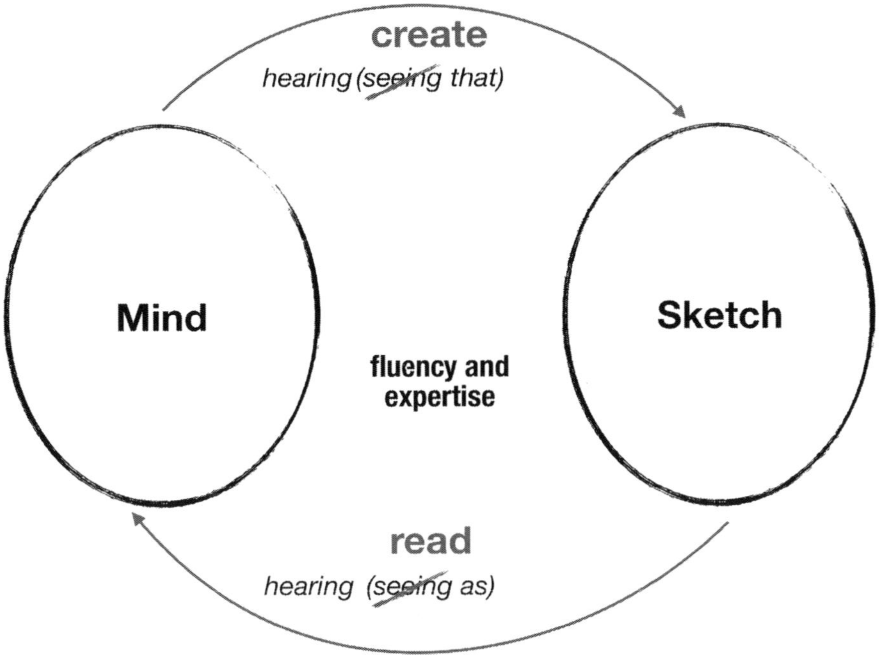


**Figure 4.4** Sonic Sketch and Human Being in Conversation.

*Source:* Adapted from Buxton 2007, p. 5, after Goldschmidt 1991.

evoking fear (see Figure 4.4). In this respect, a good sketch is primarily meant to capture ideas. Generation of new ideas comes as a consequence of the process.

A premise becomes necessary when sketching (with) sound. Visual, two-dimensional objects exist in space and time, and paper has been the natural surface and technology used to host them since the fifteenth century. Auditory objects, that is, sounds, exist in *frequency* and *time*, although they can be segregated *over* (three-dimensional) space (Kubovy and Schutz 2010). There is no such thing as a natural "surface" for sound. Sonic expressions are ephemeral and volatile. Fixed media for sound capturing are unquestionably young compared to drawing. Moreover the temporal dimension in sound capturing is linked to the duration of the sonic representation. Inspecting sound representations may take at least as long as their duration. Conversely, time in visual sketching and inspection is inherently internal to the sketcher. It follows that methods, techniques and tools that ease the capture of sonic ideas are essential for keeping the sketching process as embodied as possible (Delle Monache et al. 2018).

Scribble sketching is the quick and rough activity of collecting ideas. Scribbles and doodles are meant to be for personal use, yet depending on their detailing can be shared with collaborators as well. Essentially, they provide a starting point for idea generation and brainstorming. Diverse tools and applications can become handy to note down ideas.

### *4.3.1 Capturing Sound Scribbles*

Handy equipment for audiovisual recordings are a prerequisite for effectively capturing sound scribbles, that is, interesting sonic events that one may encounter or will want to transcribe during daily activity. Whether one is in line at the post office, or walking to the train station, waiting for the bus, or simply at home in front of the computer screen perhaps, the recording device should be always readily available to capture the moment. Visual designers learn to bring sketchbook and pen, and to scribble-sketch situations and ideas with a few lines in a very limited amount of time (a few seconds). It is not a matter of being a proficient drawer, but rather of grasping with the eyes the salient elements of that situation, and transferring them onto paper, even with very rough, yet meaningful strokes. Often, textual annotations are added to detail the figurative elements. Scribble sketching provides access to interactive imagery. Cameras can be used to sample the world around us, as well. Once more, the photograph or the video recording is not meant to convey a beautiful or well-balanced composition. They are placeholders for further reflection and thinking.

Handheld sound recorders are a good solution in sound design explorations, although they are especially meant for musical purposes. The audio quality ensures the possible reuse of the recording in more advanced stages of the creative process. On closer inspection, this possibility may even become misleading, and move the focus away from the primary objective of sound scribbling. Moreover, organizing the recordings usually requires a set of operations (e.g. transferring the sound files from the SD card) which makes the whole documentation cumbersome and asynchronous compared to scribbling on paper.

Smartphones are our shadow, and they represent an all-in-one solution suitable to sound scribbling. Audiovisual capture is readily available in any standard device on the market. Note-taking apps, as those mentioned in Section 4.2.3, support cloud computing and synchronization. The disadvantage of this solution is that the fluency of the scribbling is dependent on the mastering of the app and its opportunities.

Smart pens, such as Livescribe[18] and Neo Smartpen,[19] represent a further interesting and powerful solution. Smart pens retain the embodied quality of pinning ideas on the paper surface, and yet capture the time of sound by means of embedded microphones. In addition, notes and scribbles are digitized and shared on corresponding mobile apps for further elaboration.

As opposed to taking pictures or videos, taking audio or handwritten notes are activities that often go unnoticed, thus introducing very little perturbation in the observed environment. Audio recording can take place very close to where human action is, without actually disturbing the genuine pace of activities. This is a definite advantage of audio recording as compared to camera shooting, which often exceeds the apparent richness of the audiovisual medium.

In any case, it should be clear that choosing and mastering a capturing system does not equate with dexterity in sonic sketching.

*Exercise 3:* Go hunting for examples of bad sound designs. Pinpoint a few products or situations and record their sound. Why does it irritate you? Why do you find it inadequate? Collecting and discussing bad sound designs fosters sensitization, awareness and critique.

*Exercise 4:* Go hunting for a sound event and make the audio-only recording. Try to understand the reason(s) why that particular sound idea interests you. Go out in the wild again, and look for twenty different samples that *follow* your idea.

Now, work by contradiction, and look for sound events that evoke the opposite of your idea (e.g. dynamic-static, hot-cold, happy-sad, energized-exhausted, textural-gestural, informative-confusing, and so on).

Repeat the same exercise with audiovisual capture.

### 4.3.2 Sound Scribbling Through Vocalizations

The vocal apparatus is the primal, embodied sound scribbling tool available to access sound imagery interactively. Its effectiveness is rooted in the natural ability of humans to produce vocalizations of *non-verbal everyday, abstract sounds*. You practiced vocal scribbling during childhood, perhaps imitating the sound of a car passing by, gunshots or lightsaber sounds. *Vocal imitations* are semantic representations, directly ascribable to a person's idea of a given sound (Lemaitre et al. 2016).[20] They are an effective communication device, more than the verbal description of the same sound idea, especially when the (source of the) referent sound is not known (Lemaitre et al. 2013). They are *sketches* of auditory concepts, in the same way freehand drawings represent visual objects.

Vocal imitations rely on subtle sensory processing mechanisms, that is, the selection of the most relevant acoustic features, for the identification of a given sound, that *can be produced with the human voice*. For example, the human voice can reproduce the pitch, rhythm and sharpness (i.e. spectral centroid) of a given sound. Likewise, when vocal imitations are elicited through imagery, they seem to rely on some sort of iconic similarity with the thought sound, that is, a few salient characteristics picked up and shifted in the imitator's vocal register. In this respect, vocal imitations act much like sound *caricatures*.

Imitations can be *wild* or *tame*. Tame vocal imitations are constrained by the linguistic system they belong to, such as onomatopoeias (e.g. crash, word imitating the sound) and ideophones (e.g. sparkle, word eliciting multisensory representations), and as a consequence, they are also constrained by what they refer to. Instead, sound scribbling through vocalization refers to the exploitation of wild imitations. Wild vocal imitations are commonly used by humans when they run out of words, or simply for economic reasons within the flow of a conversation.

*Exercise 5:* Start paying attention to the use of wild and tame vocal imitations around you. Observe when you use wild vocal imitations to describe some situations and ideas to your interlocutor(s). How did you use them, and why? Observe your interlocutors, as well. Try to register the imitation strategy and the articulation of your voice. Then, focus on the concept(s) again. Could you have articulated them in other ways?

Wild vocal imitations imply two main mechanisms of sound production: myoelastic vocal fold vibrations producing tonal vocalizations, and various constrictions of the vocal tract that are used to imitate turbulences in noisy sounds. Professional imitators and artists are able to control vocal production to impersonate characters and produce complex sound effects, faithfully.

Fluency in vocal scribbling is a skill that can be honed through experience and practice. In this respect, *Mouthsounds* is a valuable resource to start learning and practicing techniques for the production of wild vocal imitations (Newman 2004). The book is organized around the increasing difficulty of the techniques, and a pragmatic taxonomy of vocal imitations that almost mirror the categories of sounds introduced in section 4.2.1. This rationale is reflected in corresponding articulatory characteristics of vocal production. With a few basic techniques and their virtuoso combination, it is possible to cover a large variety of sound categories: frictions, impacts, turbulences, machines, artificial sounds and so forth.

However, as for drawing in visual sketching, you do not have to be a virtuoso imitator to produce effective vocal scribbles.

*Exercise 6:* Vocal scribble what you hear. Pick a sounding object from your collection. Listen carefully to its sound. How is the sound produced? Which are the interacting materials? Are there frictions, impacts, liquids or airstreams involved? How do they evolve in time? Now try to imitate the sound by exploring the vocal articulation and finding the one that suits you best. Record the imitation, listen to it and compare it to the referent sound. Are they similar? Is there anything missing? Is the sound compound? Is there any apparent hierarchy of sound events in the compound? Now try to reproduce that sound as faithfully as possible by imitating the layers composing the sound. Record and edit them.

*Exercise 7:* Sketch from memory. This exercise involves two or more participants and is based on guessing and turn-taking. Make a written list of sounding objects, machines and basic mechanical interactions. Choose a verbal description, vocal scribble its sound and make the other participant(s) guess the sounding object at hand. Do not take much time thinking about how to produce the sound, just do it, iteratively. It's fun! When the verbal description is discarded, the turn passes to the next participant. Tame imitations, onomatopoeias and ideophones, are not allowed.

*Exercise 8:* Sound scribble emotions. Sonic concepts may evoke specific emotions, mood and abstract meaning. Like the previous exercise, this involves two or more participants, and is based on guessing and turn-taking. Make a list of moods and emotions. Pretend to be a robot and try to vocalize its mood this time. No speech or onomatopoeias are allowed.

*Exercise 9:* Scribble variations. Go back to Exercise 3 and recover the audiovisual capture of your idea. Try to produce twenty variations that follow your idea by replacing the original audio with your vocalization.

## 4.4 The Collaborative Sketch

As discussed in section 4.2, sketches can have diverse functions: to support individual reflection (*thinking*), to communicate ideas to peers (*talking*) and stakeholders (*prescriptive*) and to freeze ideas for later inspections (*storing*). Teamwork in contemporary design has become a common habit, if not a norm. The design process is carried out by teams of designers with different backgrounds in diverse domains. Yet, current practices in sound design confine the sound designer in her studio, immersed in a multitude of software tools and plugins. Typically, meetings with stakeholders and peers are limited to the analysis and discussion of the design brief, the presentation of advanced sound proposals and their validation. In this section, we focus on the production of sketches that are meant for collaborative interaction with other team members.

### 4.4.1 Group Interaction Around the Sonic Sketch

A variety of methods and techniques are available for collaborative sketching in the visual and, especially, interaction domains. Team composition and individual role-taking may affect team dynamics in various ways, making the design process more or less coherent and smooth. In this respect, collaborative sketching is about sharing mental models, that is, internal representations that are in the mind of individual designers. Collaborative representations allow designers to identify and negotiate shared aspects across mental models. However, as a general rule for effective group interaction, it is essential to have the sketching surface shared and accessible to all the members. Group interaction may happen around a table by using shared sheets of paper, pens, sticky notes and physical props, or standing in front of a whiteboard. It is important to provide an environment that facilitates both individual thinking and collaborative discussion.

Therefore, contrary to current sound design practices, mostly based on verbal interaction, when designing cooperatively with sound, it is beneficial to have externalized representations of sound concepts to be readily shared. As effective sketching relies on embodied interaction, it is straightforward to prepare the design environment with appropriate sounding materials and recording tools that would not divert the design flow. Collecting and sharing conceptual sounding objects is a viable way to drive sound scribbling. In addition, as cooperative sketching is inherently multimodal, the meaning and the specification of the sound design entity at hand emerge in a twine of talking, gesturing, drawing, manipulating objects and audiolizing.

*Exercise 10:* Acousmatic narratives. Vocal sketching is prone to improvisational and collaborative performance. The exercise devises a warm-up tool for team-building when designing with sound. The exercise was originally conceived by Ekman and Rinott (2010), and challenges the team to produce polyphonic sketches of fictional machines, in a constrained amount of time.

*Brief*: Design a fictional machine character that produces a living creature from sawdust.

Describe the process through sound only (input → transformation → output).

Sound design themes are (1) eighteenth-century steam engines + magic; (2) sci/fi futuristic + intelligence; (3) organic life form + horror.

You have twenty-minutes to design what is actually happening in the process, sonify the process through vocal sketching, rehearse the performance and perform it live. Constraints: no speech, onomatopoeias or gestures are allowed. Film the performance, then review and discuss it together. Does it convey your idea?

### 4.4.2 *Use of Gestures in Cooperative Sound Design*

In collaborative sketching, static forms of representation are complemented by active engagement and performative actions. Gestures are an important part of nonverbal communication. They are effective visuospatial devices that designers exploit to catch and convey explanations. Observe the way you and your peers make use of gestures to support reasoning around a conceptual sketch. You may find yourself gesturing to punctuate the discourse (*beats*), to highlight a design element by pointing at it (*deictic*), to mimic some situations and behaviors (*iconic*, e.g. knocking gesture to mean knocking at a door) or to represent abstract concepts (*metaphoric*, e.g. shapes or some other structural properties) pertaining to the design at hand (Tholander et al. 2008; Tversky et al. 2009).

Lemaitre et al. (2017) studied the use of gestures as visuospatial means supporting the communication of sonic concepts through vocal imitations, and they found that gestures are used by humans to encode some information pertaining to sound. As discussed in previous sections, visual and aural depictions are selective and bound to some salient features of the referent percept, which are perceptually relevant to the sketcher. Sonic gestures are visual signals capable of communicating auditory sensations and causal representations. In this respect, they act much like *metaphors*. Causal representations show as pantomime gestures mimicking the sound source, when it is identified (e.g. closing the door abruptly, for a slamming sound). Conversely, when the source is not known, sound tracing gestures depict the spectrotemporal evolution of the sound (see Lemaitre et al. 2017 for a comprehensive overview). The idea of the sound is thus embodied in the action.

Gestures representing auditory sensations are instead based on spatial metaphors. When combined with vocalizations, these gestures are used to complement one salient feature of the represented sound because it is either difficult or impossible to produce with the voice, such as when rapidly shaky gestures are used to represent noisy components or high density events (e.g. the rumble of the refrigerator). Small spatial trajectories, such as pinching, are used to represent the tonality of depicted sound.

Gestural interaction represents a significant part of collaborative sketching. Do not undervalue the role of gestures when explaining a sound design concept. Engaging in bodily expression makes the communication and sharing of mental models more accessible and fluent. Warm-up exercises and theatrical methods may help to overcome the natural social discomfort of exposing one's own body.

*Exercise 11:* Go back to Exercise 10 and recover the sound design proposed. Reenact the performance by resorting to vocal and gestural sketching this time. Film the performance and compare it with the previous vocal-only video recording.

## 4.5 Hybrid Sound Sketching (Getting Physical)

Embodied sound design is a process of sound creation that extensively involves the designer's body, from the early generation of ideas to the detailing of sound specifications in embedded media. We contend that the human ability to embody proximal auditory sensations and distal causal representations in vocal and gestural depictions forms the fundamental stage to access sound imagery, interactively. Why should you bother learning sketching skills, then? Learning to produce rich and malleable representations facilitates mental processing, planning and evaluating. Developing sonic sketching skills (*hearing that*) prevents mismatches with the mental, internal representation of the design entity (*hearing as*). In turn, information processing abilities, that is, interpreting and depicting representations in the mind, are gained through experience and practice.

In a rather evocative way, descending the sound design funnel (see Figure 4.3) is about progressively distilling and embodying the initial concept(s) existing in one's own body and mind only into structural properties and configurations of the design entity under scrutiny. Horizontal wanders and explorations of alternative sound ideas are kept internally towards the designer's body, while vertical developments and detailing make a specific sound design entity progressively coming to existence *per se*, which may provide a clue to the expression "to give a birth to an idea."

Diverse techniques and sound sketching strategies reflect this degree of sound embodiment centered on the sketcher or the sketch (see Rocchesso et al. 2013 for the pedagogical implications).

*Vocal and gestural sketching* is the most immediate way to enact interactive sonic representations. It is intuitive, directly available and tightly coupled to imagery in capturing and manipulating the temporal aspects of the sound design inquiry. Cons are the inherent ephemerality of the medium and the timbral limitations of the vocal apparatus (e.g. inharmonicity and polyphony are largely precluded). Computing systems that convert vocal imitations into configurations of synthetic sound models, that can be further controlled with the voice, are making progress[21] (Delle Monache et al. 2018). Vocal sketching represents an effective means of physically situated brainstorming.

*Bodystorming* (Schleicher et al. 2010) and vocal sketching are closely related, as they both involve the designer's body. More generally, bodystorming is a fast concept generation technique, widely used in interaction design. This method makes use of living personas and props to act out a design scenario. The designer can either observe the "actors" playing out the interaction or play it herself to gain insights from the emerging experience. For example, one actor may play the user, and another one may play the interface by vocalizing or by manipulating sounding objects. Bodystorming and vocal sketching clearly fulfill a *thinking* function in cooperative creation.

*Building provisional physical mock-ups* (Kemper and Hug 2014) represents the first step towards the embodiment of generated concepts in an autonomous design entity. Mock-ups do not need to be refined; they are still conceptual placeholders, and yet external to the designer's body. They can be made out of paper or other easily manipulable material. The techniques of pop-up book construction can be repurposed to create sound-augmented movables that afford performative human manipulation and compelling first-person experiences[22] (Delle Monache et al. 2012). The emerging behaviors can also be inspected and analyzed from an external point of observation. At this stage, vocal or Foley sounds can be replaced with recorded sounds. Incompleteness and ambiguity are a resource, and alternative ideas can still be produced, without compromising the creative process in some forms of design fixation.

*Video-prototyping and sonic overlay* are rather refined techniques to fix *talking* and *prescriptive* sketches while advancing in the design funnel. They consist in filming the interaction and adding recorded sounds at a later stage.[23] Here, more complex editing tools come into play, which may loosen the closed loop sketch-thinking. Therefore, it is important to not overload the cognitive activity of designing. On the other side, sonic overlay affords the production of a variety of rather refined demonstrations and solutions that can be compared and effectively communicated, without delving yet into fully working prototypes and technological issues.

*Exercise 12:* Consider the Ferrari Dino 206 Competizione in Figure 4.1. How would it sound today if it were electric? Produce different sonic overlays by vocal sketching first, and creating more refined sounds later.[24]

There comes a point in the design process when *physical and sound computing* solutions become necessary to embody concepts in most advanced mock-ups and prototypes. Computing technologies are not only tools, but also proper design materials. Sensor and actuators can be physically embedded in designs and provide perceptual and expressive means to manifest computed effects on the environment. Sketching by computing concerns both the control/display dimension and the rough crafting of the electronic circuitry. The concept is fully embodied in the sketch, and requires appropriate and malleable digital sound models that afford a sensible fitting with the overall behavior of the designed artifact[25] (Rocchesso et al. 2009; Baldan et al. 2017).

## 4.6 Conclusions

This chapter provided an overview of methods and practices for sketching sound in interactive contexts. We proposed an account of the sketching activity centered on the designer's bodily engagement with representations and displays. We presented several design approaches to the embodiment of sound in concept generation, along the design funnel. For additional online resources related to this chapter, please visit http://bit.ly/SonicSketching.

## Notes

1 www.wordreference.com/definition/sketch, accessed online on January 16, 2018.
2 www.uisoftware.com/MetaSynth/, accessed online on April 19, 2018.
3 www.sound-ideas.com/, accessed online on January 20, 2018.
4 www.hollywoodedge.com/, accessed online on January 20, 2018.
5 www.bestservice.com/blue_box_16_cd-set.html, accessed online on January 20, 2018.
6 www.crowdsourcesfx.com/, accessed online on January 20, 2018.
7 https://freesound.org/, accessed online on January 20, 2018.

8 https://labs.freesound.org/apps/freesound-explorer.html, accessed online on January 20, 2018.
9 www.getsoundly.com/, accessed online on January 20, 2018.
10 www.sound-ideas.com/Page/metadigger-data-management-software, accessed online on January 20, 2018.
11 www.baseheadinc.com/, accessed online on January 20, 2018.
12 http://store.soundminer.com/, accessed online on January 20, 2018.
13 https://evernote.com, accessed online on January 22, 2018.
14 www.onenote.com/, accessed online on January 20, 2018.
15 https://paper.dropbox.com/hackpad/, accessed online on January 20, 2018.
16 https://soundcloud.com/, accessed online on January 20, 2018.
17 https://pinterest.com, accessed online on January 20, 2018.
18 www.livescribe.com/, accessed online on January 20, 2018.
19 www.neosmartpen.com, accessed online on January 20, 2018.
20 https://vimeo.com/125024731, accessed online on April 10, 2018.
21 miMic: the microphone as a pencil, https://vimeo.com/142351022, accessed online on April 10, 2018.
22 www.youtube.com/watch?time_continue=2&v=msvTfA5uHis, accessed online on April 10, 2018.
23 https://vimeo.com/12549217, accessed online on April 10, 2018.
24 A video of the car model running is available at www.youtube.com/watch?v=LHGOFPGgDl0 - *YouTube Standard License*, accessed online on January 20, 2018. An example of the exercise is available at https://vimeo.com/128886746, accessed online on April 10, 2018.
25 Sound Design Toolkit (SDT), a collection of physically-informed sound models for procedural audio, video tutorials available at https://vimeo.com/album/2105400, accessed online on April 10, 2018. The SDT is available for download at http://soundobject.org/SDT/, accessed online on April 10, 2018.

## References


Baldan, S., Monache, S. D., and Rocchesso, D., 2017. The sound design toolkit. *SoftwareX* 6, pp. 255–260. doi:https://doi.org/10. 1016/j.softx.2017.06.003.

Buxton, B., 2007. *Sketching User Experiences: Getting the Design Right and the Right Design*. San Francisco: Morgan Kaufmann.

Delle Monache, S., Rocchesso, D., Bevilacqua, F., Lemaitre, G., Baldan, S., and Cera, A., 2018. Embodied sound design. *International Journal of Human Computer Studies* 118, pp. 47–59. https://doi.org/10.1016/j.ijhcs.2018.05.007.

Delle Monache, S., Rocchesso, D., Qi, J., Buechley, L., De Götzen, A., and Cestaro, D., 2012. Paper mechanisms for sonic interaction. *Proceedings of the Sixth International Conference on Tangible, Embedded and Embodied Interaction*. New York: ACM, pp. 61–68. doi:https://doi.org/10.1145/2148131.2148146.

Ekman, I., and Rinott, M., 2010. Using vocal sketching for designing sonic interactions. *Proceedings of the 8th ACM Conference on Designing Interactive Systems*. New York: ACM, pp. 123–131.

Font, F., and Bandiera, G., 2017. Freesound explorer: Make music while discovering freesound!. *Proceedings of 3rd Web Audio Conference*. London: Queen Mary University of London, August 2017.

Font, F., Roma, G., and Serra, X., 2018. Sound sharing and retrieval. In: Virtanen, T., Plumbley, M., and Ellis, D. (Eds.) *Computational Analysis of Sound Scenes and Events*. Cham: Springer, pp. 279–301.

Goldschmidt, G., 1991. The dialectics of sketching. *Creativity Research Journal* 4, no. 2, pp. 123–143.

Goldschmidt, G., 2003. The backtalk of self-generated sketches. *Design Issues* 19, no. 1, pp. 72–88.

Goldschmidt, G., 2017. Manual sketching: Why is it still relevant? In: Ammon, S., and Capdevila-Werning, R. (Eds.) *The Active Image. Philosophy of Engineering and Technology* 28, pp. 77–97. Cham: Springer.

Greenberg, S., Carpendale, S., Marquardt, N., and Buxton, B., 2011. *Sketching user experiences: The workbook*. Boston, MA: Morgan Kauffman.

Hug, D., 2010. Investigating narrative and performative sound design strategies for interactive commodities. In: Ystad, S., Aramaki, M., Kronland-Martinet, R., and Jensen, K. (Eds.) *Auditory Display. Lecture Notes in Computer Science*, Vol. 5954. Berlin: Springer.

Keller, A. I., Pasman, G. J., and Stappers, P. J., 2006. Collections designers keep: Collecting visual material for inspiration and reference. *CoDesign: International Journal of CoCreation in Design and the Arts* 2, no. 1, pp. 17–33.

Kemper, M., and Hug, D., 2014. From foley to function: A pedagogical approach to sound design for novel interactions. *Journal of Sonic Studies* 6, no. 1. Viewed January 20, 2018 <http://journal.sonicstudies.org/vol06/nr01/a03>.

Kirsh, D., 2013. Embodied cognition and the magical future of interaction design. *ACM Transactions on Computer-Human Interactions* 20, no. 1, Article 3. doi:https://doi.org/10.1145/2442106.2442109.

Kubovy, M., and Schutz, M., 2010. Audio-visual objects. *Review of Philosophy and Psychology* 1, no. 1, pp. 41–61.

Kulkarni, C., Dow, S. P., and Klemmer, S. R., 2014. Early and repeated exposure to examples improves creative work. In: Leifer, L., Plattner, H., and Meinel, C. (Eds.) *Design Thinking Research. Understanding Innovation*. Cham: Springer.

Lemaitre, G., Houix, O., Voisin, F., Misdariis, N., and Susini, P., 2016. Vocal imitations of non-vocal sounds. *PloS One* 11, no. 12, e0168167. https://doi.org/10.1371/journal.pone.0168167.

Lemaitre, G., Scurto, H., Françoise, J., Bevilacqua, F., Houix, O., and Susini, P., 2017. Rising tones and rustling noises: Metaphors in gestural depictions of sounds. *PloS One* 12, e0181786. https://doi.org/10.1371/journal.pone.0181786.

Lemaitre, G., Susini, P., Rocchesso, D., Lambourg, C., and Boussard, P., 2013. Non-verbal imitations as a sketching tool for sound design. *International Symposium on Computer Music Modeling and Retrieval*. Cham: Springer, pp. 558–574.

Lemaitre, G., Voisin, F., Scurto, H., Houix, O., Susini, P., Misdariis, N., and Bevilacqua, F., 2015. *A Large Set of Vocal and Gestural Imitations, Deliverable of Project SkAT-VG*. Paris: Ircam—Institut de Recherche et Coordination Acoustique/Musique. Viewed January 23, 2018 <http://skatvg.iuav.it/?page_id=388>.

Leman, M., 2008. *Embodied Music Cognition and Mediation Technology*. Cambridge, MA: MIT Press.

Löwgren, J., and Stolterman, E., 2004. *Thoughtful Interaction Design: A Design Perspective on Information Technology*. Cambridge, MA: MIT Press.

Newman, F., 2004. *MouthSounds: How to Whistle, Pop, Boing and Honk for All Occasions . . . and Then Some*. New York: Workman Publishing Company.

Özcan, E., and Sonneveld, M., 2009. Embodied explorations of sound and touch in conceptual design. *Proceedings of Design Semantics of Form and Movement '09*, pp. 173–181. Taipei: Koninklijke Philips Electronics N.V.

Özcan, E., Van Egmond, R., and Jacobs, J., 2014. Product sounds: Basic concepts and categories. *International Journal of Design* 8, no. 3. Viewed January 23, 2018 <www.ijdesign.org/index.php/IJDesign/article/view/1377/655>.

Purcell, A., and Gero, J. S., 1998. Drawings and the design process: A review of protocol studies in design and other disciplines and related research in cognitive psychology. *Design Studies* 19, no. 4, pp. 389–430.

Rocchesso, D., Polotti, P., and Delle Monache, S., 2009. Designing continuous sonic interaction. *International Journal of Design* 3, no. 3. Viewed January 23, 2018 <http://ijdesign.org/index.php/IJDesign/article/view/620/271>.

Rocchesso, D., Serafin, S., and Rinott, M., 2013. Pedagogical approaches and methods. In: Franinovic, K., and Serafin, S. (Eds.) *Sonic Interaction Design*, Chapter 4. Cambridge, MA: MIT Press, pp. 125–150.

Schleicher, D., Jones, P., and Kachur, O., 2010. Bodystorming as embodied designing. *Interactions* 17, no. 6, pp. 47–51. New York: ACM.

Tholander, J., Karlgren, K., Ramberg, R., and Sökjer, P., 2008. Where all the interaction is: Sketching in interaction design as an embodied practice. *Proceedings of the 7th ACM Conference on Designing Interactive Systems*. New York: ACM, pp. 445–454.

Tohidi, M., Buxton, W., Baecker, R., and Sellen, A., 2006. User sketches: A quick, inexpensive, and effective way to elicit more reflective user feedback. *Proceedings of the 4th Nordic Conference on Human-computer Interaction: Changing Roles*. New York: ACM, pp. 105–114.

Tversky, B., Heiser, J., Lee, P., and Daniel, M.-P., 2009. Explanations in gesture, diagram, and word. In: Coventry, K., Tenbrink, T., and Bateman, J. (Eds.) *Spatial Language and Dialogue*, Chapter 9. New York: Oxford University Press, pp. 119–131.

Van der Lugt, R., 2005. How sketching can affect the idea generation process in design group meetings. *Design Studies* 26, no. 2, pp. 101–122.

5

# Bringing Sound to Interaction Design

## Challenges, Opportunities, Inspirations

**Daniel Hug and Simon Pfaff**

## 5.1 Introduction

Interaction design (IxD) is a relatively young discipline that has its roots in human-computer interaction (HCI) and links to various other design disciplines, such as industrial design or architecture. Susanne Bødker (2006) coined the term "third wave HCI" to describe how the scope of HCI design moved from work and task orientation towards everyday lives and meaning making. This is complemented by the notion of human action often being opportunistic and situational, rather than based on the execution of formalized plans (Suchman 1987). This approach to design encountered the understanding of human experience as essentially embodied (Dourish 2004) with a new focus on bodily experience, gestural and spatial interaction and multisensory design. Among other aspects, this forms the conceptual background for a new consideration of the possibilities and potentials of using sound in interaction design.

Auditory display design literature posits that sound offers several benefits for interaction design. Concepts and guidelines in relation to notification and warning sounds, design concepts such as earcons and auditory icons, and sonification strategies for the representation of data through sound have been widely discussed and investigated (Neuhoff 2011). The relevance of sound for interaction design has also been recognized at the Interaction Design Department (IAD) of the Zurich University of the Arts (ZHdK), and sound design has been a fixed part of the curriculum since 2005, with students, lecturers and researchers equally contributing to the emerging field of Sonic Interaction Design (Franinovic and Serafin 2013). But in interaction design education, which is multidisciplinary in nature, sound competes with many other areas of knowledge and skills such as user interface design and programming, user experience and service design, as well as electronics, digital fabrication and many more.

The aim of this chapter is to illuminate the approach to sound design activity from an interaction design perspective. This implies that sound is seen as material of design equivalent to other sensory dimensions. We will first outline the cornerstones of the interaction design process as it is implemented at IAD in education and projects. We will then report our approach to educating interaction design students in sound design in the context of the Sonic Interaction Design courses held at the ZHdK. Finally, we will present the approaches and discourse related to the exploration of design solutions from projects that aim at using sound as the principal modality in interaction processes. Retracing the design processes of three workshops carried out between January 2016 and January 2018, we will describe the prevalent motivations for using sound and the scenarios that emerge from the perspective of using sound for interactive artifacts. We will investigate how our design students approach and justify the potentials of sound in interaction design and where they see the limits and problems associated with using sound. We will describe how these concepts materialized in their initial sound ideas and the resulting design approaches. This includes the progression from first results via initial evaluations to subsequent revisions. We will also investigate how the designers interpret results of evaluations, how sound ideas (or issues) influenced the decision-making in terms of interaction design and how this led to the end product.

Central to our approach is the assumption that the resulting projects and artifacts are—and should be—primarily motivated and informed by an interaction design perspective. This results in a relatively pragmatic approach to leveraging "sonic potentials." The motivation of using and designing sound stems from the recognition of their added value from a functional and aesthetic point of view, rather than from a genuine interest, intrinsic motivation and mere joy of creating sounds, which students in a sound or music-related discipline might exhibit.

## 5.2 Beyond Beeps and Clicks: The Ethos of Sound Design

While the potential of sound for products and design in general is increasingly recognized, the actual use of "intentional" (Langeveld et al. 2013) or functional (Spehr 2015) electroacoustic sound, often in the form of warnings, notifications or input confirmations, has been limited to a few dominating design paradigms, namely auditory icons and earcons (Hug 2017). Only recently has the relatively new field of Sonic Interaction Design (Franinovic and Serafin 2013) extended these two central concepts by the notion of continuous sounds (Rocchesso et al. 2009) which make use of generative or procedural methods for sound design (Delle Monache et al. 2010; Farnell 2010).

Still, the majority of functional sounds encountered in auditory displays of everyday products are more or less refined beeps, short melodies or noises. If we compare this current situation to the sounds of the "fictional everyday" from sci-fi and fantasy film, it is striking how elaborate, original, affective and effective—yet still plausible—these turn out to be (Hug 2008). If we consider the effectiveness of these sounds in establishing elaborate visions of a potential future to a degree where the sounds are perceived as natural and highly functional in terms of helping the audience making sense of the interactions depicted on screen, we can also consider them an inspiration—and benchmark—for sound design today, in particular as this future partially has become reality with ubiquitous computing, gestural interfaces and smart speakers.

In a study conducted by Hug and Misdariis (2011), sound design experts emphasize the importance of achieving sonic quality and elaboration. They report, that a common strategy is to play with familiarity, or "naturality," of sounds by carefully processing them in order to create sounds that are novel and innovative, yet approachable. Furthermore, sound for them is never just an optional embellishment or simple "feedback." Emotional and expressive qualities play a central role. As a consequence, sound designers usually strive to combine several sonic elements rather than directly translating a parameter to a single sonic change. This attitude towards sound design is particularly evident in sound design for film, where sounds associated with interactions not only inform about an artifact, but also contribute to its unique emotional quality and role in the narrative (LoBrutto 1994; Hug 2008). However, this design attitude (and ethos) seems to be largely absent in the area of auditory display and product sound design.

While there may indeed be technical reasons for the limited sonic spectrum of most of today's (interactive) products, this is certainly no longer the limiting factor. Also, a clear separation between the experience and interpretation of sounds encountered in "reality" and those encountered in films seems to be missing: studies of interpretational processes related to sounds in the context of using prototypes of interactive artifacts have shown that those interpretations often are based on sounds familiar from film (Hug 2017). This "filmic listening mode," which gets its power from the specific normative (sonic) narratives of mainstream film, can be seen as complementing other listening modes,[1] and could be more actively exploited in sound design.

Thus, it might be beneficial to bring the "filmic" sound design attitude and ethos to interaction design practice. We should not forget that (interaction) designers who use sound contribute to the process of defining possible future experiences of sound in interactive artifacts and what they might sound like. From an optimistic standpoint, the situation can be compared to the time of the "New Hollywood" movement in 1970s cinema, where entirely new sound aesthetics were introduced, which today belong to a—partially globalized—cultural mainstream. In order for this to happen in interaction design, a deeper understanding of the interaction design process and its relation to sound design practice is required. This will be outlined in the following section, focusing on the specific implementation and practice within the IAD of the ZHdK.

## 5.3 The Interaction Design Process as Challenge for Sound Design

Of central concern in the practice of interaction design is an idealized model and methodology of a design activity that is often not only aimed at "problem solving" or changing existing situations into preferred ones (Simon 1996) and goes beyond merely creating the interface between human and artifact (Bonsiepe 1996). Rather, it is understood as a more universal and pervasive creative activity in all possible contexts of human life (Dorst 2006; Löwgren and Stolterman 2004), often dealing with "wicked problems" (Rittel and Webber 1973) or the need to identify and exploit new and unknown design potentials. Common to all design activities is the notion of iteration and a more or less systematic progression through a creative process. While there are relatively formalized notions of the ideal design process, for instance in the context of the "design thinking" movement (Brown 2009), there is in fact a plethora of different models of the design process (Dubberly 2004, Figure 5.1). What most of them share, however, is a rather open, explorative phase in the beginning, with a gradual process of iterative prototyping and refining towards a certain goal. This goal is not known in the beginning of the process and there is no initial and more or less definite requirement analysis and specification. This distinguishes the design process from an engineering approach, which, in this respect, is comparable to the traditional sound design process from film, where a relatively clear functional and formal-aesthetic goal is formulated at a very early stage.

Another important aspect of the design process is that it is supposed to provide a defined space for systematic creativity and innovation using methods to generate and synthesize new ideas, and through related experiment and testing. This implies various levels of prototyping, from sketching (Buxton 2010) or bodystorming (Oulasvirta et al. 2003) to screen prototyping, functional prototypes and tech demos (Lim et al. 2008). Prototyping is used as means of communicating with both members of a design team and the "outside world," be they clients or potential users. It

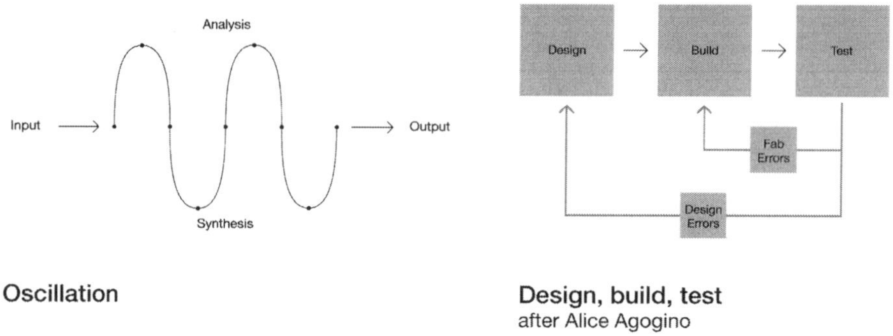


**Figure 5.1** Examples of a Design Process.

*Source:* Dubberly 2004.

thus relates to another topic of our interaction design education, namely, narration and externalization, which serves to convey a design concept with storytelling using personas, scenarios and dramaturgy (Quesenbery and Brooks 2010; Carroll 2000). Besides rational considerations, at IAD it is also important that the design process is informed by sensory experience and abduction (Kolko 2010), which results in the increasing relevance of "experience prototyping."[2]

This openness to sensory experience, storytelling and narration is certainly shared with the sound design discipline, which also thinks in terms of narratives and sonic character design. But apart from this aspect, the topics of innovation and prototyping seem to be a great challenge in sound design. This often manifests itself in the interaction with potential users or clients. For instance, there is no straightforward analogy to visual sketching in sound. As opposed to a sketch, which is able to indicate a design potential, and offers room for imagination, a rough or simplified sound usually just sounds bad.[3] As a consequence, the sound design process is often associated with hours of secluded tinkering and honing of sonic details before a result will be presented to an audience or a client.

An important aspect in interaction design is its reliance on computation as the enabler of the actual interaction between human beings and artifacts (Moggridge and Atkinson 2007). This dialogical principle is important to understand for a sound design process in the context of interaction design. Moreover, interactive computational technology also leads to questions revolving around sound generation and control. The design of sonic interactions takes place in a nonlinear, potentially generative, open-ended setting, which has a strong impact on the actual sound design process. This means that in some use cases it might be necessary to realize the design of sounds using algorithms and control conditions for synthesizing sounds in real time (Farnell 2010), which is very uncommon in the context of the traditional linear sound design process found, for example, in film.

Finally, interaction designers usually explore a more or less unfamiliar domain and often have to justify or substantiate their design decisions vis-à-vis clients. Research, therefore, is an important cornerstone of the design process, helping to gather information about the artifacts, potential users, environments and other aspects of an interaction setting. Often, ethnographic field research methods are adopted, and combined with designerly interventions, such as "probes" (Boehner et al. 2007). Formal-aesthetic analysis and comparative approaches are also often used, and finally, prototypes are evaluated with focus groups or through user tests with methods adopted from psychology to evaluate usability and user experience (Hassenzahl 2013).

This level of detail in research and argumentation is very uncommon in sound design. Usually, the sound designer's artistic intuition still dominates the design process and she or he is required to deliver the potential final product, with the sounds blending into the overall experience of a film.[4] Even in the context of sonic branding and corporate sound, where there is a growing demand for systematic approaches and structured processes to cope with increasingly complex design problems, a systematic, even scientific, evaluation only occurs after significant effort has been made to finalize a sound design (Hug and Misdariis 2011).

## 5.4 Approaches to Sound From an Interaction Design Perspective

In this section, we will discuss the motivations to use sound and the approaches to sound design from an interaction design perspective as it has manifested itself in our work with interaction design students. This is based on a thorough analysis of a series of classes, which were conducted during the third semester of the interaction design program at IAD ZHdK. During the classes, the students were introduced to various methods of creating and representing concepts for using sound, such as graphical scoring, sonic mood boards or bodystorming. Additionally we sensitized the students regarding the relevance and meaning of timbre, composition, temporal development and gesture-sound relationships through a series of short exercises. This was supposed to minimize the students' tendencies to overly rely on pitch and volume in their sonic creations. Such introductory sessions took place in the first two days of the workshops. The students were then tasked to explore the opportunities of using sound in interactive scenarios that were to be developed in small teams. In order to provide initial directions and to make sure that a relevant and broad range of design challenges was addressed, five design assignments were presented as starting points for the students' project work. The design assignments covered various configurations of social and spatial situations, artifact-user relationships, task and goal properties and narrative or emotional aspects.[5] Throughout the classes we made use of responsive, flexible design methods and tools that are open to improvisation and variation, also when used by non-experts in sound design. At the same time, these tools and methods enable one to work with sound from the very beginning of the creative process. The core method used in the workshops was the "electroacoustic Wizard-of-Oz" method (Hug 2010; Hug and Kemper 2014), which enables the quick creation of compelling sonic experience prototypes without the need to code or implement fully operational sensor systems. This set the focus on sound design as much as possible, which was a primary concern, given the amount of other topics related to interaction design and the short duration of the workshop. This way, the involved students would acquire all necessary knowledge and skills required to use sound in multimodal design projects later during their studies.

Due to the explorative and somewhat speculative nature of sound design in the context of interaction, the students were also obliged to incorporate research elements in their projects. After initial concepts had been formulated, prototypes were created to evaluate related design hypotheses. This was achieved using various methods, such as ranking experiments or semantic differentials, questionnaires or think-aloud sessions. The classes were thus transformed into small design research labs, which enabled a guided exploration and provided opportunities to share experiences and discuss sound design approaches, in turn, contributing to the appreciation of the benefit of well-crafted sound design in interaction design.[6]

The following discussion presents an aggregated overview over the prevalent design phenomena and patterns, and is based on our mentoring notes and the process documentation the students were required to produce. In our discussion of the conceptual approaches and solutions provided by the interaction design students, we will cover the motivations and scenarios inspired by the discovered potentials of sound in interaction design. We will also look at how they were implemented throughout the iterative process of formulating and evaluating sound design hypotheses up to the final experience prototypes. Finally, we will discuss emerging patterns and paradigms that underlie the shared practices of sound design and interaction design and propose ways how the two disciplines can profit and learn from each other.

**Figure 5.2** Students Investigating the Relationships Between Sound and Location.

### *5.4.1 Motivations for Using Sound*

A very common motivation for using sound was to indicate that *a device is active, working*, both in terms of being switched on or "running." The former was motivated by the example of a mechanical switch where the mechanical process of activation is intrinsically linked to a sound. The use of such sounds emerged from the desire to indicate when the interaction actually starts (activation) and ends (deactivation). In the latter case, of indicating an ongoing process, we are familiar with the sonic phenomena associated with running engines or mechanical machinery.

In many cases, these on-off sounds were sonically isolated from the remaining sounds. For instance, while most of the sound design would be subtle noises, the on-off switch would be a classic synthesized earcon. Using rising intervals for activation and falling intervals for deactivation was common. Activation and deactivation sounds based on related patterns from natural processes were more rare.

An extension of this design approach was the *indication of different states* in a finite state machine, motivated by the need to indicate the transition from one state of a system to another (Figure 5.3). Most often this seemed necessary for devices with some sort of data recording function (e.g. "device ready," "start scanning"), for games or functions related to

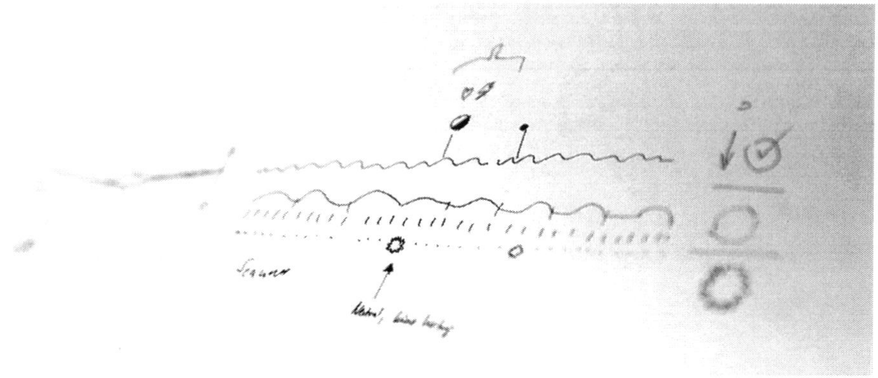

**Figure 5.3** Visual "Score," Representing State Changes.

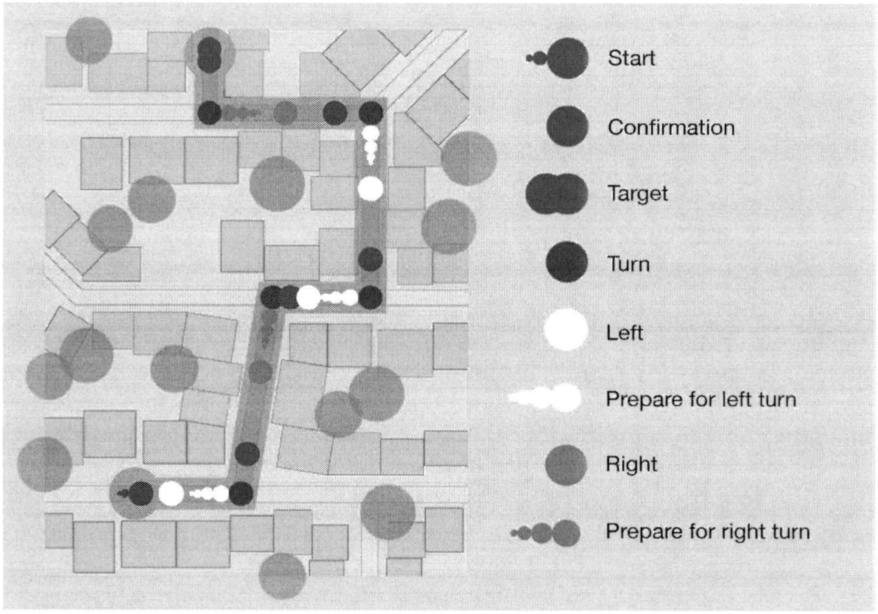


**Figure 5.4** Visual Representation Depicting Several Motivational and Attention Grabbing Sound Events.

gamification (e.g. "turbo-boost ready") or to indicate the transition from one physical area of interaction to another in a spatial navigation scenario.

Another common motivation for using sound was to *express a continuous process or give feedback regarding a continuous operation*. The goal here often was to give the user a sense of progress within a sub-task or a sense of distance/position within a fixed one-dimensional linear space. Both the mapping of physical (e.g. space, Figure 5.4) and metaphorical (e.g. distance from an "ideal" state) parameters were very common. Occasionally, several individual one-to-one mappings were combined in order to give linear feedback about a multidimensional system.

In several cases, there was the motivation to use sounds for their power to *create or focus attention*, which is also one of the fundamental perceptual effects of sound in everyday life. Sounds notified the users about a potential interaction with either themselves, other people or the environment. Other usages included the indication of danger and its source. Interestingly, such sounds were almost always used to communicate an issue with a negative connotation, or a deviation from some ideal value. This may be related to the convention that sounds are more often used for some kind of warning function rather than as positive reinforcement of an interaction.

Related to the uses of sounds for creating attention was their use to *motivate user input*, as a "call to action," so to speak. Asking for user input at specific points during an interaction is a common function in interaction design, where a system operates solely based on a user's input actions.

In the process of interaction, the sonic *motivation of the user to reflect, adapt or change behaviors* was also apparent from the resulting prototypes. Typical use cases for such sounds were apps used to micromanage our daily routines, often related to health. As such apps are meant to become part of everyday routine and habits, the motivation to use sound was also to reduce the need to look at the phone all the time.

Many of these interaction scenarios have in common that the interaction between device and user is rather one sided, the device prompting the user for action, either because he/she does not exhibit the desired behavior or has to interact with the device or his/her environment. As a consequence, that sound design often had a "commanding" quality (i.e. short, clearly audible and foregrounded), and if there was no "correct" action, the interaction process would stop.

In some cases, however, the goal was to use sounds to *interact and communicate with the user in a dialogical way*. This approach emerged from the notion of "smart assistant" systems, becoming increasingly common for instance in the form of "smart speakers," which are able to act as a conversational agent in a broader sense. These systems not only order the user to do something, but may negotiate such actions considering related trade-offs (e.g. reduced energy consumption versus brightly lit flat), making such "assistants" more socially acceptable. In terms of sound design, this required a systematic approach to conveying various degrees of appreciation, rejection or urgency sonically.

Lastly, a few examples used sound to *express or communicate emotions* in an interaction between humans. The underlying incentive was in

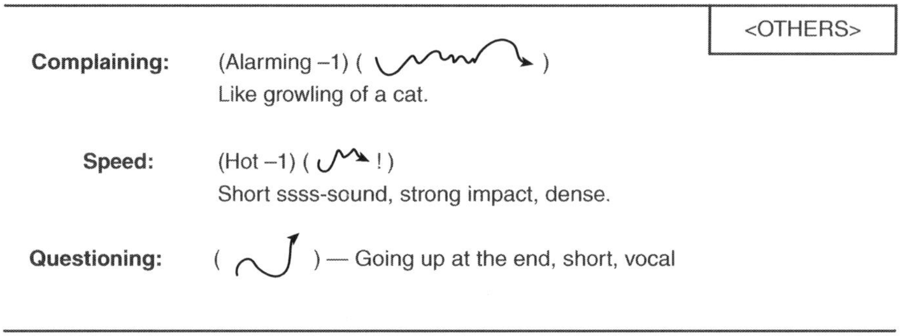


**Figure 5.5** Analysis of Vocal Patterns to be Used in a Robot.

most cases the observation that humans often have difficulties expressing their emotions towards work colleagues or towards less familiar acquaintances. In these cases, the product would act as a sonic mediator. In one design case, for instance, the users could communicate their emotions using expressive gestures that modulated their (neutral) voice. The inspiration for the related sound design often was either the human voice and its emotional variations, or robots familiar from films such as *Star Wars* or *Wall-E*, featuring nonverbal expression of emotions. Figure 5.5 shows an illustration for sonic movements of the utterances of such an object.

In some design cases sounds were used to actually *create emotions or motivate affective behavior*. Related examples were a navigational device whose purpose was to make certain routes or places either appealing or repelling through sound. It was evident in these cases that students tended to focus on use of sound that would cause negative emotions rather than positive ones.

## 5.4.2 Emerging Design Approaches and Related Phenomena

### 5.4.2.1 Sound Follows Function

While the goal in interaction design may be to enable relatively open interaction experiences, many design approaches still boiled down to offering a specific set of functional components. This related to the need to give the user some kind of feedback about the availability of a function and then about its execution. By consequence, the use of sound in the projects, and by consequence the related design considerations, often were largely subjected to improving functionality and ease of use. Hug and Kemper put it in a nutshell when they observed that "sound in interaction design is often

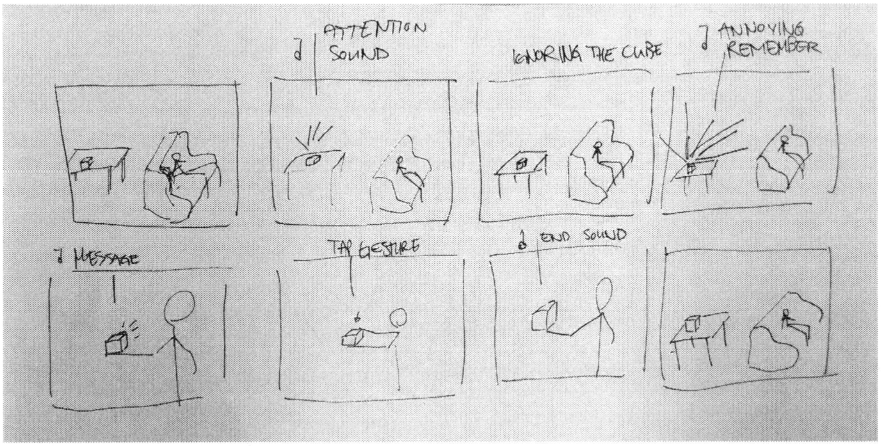


**Figure 5.6** A Storyboard Indicating the Functions to be Associated With Sound.

understood as design element that has the right to exist if, and only if, it offers a specific functional benefit—otherwise it is superfluous." (Hug and Kemper 2014).

A typical example would be sounds that worked as symbols for a function, similar to a visual icon. The motivation to elaborate such sounds was relatively small: as soon as it served its purpose as a sonic symbol, further development of the sound design would stop. In this "functionalistic" and utilitarian approach to sound, an isolated sound parameter is changed only in order to convey specific information, and the resulting sonic quality is secondary. The problem with such sounds is that while the may work "technically," they soon become annoying, in particular in repetitive tasks.

### *5.4.2.2 The Power of the Design Concept*

We have mentioned earlier that design often is strongly concept-driven, in particular when dealing with clients. This could also be observed in many projects from the workshops where sound design decisions were driven by an underlying (theoretical) concept. In some cases, those concepts made use of knowledge about auditory perception, but it turned out that designers often seemed to trust the theory more than what they could perceive themselves. For instance, sounds with a low amount of high frequencies may in theory work better as background sounds, but if they have a "boomy" resonance, and a strong indexical quality (evoking a specific sonic event familiar from other contexts), their effect will turn into the opposite and become very disruptive. A gentle, "whispery" sound on the other hand may contradict the perceptual "rule," that higher pitches are generally more attention grabbing.

### *5.4.2.3 The Impact of "Visual Thinking"*

It was striking how the visual paradigm pervaded design concepts and practice. First of all, this "dominance of the visual" (Welsch 1996) often surfaced in discussions about whether sound design makes sense, as everything supposedly is solved already—and better—with visual solutions. While in many cases this is a fair objection, the conclusion that there are no useful design scenarios for sound would certainly be disastrous. The dominance of "visual thinking" is maybe not very surprising, since in interaction design, visual aspects of, and visual means for design, are both of great importance.[7] As a consequence, the students came up with various visual representations or scores to describe the sound itself or their occurrence over time. The representations ranged from visual scores (such as Figure 5.7) to more complex representations of single sound events (such as Figure 5.8) to figurative descriptions of the situation at hand and the use of sound therein. For instance, in some cases, there was a simple "start"

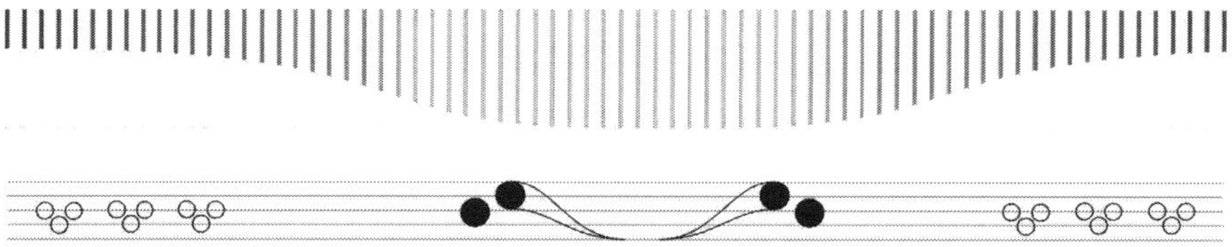

**Figure 5.7** A Visual Representation of Start and End Phases and Intermediary Transition, Related to Linear Mapping of Movement.

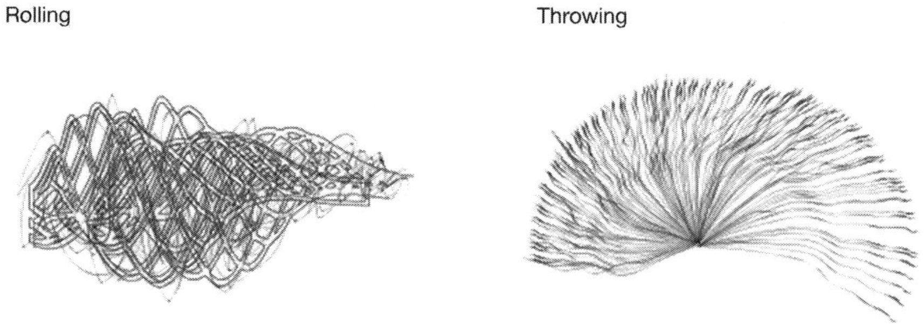


**Figure 5.8** Drawings Illustrating Complex Movements as Means to Visualize Desired Sonic Qualities.

and "end" signal, with another indicator sound in between. This design directly emerged from a visual representation of the process (Figure 5.7). The visual approach in general seemed to foster a tonal, earcon-like design approach, as pitches can be easily represented visually.

In a few cases the students also employed visual strategies to frame the desired sonic qualities. In this case, drawings were used to express the desired aesthetics of more complex sounds which are not mapped directly to any interaction parameter (Figure 5.8).

Many design solutions were close to visual representations also on the level of interface design, using buttons knobs or sliders, which embody binary states or linear parameter changes (with both limited and unlimited range). This was often translated directly to the sound design, even if the interface was translated to a scenario of, for instance, navigating space or finding a point in space (Figure 5.9).

In many cases, the limitation of the two- or three-dimensional visual representation of both interaction processes and sound morphologies in terms of capturing multidimensional aspects was evident, for instance, in the attempt to represent the complex qualities of emotions with specific sound qualities (e.g. "darker" or "brighter"), or isolated effect parameters, such as changing the LFO speed of an amplitude modulation. While acoustically produced sound is often highly dynamic and complex in its structure, the visual thinking resulted in sound being treated as a static, homogenous entity. Sounds would often be clearly associated with the event that caused it, and not change upon repeated triggering of the event, thus behaving very much like a visual notification.

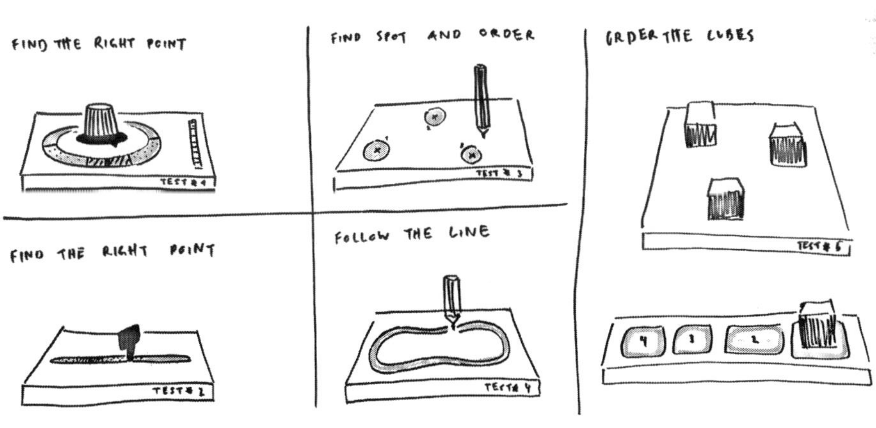


**Figure 5.9** Example of a Series of Interfaces Exploring Associations Between One- or Two-Dimensional Input and Sound.

### *5.4.2.4 Managing Additive Design*

A consequence of the understanding of sound as isolated entity linked to a specific function in the interaction process was what could be called "additive sound design," where several (individually tested and working) sound events are lined up in a sequence. In this case, each sonic "building block" served a specific purpose (e.g. as a call to action, to inform about a certain state, or to provide some kind of input confirmation) and transitions were often rather abrupt. This conceptual approach to sound also relates to the use of sound to represent specific states in an interaction or system and lends itself well to the representation of interaction sequences as flow diagrams (Figure 5.10). The additive approach to sound design was particularly problematic when the types of sounds used were aesthetically unrelated, for instance, when sequencing noisy, vocal, tonal or indexical sounds, or when design strategies such as auditory icons and earcons were arbitrarily combined.

In this context it turned out that the take on the method of sonic layering, which is a fundamental and established strategy in sound design, also was influenced by visual thinking. If sonic layers are carefully composed, they will result in a coherent sonic composition. But if layers are understood visually as individual "tracks" or even static "icons" with their own

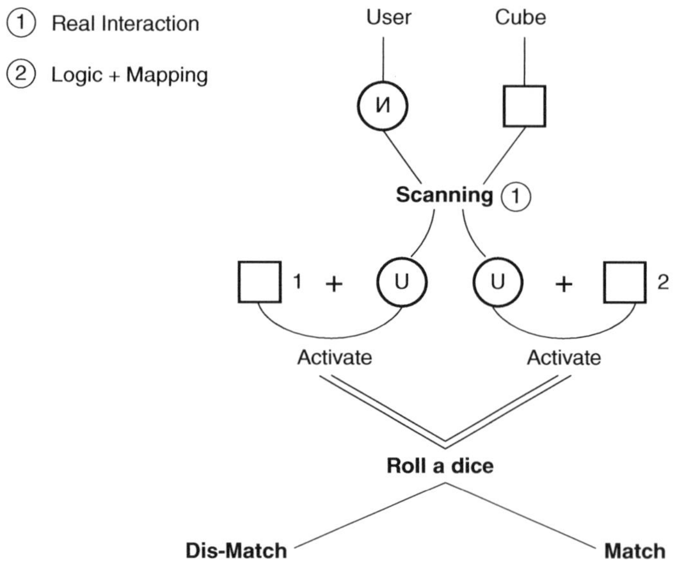


**Figure 5.10** Flow Diagram of an Interaction as Sequence of States, Encouraging the Design of Individual, Unrelated Sound Events.

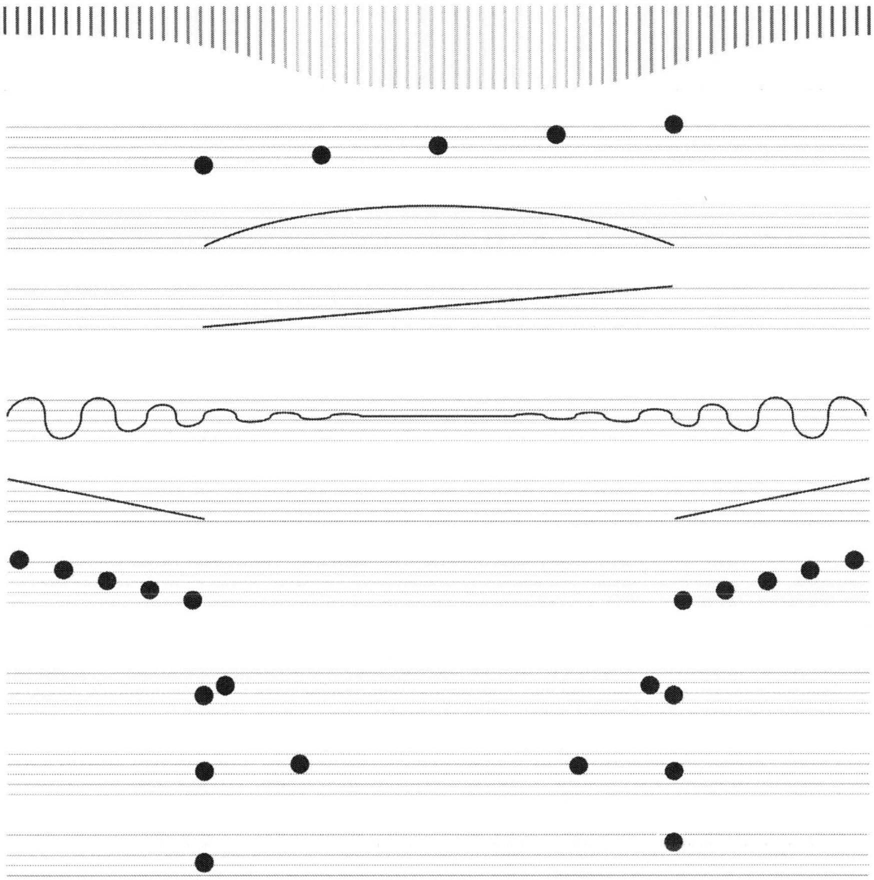

**Figure 5.11** Additive Design Pattern Showing Various Functional and Sonic Layers.

structure and sonic quality, they will not blend and the sound design will be incoherent. This effect is aggravated if the layers are designed to represent isolated, simple parameters such as a pitch modulation or intermittent beeps.

It is striking that design students in principle are very familiar with the need to layer, combine and refine in the visual domain and know about proportions and composition, coloring, or blending elements into a coherent whole. Also they can deal with material qualities such as stiffness, viscosity and roughness. However, while there seemed to be an interest in blending several "informative parameters" into one "sonic movement," the participants often struggled with transferring

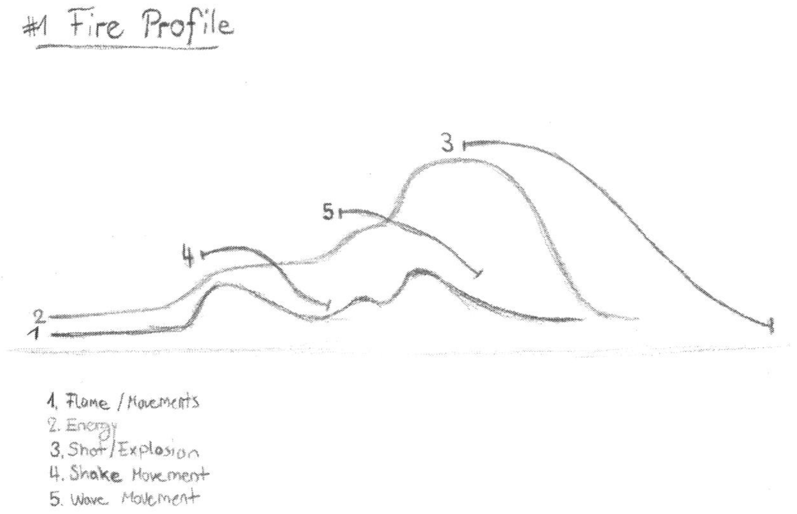


**Figure 5.12** Visual Score Showing the Intended Layering of the Sound Events.

these competencies to the sonic domain. One reason might lie in the fact that dealing with material and visual qualities is already a topic in school whereas sound-related education is restricted to conventional musical education, which usually does not cover critical listening for spectral qualities and abstract sonic properties beyond pitch, volume and rhythm.

However, it also turned out that additive design as such is not a generally "bad" or "wrong" approach to sound design in an interaction design context. Sometimes it was the most efficient approach, and a function centered design may even be a necessity if the functionality is key to a successful user experience. Thus, the aim is to overcome the inherent issues of additive design, for instance, by exploiting the power of temporal sequencing and composition (Hug 2017, Chapter 27). For instance, individual elements or states could be connected by a transitional sound, which can help to link the end of one state with the beginning of the subsequent state. Individual sounds could moreover be aesthetically linked or unified using a common processing method, such as a common reverberant space or a general timbral property.

### 5.4.3 *The Challenge of Sonic Invention*

#### 5.4.3.1 *Keep it Simple, Stupid? Trade Off Between Simplicity and Complexity*

As mentioned above, in many design cases, the main sound parameters used were pitch, volume or panning. Some projects additionally made use of rhythm. The mapping of multiple parameters of a layered sound to multiple parameters of a system or the attempt to sonify the outcomes of complex (e.g. movement) patterns was very uncommon. These considerations reveal one central line of conflict between ideals and motivations of sound and interaction designers: The diverging relevance and approach to simplicity and complexity.[8] Students often referred to the design of "simple" sounds as one of their goals, matching the intended simplicity and ease of use of their application scenarios. In these cases it was ignored that this striving for simplicity in sound can often be counterproductive. While visual simplicity is usually regarded as noble goal, what is clarity in the visual world quickly becomes monotony in the auditory domain.[9] A "sleek" appearance of a product does not have to be underlined by equally "sleek" sounds.

Another, closely related, reason why the interaction design students strove for simplicity was that it was considered a correlate or even synonym for "intuitive" and "easy to use." Many of the students' test setups aimed at achieving an intuitive understanding of a sound's meaning. This may be appropriate for many applications focusing on state changes or one-way communication with a user through notifications. But in more complex and dynamic situations a simple sound actually can lead to confusion. Furthermore, striving for clarity often is motivated by the desire to avoid ambiguity. But ambiguity and the "instability" of interpretation are inherent in the experience of signs in both their representational and presentational function (Hug 2008), and it is advisable to see ambiguity as a resource rather than a threat (Gaver et al. 2003; Hug 2017).

#### 5.4.3.2 *Who's Afraid of the Unknown Sound? Tradeoff Between Discreetness and Novelty*

Another line of thought concerns the question of openness for new sonic solutions and innovation. In film, inventive sound design lets a simple prop become a futuristic device, like the prevalent example of the salt shaker prop in *Star Trek*, which is transformed into a complex medical scanning device purely through sound. Another well-known example would be the lightsaber from *Star Wars*, which only through sound gains qualities such as heat and cutting power. In the real world, however, even highly advanced appliances may sound like a generic microwave oven. Also, the design solutions presented by the participants often resorted to seemingly familiar sounds assuming they would contribute to an easier understanding. It is certainly a useful strategy for interaction designers to approach sound design using existing everyday sound experiences. However, it has to be taken into account that along with the benefits of this approach comes its flaws. For instance, the parking assistant paradigm for approaching something, which was used in several cases as a design pattern, suffers from a lack of information and detail in the final, most crucial part of the interaction, when the remaining distance to another car or an obstacle can no longer be estimated visually. Here, the intermittent beeps turn into a continuous tone, which in addition is highly annoying. Another example is the exclusive and uncritical use of conventional earcons and auditory icons, whose design limitations have been reported in detail by Mustonen (2008) and Hug (2017).

#### *5.4.3.3 Aiming to Please: The Fear of Being Annoying*

In the context of the "function centered" design approach described above, it was notable that the devices' functions were often intended to improve something or do something in a "correct" or "better" way. The participants often stated that their goal was to communicate such indications in a "pleasant," "not annoying" way.[10] This can be interpreted as an attempt to make normative and "educational" aspects of interactions more acceptable. As a consequence, many of the students' evaluations focused on the assessment of how "annoying" a sound was. Related to this was the concern that sound could disturb, which also led to a certain level of self-censorship in the designers' decision-making. On the other hand, it was often the rather simple, sometimes tonal approaches in the designs reported above that resulted in rather "attention grabbing" sounds, automatically increasing the potential to annoy. As a variation on this issue, it was common to design "neutral," even "boring," information sounds that were meant to stay in the background, again resorting to simple sound timbres that—contrary to the designer's intention—tended to grab attention because of their salience. This issue is also related to the problem of framing sonic design by visual metaphors and methods, as sonic and visual "backgrounding" work differently.

Of course, there are many good reasons for the quest to reduce the annoyance of sounds. We are surrounded by so many little contraptions that beep, whistle and hum to get our attention. It is thus necessary to consider the impact on our soundscape when designing sounds, in particular if they have a prominent role in our everyday life. In this case, unobtrusive, subtle sounds are obviously much more valuable than sounds designed to grab the attention of anyone within earshot.

#### *5.4.3.4 User Studies Between Conservative Confirmation and Inspirational Impulse*

As outlined above, systematic evaluation was an important element in the iterative design process used during our workshops. A common strategy employed by the students at earlier stages of the design process was to design a set of sounds for predefined generic functional categories, such as "warning" or "alert," or for specific moments in an interaction process, such as "friendly welcome," or "activation," and then ask participants to attribute them to the predefined categories in order to verify whether their attributions would match those of the designers. Rarely there were sustained efforts to develop a sonic solution for a specific category through variation and iteration. Only occasionally, the sound design parameters were operationalized in a way that would help to define either the "design guidelines" for a specific category of sounds or their implementation in consistent sound variations. This resulted in flawed findings resting on a lack of systematic design alternatives, or a forced selection from flawed initial design examples.

Among the evaluation methods introduced and used by the students was a variation of the threshold test, which helps to identify the point

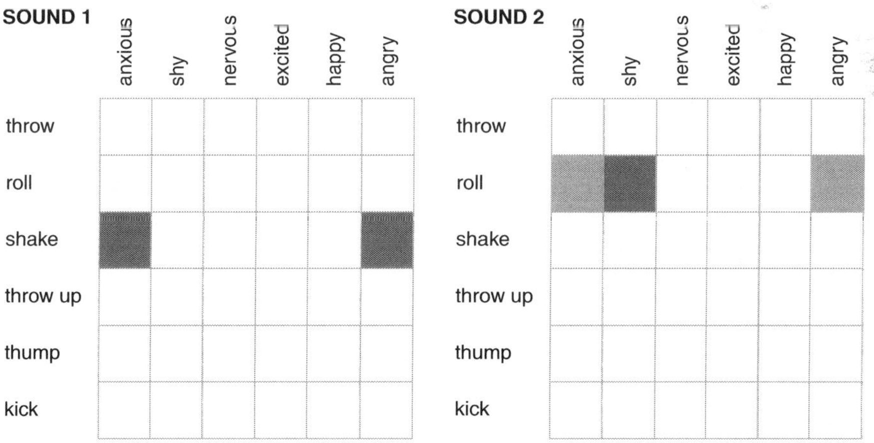


**Figure 5.13** Result of Ranking Experiment Used to Determine Whether Sounds Would Fit Into Predefined Categories.

when a linear change in a sound becomes noticeable or motivates a particular association or reaction. This method is particularly useful for less experienced sound designers; it is easy to implement and was used more often than anticipated. But it turned out that it also contributed to a rather conservative and reductionist approach to sound design as it motivated the use of rather obvious sound parameters, such as pitch, volume or rhythm, which allow for measuring and establishing clear correlations. This issue was further aggravated as the related tasks for such setups often had to be rather simple. Finally, in the case of a successful outcome of evaluations of individual sounds, or simple sound modulations, it often was forgotten that in the end the sounds had to work together, mix well and exhibit aesthetic coherence.

The issue can also be looked at from the opposite direction: If the sound design would require a certain level of complexity, but does not achieve this in a way that sounds good, it will inadvertently result in negative evaluation results. One interpretation of such tests was that only simple tasks can be expressed sonically, but more complex interactions were impossible for users to perform.

Even if user tests in design were often deployed for an inspirational purpose rather than as proof in a scientific sense, they could be counterproductive.[11] Often, design decisions would be based on user statements, such as "sound x worked" or "users confirmed the sound design," rather

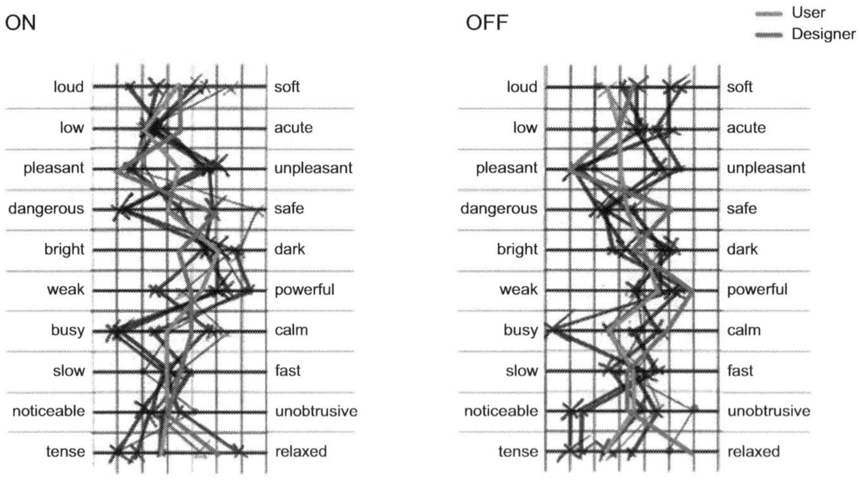


**Figure 5.14** Semantic Differential Used to Evaluate the Judgments of Two Sounds by Users (Gray Line) and Designers (Black Line).

than involving an aesthetic judgement from the designer's perspective.[12] Although many sounds were explored, only a few trivial or familiar sounds may "survive" the evaluation procedure, which, ironically, may often lead to a more annoying design, as we have argued above. Also once an initial result was achieved, the creative exploration would end. While generally being valuable and highly relevant, user studies thus tended to prevent the exploration of sonic solutions early in the process. Initial design approaches would either be frozen and no longer questioned, or dismissed completely—even if the approach was interesting or promising. This prevented further questioning of the sound design and, consequently, the emergence of new ideas and original approaches. The question may be asked: what would have happened if Ben Burtt had relied only on user studies for evaluating the lightsaber sounds he used in *Star Wars*?

## 5.5 Towards a Fruitful Integration of Sound and Interaction Design

### 5.5.1 *Bringing Interaction Design Thinking to Sound Design*

We have argued that from a sound designer's perspective, the current mainstream use of sound in interaction design leaves a lot to be desired. But the experience of working with our students on sound design solutions and exchanging with them about their motivations and solutions resulted in a deeper understanding of the intricacies resulting from bringing sound design to interaction design. Both disciplines—and they still can be considered distinct disciplines, at least until a genuine "Sonic Interaction Design" discipline has been formed—can and should learn from each other.

Firstly, interaction designers acknowledge the importance of the design process and its dynamic iterative and somewhat nonlinear nature. They are familiar with being confronted with "the world out there" through field studies and user-centered evaluation procedures, and they know about the power of multiple iterations and prototyping from "quick'n'dirty" to functional implementation. The quality of experience and functionality is paramount and all means of design have to contribute towards this goal. This also applies to any sound being produced during an interaction. As a consequence, sounds also have to be iteratively developed together with the functionality and the envisioned interaction scenario. The sound design process has to open up, become more transparent and is subjected to many influencing factors—many more than in film, the "original medium" of sound design. Furthermore, interactivity requires us to design sound with coding in mind, as nonlinear, interactive ad hoc compositions, in order to be able to build functional, interactive prototypes and products. As a consequence, the sound design often has to stay open in terms of finalization and composition, which is a challenge for many sound designers.

Also, the use of verbal and visual representations, the general use of externalizations in visual and material form and the related analytic competence in terms of visualizing stages, progressions, relationships and so forth, is a valuable contribution to sound design. By using sonic mood boards in combination with visual and verbal representations of sound spectra and sound dramaturgies, we can combine the strengths of visual and auditory thinking. The notion of spectromorphology (Smalley 1997) of sound as formally distinct development of audible spectra over time can serve as a useful conceptual and creative tool here.

Sound design for interactions essentially is a process involving not only sonic composition, but the composition of sound in the context of actions, objects, functions, users and spaces. This compositional activity integrates language, visual means, sounds and bodily movements into one system of thinking and doing.

### 5.5.2 Bringing Sound Design Thinking to Interaction Design

On the other hand, in order to get most out of sound, interaction designers need to accommodate principles and approaches from sound design, which also means adopting sound into the design process from the beginning.[13] This implies that methods such as brainstorming, scenario writing, bodystorming, mind mapping and creating mood boards have to be adapted to accommodate sound.

Also, the successful integration of sound into the interaction design process requires that one possess a thorough understanding of the interactive, multidimensional nature of sound, and seek to find solutions that work not only "functionally" but also "sonically." This is also related to the need for appropriate tools and methods to prototype sonic interaction experiences. Rather than focusing on avoidance of annoyance, the attitude towards sound has to be the same as towards visual and material qualities: to strive for aesthetic refinement and elaboration.

The "parking assistant paradigm" for indicating changing distances, which was quite popular in the students' projects, is a good example to illustrate this. From a functional point of view the underlying principle of sonic representation works, is well-established and easy to control. However, the resulting sound and its transformation is very linear and annoyingly simple to our ears, and the mapping of sound modulation with distance exhibits the functional flaws mentioned earlier. With the introduction of nonlinear modulations and interacting parameters, sounds become richer, more interesting and potentially more functional. And this is where the strength of sound design thinking comes in, which always aims at combining several parameters in a multidimensional and nonlinear transformation of sound.

It might be valuable at this point to look at product sound design as a kind of middle ground between film sound design and interaction design process. Langeveld et al. (2013) proposed a general structure of sound design for products, composed of the steps "Problem Analysis," "Conceptualization," "Embodiment" and "Refinement." This understanding is comparable with the general understanding of a design process outlined above. While it is still rather linear and follows more of an "engineering" approach, it still is a good example of a design process that integrates elements of sound design with elements of (interaction) design.

But, while in principle many aspects of the creative process can be shared—both are design disciplines in the end—there are many aspects that require a specialized skill set and a substantial amount of experience. Firstly, for sound design, a sound-specific analytic competence is required, not only in a qualitative domain but also on the level of basic acoustics. This needs to be complemented by a descriptive competence in relation to the desired qualities a certain sound should have with a precision comparable to that of a visual sketch.[14] Then, the designer has to understand the effect of the sound on the situated perception, action and overall experience of the user. Finally, it is important to understand sound design as the construction and composition of spectromorphologies, where temporal dramaturgies are sonically expressed. Every sound is composed of several elements that give it its unique signature. The interplay (both timewise and frequencywise) between these elements forms the unmistakable character of a sound and its impact on the listener. This is also the case if a design builds on a sequence of individual sound events, as is often the case in many interaction design situations, where a continuous mapping would not make much sense. All these individual events have to be understood as an aesthetic progression, where a sound is always taking place in the context of its sonic predecessor, as well as its sonic successor. To successfully design a sound for any application, be it product or film, the understanding and skillful use of such "building blocks of sound" are essential.

## 5.6 Coda

Throughout this chapter, we have explored approaches to sound design from an interaction design perspective, and we have indicated our appreciation for the innovative power, artistic ethos and attitude encountered in film sound design. This design approach is based on a detailed understanding of the narrative, and the protagonists and artifacts it contains. In fact, film sound is always part of a narrative and expressive whole, which the viewers enter, willingly subjecting themselves to the suspension of disbelief.

On the other hand, in interaction design, we are mostly dealing with ephemeral everyday interactions, where ease of use is important and things have to ultimately work without further ado, which can be understood as an aesthetic and functional "app" paradigm. From this perspective, even the most elaborate sounds are but one of many little details in the complex flow of everyday life.

The space for narrative and expressive complexity is thus limited and a certain level of simplicity and reduction is required. In addition, we must consider that layers and sonic complexity also may emerge from accidental juxtaposition and co-occurrence of sounds in everyday life. This "situational mixdown," as it were, can not be fully controlled. Here a traditionally trained sound designer may have to change perspective and design approaches.

But beeps are the worst possible interpretation of simplicity. Sound design has to find ways to create sounds that are simple and accessible yet still original and with a clear profile. We must also not forget that with the arrival of smart environments, AI and autonomous robotic systems, the scope of interaction and the required complexity in communication will most certainly increase in the years to come.

Beyond all ideals, we have to acknowledge that the potential of sound and in particular the benefit of investing in sound design elaboration, is not necessarily understood in interaction design.[15] The lack of sonic competence and awareness results in a certain "deafness" for the potential of proper sound design. Also, sound will not play a central role in the professional lives of most interaction design students. Therefore, a central question is how do we ingest or stimulate interaction design ideas and solutions that take the potential of sound into account? And, on the other hand, how can the power of sound design be leveraged in an interaction design context, where the ultimate goal is to create a functional and enjoyable interaction experience?

Firstly, it is central to advocate and leverage the potential of elaborated and original sound design, also in seemingly trivial interaction situations. As long as we resort to sonic shortcuts and seemingly "working" paradigms—as long as we simple *use* sound, rather than *design* it—this is not possible. Designers have to understand the value of sound design elaboration, how sound can affect us and afford certain behaviors and actions, and they have to be enabled to understand the nature of sound as a material of design and how it can be shaped and composed. This can be achieved best by both argumentation and demonstration.

On the other hand, sound designers have to learn how to externalize ideas and prototype sonic experiences that are not isolated from actual interactions. And they have to understand their sounds as dynamic, relational, procedural phenomena that are able to coexist with a plethora of concurrent events and sounds, rather than in terms of perfectly tuned, final and isolated entities.

Finally, it is important to understand that in many cases the design of sonic interactions will not be in the hands of some kind of "jack of all trades" Sonic Interaction Designer. Rather, the complexity of the two domains may require increased and close collaboration between experts from both interaction and sound design. This means that both interaction and sound designers need to acquire sufficient understanding and communication skills in each other's disciplines and design practices. In the end, it all comes down to cross-disciplinary and cultural boundaries in a shared effort to work towards a better sounding world, in whatever way possible. And in this mission, every sound matters.

## Notes

1 The concept of "listening modes" describes ways how we perceive, experience and judge the same sound differently based on how we listen to it. The modes most commonly referred to are "causal," "semantic" and "reduced" listening (Chion 1994). Other modes, such as "critical," "empathic" or "functional" listening have been proposed by Tuuri et al. (2007).
2 Buchenau and Suri describe "experience prototyping" as "any kind of representation, in any medium, that is designed to understand, explore or communicate what it might be like to engage with the product, space or system we are designing" (Buchenau and Suri 2000, p. 425). This also means that an experience prototype does not demonstrate a technological solution, and thus can use mockup and make-believe techniques.
3 An alternative take on the notion of sonic sketching is presented by Delle Monache and Rocchesso in the previous chapter.
4 The perfect film sound design often is described as "not being noticed at all," which is somewhat misleading regarding the many examples of iconic film sounds and their memorability, leading to the "filmic listening mode" described above.
5 For details about the design challenges, see Hug (2013).
6 Further details about the workshop's structure and methods are described by Hug and Kemper (2014).
7 Even in terms of sound in HCI, terms such as "auditory display," or "auditory icon", although not intentionally, allude to visual paradigms.
8 It should be noted, that when using the term complexity, we do not refer necessarily to complexity on the level of the sound signal, but rather in terms of timbral richness and

compositional elaboration. If done well, complex sounds may indeed have a "simple" overall appearance, and are perceived as one "Gestalt." And of course, sounds may also be too complex. Most likely, there is an inverse U-shaped relationship between pleasantness and complexity: something perfectly static or repetitive is boring, while something completely random or erratic is cannot be related to at all (McDermott 2012).

9 To make the point on a more fundamental level, consider the balanced beauty of a perfect circle in comparison with the penetrating monotony of a sine wave.

10 On a side note, it was striking that despite the recurring statement that the sounds were meant to be positive, or express a positive quality, many designs used sounds rather to convey a negative message (alerts) or emotions.

11 This can be related to the general issues associated with integrating scientific evaluation procedures in a sound design process, as reported by Hug and Misdariis (2011).

12 A sound designer might be convinced of a sonic aesthetic, and then modify and enhance it until it is accepted by users.

13 It was striking that some interaction design students stated that the fact that they were working with sound as medium made them reconsider their design process. This is also due to the fact that in relation to sound there is still so much unknown, and so many things to try, at least once one decides to leave the beaten track of beeps and clicks.

14 An approach for addressing this challenge is proposed by Delle Monache and Rocchesso in the previous chapter.

15 It is striking that the Holoscreen scene of the Movie *Minority Report* (Spielberg 2002), is often featured as a case study of interaction design, but the related elaborate sonic interactions are not further considered and analyzed.

## References


Bødker, S., 2006. When second wave HCI meets third wave challenges. *Proceedings of the 4th Nordic Conference on Human-computer Interaction: Changing Roles*. New York: ACM, pp. 1–8.

Boehner, K., Vertesi, J., Sengers, P., and Dourish, P., 2007. How HCI interprets the probes. *Proceedings of the SIGCHI Conference on Human Factors in Computing Systems*. New York: ACM, pp. 1077–1086.

Bonsiepe, G., 1996. *Interface: Design neu begreifen*. Cologne: Bollmann.

Brown, T., 2009. *Change by Design: How Design Thinking Creates New Alternatives for Business and Society*. Glasgow: Collins Business.

Buchenau, M., and Suri, J. F., 2000. *Experience Prototyping. Proceedings of the Conference on Designing Interactive Systems*. New York: ACM, pp. 424–433.

Buxton, B., 2010. *Sketching User Experiences: Getting the Design Right and the Right Design*. Burlington, MA: Morgan Kaufmann.

Carroll, J. M., 2000. *Making Use: Scenario-based Design of Human-computer Interactions*. Cambridge, MA: MIT press.

Chion, M., 1994. *Audio-Vision: Sound on Screen*. New York: Columbia University Press.

Delle Monache, S., Polotti, P., and Rocchesso, D., 2010. A toolkit for explorations in sonic interaction design. *Proceedings of the 5th Audio Mostly: Conference on Interaction with Sound*. New York: ACM.

Dorst, K., 2006. *Understanding Design*. Amsterdam: Bis Publishers.

Dourish, P., 2004. *Where the Action Is: The Foundations of Embodied Interaction*. Cambridge, MA: MIT Press.

Dubberly, H., 2004. *How Do You Design. A Compendium of Models*. Viewed May 20, 2018 <www.dubberly.com/wp-content/uploads/2008/06/ddo_designprocess.pdf>.

Farnell, A., 2010. *Designing Sound*. Cambridge, MA: MIT Press.

Franinović, K., and Serafin, S. (Eds.), 2013. *Sonic Interaction Design*. Cambridge, MA: MIT Press.

Gaver, W. W., Beaver, J., and Benford, S., 2003. Ambiguity as a resource for design. *Proceedings of the SIGCHI Conference on Human Factors in Computing Systems*. New York: ACM, pp. 233–240.

Hassenzahl, M., 2013. User experience and experience design. *The Encyclopedia of Human-Computer Interaction*. 2nd ed. Viewed June 10, 2018 <www.interaction-design.org/literature/book/the-encyclopedia-of-human-computer-interaction-2nd-ed/user-experience-and-experience-design>.

Hug, D., 2008. Genie in a bottle: Object-sound reconfigurations for interactive commodities. *Proceedings of Audio Mostly, Conference on Interaction with Sound*. Pitea: Audio Mostly, pp. 56–63.

Hug, D., 2010. Investigating narrative and performative sound design strategies for interactive commodities. In: Ystad, S., Aramaki, M., Kronland-Martinet, R., and Jensen, K. (Eds.) *Auditory Display*. Berlin: Springer, pp. 12–40.

Hug, D., 2013. Barking wallets and poetic flasks: Exploring sound design for interactive commodities. In: Franinović, K., and Serafin, S. (Eds.) *Sonic Interaction Design*. Cambridge, MA: MIT Press, pp. 351–368.

Hug, D., 2017. *CLTKTY? CLACK! Exploring Design and Interpretation of Sound for Interactive Commodities*. Dissertation. Linz: University of Art and Design Linz.

Hug, D., and Kemper, M., 2014. From foley to function: A pedagogical approach to sound design for novel interactions. *Journal of Sonic Studies* 6, no. 1. Viewed March 12, 2018 <http://journal.sonicstudies.org/vol06/nr01/a03>.

Hug, D., and Misdariis, N., 2011. Towards a conceptual framework to integrate designerly and scientific sound design methods. *Proceedings of the 6th Audio Mostly: Conference on Interaction with Sound*. New York: ACM, pp. 23–30.

Kolko, J., 2010. Abductive thinking and sensemaking: The drivers of design synthesis. *Design Issues* 26, no. 1, pp. 15–28.

Langeveld, L., Van Egmond, R., Jansen, R., and Özcan, E., 2013. Product sound design: Intentional and consequential sounds. *Advances in Industrial Design Engineering*, IntechOpen. Viewed April 23, 2018 <www.intechopen.com/books/advances-in-industrial-design-engineering/product-sound-design-intentional-and-consequential-sounds>.

Lim, Y. K., Stolterman, E., and Tenenberg, J., 2008. The anatomy of prototypes: Prototypes as filters, prototypes as manifestations of design ideas. *ACM Transactions on Computer-Human Interaction (TOCHI)* 15, no. 2, p. 7.

LoBrutto, V., 1994. *Sound-on-film: Interviews with Creators of Film Sound*. Westport, CT: Greenwood Publishing Group.

Löwgren, J., and Stolterman, E., 2004. *Thoughtful Interaction Design: A Design Perspective on Information Technology*. Cambridge, MA: MIT Press.

McDermott, J. H., 2012. Auditory preferences and aesthetics: Music, voices, and everyday sounds. In: Sharot, R., and Dolan, T. (Eds.) *Neuroscience of Preference and Choice*. San Diego: Academic Press, pp. 227–256.

Moggridge, B., and Atkinson, B., 2007. *Designing Interactions*. Cambridge, MA: MIT Press.

Mustonen, M. S., 2008. A review-based conceptual analysis of auditory signs and their design. *Proceedings of ICAD*. Paris: International Community for Auditory Display.

Neuhoff, J. G. (Ed.), 2011. *The Sonification Handbook*. Berlin: Logos Verlag.

Oulasvirta, A., Kurvinen, E., and Kankainen, T., 2003. Understanding contexts by being there: Case studies in bodystorming. *Personal and Ubiquitous Computing* 7, no. 2, pp. 125–134.

Quesenbery, W., and Brooks, K., 2010. *Storytelling for User Experience: Crafting Stories for Better Design*. New York: Rosenfeld Media.

Rittel, H. W., and Webber, M. M., 1973. Dilemmas in a general theory of planning. *Policy Sciences* 4, no. 2, pp. 155–169.

Rocchesso, D., Polotti, P., and Delle Monache, S., 2009. Designing continuous sonic interaction. *International Journal of Design* 3, no. 3.

Simon, H. A., 1996. *The Sciences of the Artificial*. Cambridge, MA: MIT Press.

Smalley, D., 1997. Spectromorphology: Explaining sound-shapes. *Organised Sound* 2, no. 2, pp. 107–126.

Spehr, G. (Ed.), 2015. *Funktionale Klänge: hörbare Daten, klingende Geräte und gestaltete Hörerfahrungen*. Berlin: Transcript Verlag.

Spielberg, S., 2002. *Minority Report*. United States: 20th Century Fox.

Suchman, L. A., 1987. *Plans and Situated Actions: The Problem of Human-machine Communication*. Cambridge, MA: Cambridge University Press.

Tuuri, K., Mustonen, M. S., and Pirhonen, A., 2007. Same sound—different meanings: A novel scheme for modes of listening. *Proceedings of Audio Mostly, Conference on Interaction with Sound*. Ilmenau: Audio Mostly. pp. 13–18.

Welsch, W., 1996. *Grenzgänge der Ästhetik*. Ditzingen, Germany: Reclam.

6

# Auditory Display in Workplace Environments

Michael Iber

## 6.1 Introduction

In a highly interconnected data-driven world, monitoring and analysis of activities, processes, or system conditions have become complex and often time-critical tasks. Data of various kinds are captured by manifold types of devices and applications including sensors, websites or surveillance cameras (Höferlin et al. 2012), to name just a few. Huge amounts of data are streamed or stored waiting to be interpreted in real time or in retrospect, hence raw "data without context is incomprehensible" (Roddy and Furlong 2014). Sense-making is the keyword for conveying data from their actually futile states into viable pieces of information. Many sense-making procedures run autonomously at the device level (e.g. the automatic shutdown of an engine because the measured temperature exceeded a predefined threshold, or an emergency turn of a self-driving car because some deep-learning algorithm detected obstacles by the analysis of multivariate sensor data). Despite recent enormous progresses with regard to machine learning and automation, human activity and interaction has by no means lost its eligibility for the surveillance of processes or the analysis and interpretation of data in order to derive appropriate measures. Especially in real-time scenarios, decision-making processes are primarily triggered by human sensual perception. Generally, any of the five senses including vision, hearing, taste, smell and touch are suitable to contribute to these processes, however, the indisputable perceptual focus lies in the visual channel. In conjunction with cognitive sciences and research in user experience (UX), scientific fields such as visual analytics or information visualization have received increased attention and provide tailored methods and tools for meaningful displays of complex data scenarios (Aigner et al. 2015). However, in practice, an increase in visual information to be displayed usually leads neither to the new design of an adjusted visual display nor to an involvement of available attentional capacities provided by other senses, but is "generally addressed by adding more computer screens" (Roginska et al. 2006). Kaiser and Fuhrmann (2014) confirmed that additional information to be monitored is usually "implemented by adding additional interfaces" without actual integration into a complementary system.

While the number of approaches to data interpretation based on tactile (Hogan et al. 2017), olfactory and gustatory sensations are rather rare (Sanders and McCormick 1987), there has been vast research on "auditory display"—as opposed to "visual display"—involving the human auditory channel to provide information. Due to the volatile—or as Watson and Sanderson (2007) say, "transitory"—character of sound, implementations of auditory displays are in most cases multimodal, comprising visual and auditory senses either redundantly or complementarily.

## 6.2 Auditory Perception and Cognitive Attention

From an evolutionary perspective, monitoring the acoustic environment with our ears is one of the most fundamental functions contributing to the sustainability of human existence. As early as millions of years ago, hearing provided our ancestors a permanent and omnidirectional awareness of possible threats and dangers in their vicinity both during the day and especially at nighttime in the dark, when their visual perception was rather limited (cf. Carlile 2011). Also, in everyday life today, we permanently utilize our auditory channel to monitor our ecological[1] system. We heavily depend on our ears in order to become aware of an incoming call on our mobile phone or to decide on the appropriate moment to change gear while driving our car. We also become immediately suspicious that something might go wrong when the sound of its combustion engine even slightly deviates from the one to which we are accustomed. This self-evident perceptual behavior also plays a major role in industrial production, denoted as "working knowledge" (Berner 2008) or "tacit knowledge" (Reeves and Shipman 1996), where workers' auditory perception is regarded as a major contribution to an experience-based understanding of processes in work. Although major efforts in vibration analysis (Renwick and Babson 1985) and condition monitoring have been made using the information content of sonic process emissions for machine-based measuring and analysis procedures (Martins et al. 2014), (not only) sound-related working knowledge remains a significant element in terms of functioning human-machine interaction in industrial environments. While these acoustic emissions are sounds inherent to the machines and processes concerned, auditory displays[2] are explicitly designed sonic systems that provide acoustic information to "operators, engineers, maintenance personnel, and managers"[3] (Johannsen 2004). Both acoustic machine emissions and auditory displays have in common that they deploy the auditory channel for monitoring and analysis purposes. They both function eyes-free, allowing operators to focus on their primary tasks. The advantage of auditory displays is that only functional and acoustically shaped sounds are chosen to convey relevant information (Gaver et al. 1991). However, since acoustic emissions at a plant shape operators' understanding and knowledge about processes, these sounds are very suitable to be used as sonic bases for the design of auditory displays (Alexanderson and Tollmar 2006; Johannsen 2004).

Salient features of auditory displays and sonifications (i.e. "the use of nonverbal audio to convey information"; Kramer et al. 1997) are summarized by Kramer (1994a). The systematic and reproducible transfer from data domains to sonification domains is denoted as "data mapping." Before dealing with aspects of sound design and generation, sonification designers face the essential challenge to define appropriate mappings. Also, considering auditory perception, including (ecological) psychoacoustics (Neuhoff 2004; Walker and Kramer 2004) and Gestalt psychology (Bregman 1994), is critical to the development of auditory displays that present meaningful information without causing listeners' annoyance. Literature refers to terms such as "alarm fatigue" (Paterson et al. 2016), frequently caused by a high rate of "false positive" alarms (Hildebrandt et al. 2016), namely sounding alerts without evidence.[4] "Alarm flooding" (Viraldo and Caldwell 2013) denotes, for instance, the risk of causing users to interrupt their primary tasks to silence alarms instead of solving the causing problem (Xiao et al. 2000), while "alarm showers" (Johannsen 2004) consist of a sequence of interdependent errors causing multiple alerts that overburden operators on duty (Hearst 1997). Frequently, auditory alarms are considered to be "redundant to the operators' work" with the consequence of being ignored or even switched off (Li et al. 2012). Brock et al. (2005) mention even the clipping of wires in order to make buzzers keep silent. Inattentional deafness (Chamberland et al. 2017)—the failure of noticing sounds—is another risk of auditory overloads caused by inappropriately designed and implemented sonifications neglecting aspects of auditory perception and working environments.

## 6.2.1 *Types of Monitoring and Listing Strategies*

Concerning auditory perception, there is a strong relationship between modes or strategies of listening as they have been defined by several authors such as Chion et al. (1994), Gaver (1994), or Schaeffer (2017) and types of monitoring as described by Vickers (2011). Whereas listening modes denote attitudes of hearing in common situations on a more general level, monitoring types more specifically refer to task- or work-related scenarios. Truax (2001) distinguishes three modes of listening, each of them characterized by a different level of attention. The first kind, "background listening," is entirely passive and refers to listening that is not directed at achieving any practical purpose. By comparison, the second mode "listening-in-readiness" (for instance the practice of recognizing a vehicle by its sound) is more active, while "listening-in-search" (exemplified by a ship captain whistling and using the echo for the purposes of orientation) requires the highest activity. Classifying three types of monitoring, Vickers (2011) relates aspects of Truax's categorizations to the auditory monitoring of processes. "Direct monitoring" corresponds to "listening-in-search," requiring the user to exclusively engage "with the system being monitored." This type of monitoring requires high activity and, in most cases, exclusive focus by operators. The primary focus of "peripheral monitoring" lies "elsewhere" with "attention being diverted to the monitored system either on your own volition at intervals by scanning the system . . . or through being interrupted by an exceptional event signaled by the system itself or through some monitor (such as an alarm)." Its allocation to "listening-in-readiness" or "background listening" is situation dependent. However, in view of process monitoring, we can assume that in the majority of applications, "listening-in-readiness" applies. Mynatt et al. (1998) coined the term "serendipitous information" for hints that may give notice to office workers about incoming emails and other useful but not necessarily indispensable messages. Vickers (2011) includes "serendipitous-peripheral monitoring" as a monitoring type for situations, in which additional information is "useful and appreciated, but not strictly required or vital either to the task in hand or the overall goal."

### 6.2.2 Levels of Attention

The Skills, Rules, Knowledge (SRK) framework by Rasmussen (1987) extends these perception- and attention-related aspects of listening and monitoring by considerations about cognitive control of human behavior in complex work scenarios. Rasmussen distinguishes between three different levels of psychological processes: "Skill-based behavior" stands for sensorimotor activities executed without conscious attention based on "smooth, automated, and highly integrated patterns of behavior." Riding a bicycle may serve as an example for such an activity requiring little attention once the skill is learned. The higher-level "rule-based behavior" refers to executing tasks according to given instructions or empirically derived rules without necessarily understanding the underlying context, that is, "although the cues releasing a rule may not be explicitly known." Following the recipe of a cookbook gives a representative example for rule-based behavior. In novel and unexpected situations for which no rules derived from former experiences are known "control must move to a higher conceptual level," knowing and understanding the functionality of a system in order to formulate an explicit target based on current analysis of the situation. This "knowledge-based behavior" typically requires higher cognitive workload (Hearst 1997) than the former categories of behavior including "dynamic interaction between the activities at the three levels" (Rasmussen 1987).

Johannsen (2004) broaches the issue of "human errors possibly triggered or avoided by means of auditory display" and attributes several categories of failures to the three levels of the SRK framework. While "slips and lapses" are typical errors within skill-based behavior tasks that can be minimized by appropriate sonification design, mistakes on the rule- and knowledge-based levels are either caused by the "misapplication of good rules and the application of bad rules" or by "bounded rationality or an incomplete or inaccurate mental model of the problem space" (Johannsen 2004). Johannsen concludes that the design of auditory displays at rule- and knowledge-based levels requires "well-organized user participation" and strict orientations towards the defined tasks and targets. By its transitory nature, the main application of auditory displays would therefore be "to minimize human error at the skill-based level" (Watson and Sanderson 2007) shifting—especially in visually overloaded situations—parts of the attention to the auditory domain. Thus the cognitive workload on the visual domain can be supported (Hearst 1997). This conclusion conforms to the current state of in situ implementations in industry and other control room environments focusing on skill-based monitoring behaviors, particularly because auditory displays apparently outperform visual presentations in monitoring tasks requiring sustained vigilance (Szalma et al. 2004). In high-stress monitoring environments, auditory displays are therefore most suitable to operate in the background as a secondary channel facilitating operators to act "eyes-free" focusing on primary activities, for instance, during surgery or machine operation (Watson and Sanderson 2004).

### *6.2.3 Spatial Auditory Displays and Stream Segregation*

Human auditory perception is particularly sensitive to changes within an acoustic environment and the directivity of sound sources (cf. Blauert 1983), which make audio signals suitable as alerts in critical situations, giving evidence not only of the state of urgency and the information content, but also of the local origin of the causing event. Besides supporting localization, acoustic spatialization is also useful for the segregation of simultaneous streams of information. Our perceptual ability for auditory stream segregation (Bregman 1994) facilitates discrete identification of simultaneous sonic events and their assignation to corresponding incidents. Shinn-Cunningham and Ihlefeld (2004) showed that overall performance improved when sound sources were perceived from different locations, both for selective—focusing on one specific auditory stream—and divided attention-tasks.

Sound spatialization can manifest itself in various forms: by discrete speakers for each information stream (Sirkka et al. 2014) or by using spatialization algorithms, such as vector-based amplitude panning (VBAP)[5] and Ambisonics, in combination with multiple speaker setups. The latter require appropriate and calibrated monitoring conditions making them ineligible for most existing control room environments. Solutions for headphones including head tracking and binaural encoding by head-related transfer functions (Arrabito 2000; Roginska et al. 2006) are much more feasible, since they are comparatively cheap and easy to implement. The drawback of wearing headphones is that they easily can be sensed as physical obstacles impeding natural behavior and perception. Head-related transfer functions (HRTFs) are a set of filters that compensate for the directional spectral functions of our outer ear when we are listening to an audio signal via headphones. Processing sounds via HRTFs usually includes head tracking in order to identify their directionalities and to apply correspondent frequency responses. Using HRTFs, our natural localization cues can be modeled to provide us with realistic soundscape environments. Since we experience the emerging acoustic image provided by HRTFs outside of our head, this perceptual phenomenon is also called "externalization." Using headphones without HRTF processing, we perceive the location of sounds only inside our heads ("internalization"). Roginska (2012) compared the accuracy and response times of stimuli perceived inside and outside the head and concluded that internalized presentations outperformed externalizations in both aspects investigated. She also observed that externalized sounds cause less fatigue since they are "less 'attention grabbing.'" The implementation of 3D spatialization does not necessarily increase the discrimination rate of parallel auditory streams significantly. A comparison of stereo with 3D sonifications resulted in only a slightly increased number of correctly identified call signs by fire and rescue command operators in high cognitive workload situations (Carlander et al. 2005).

## 6.3 Auditory Display for Process Monitoring and Analysis Tasks

Kramer (1994a) differentiates between monitoring and analysis tasks for the design of auditory displays. While the available definitions of sonification explicitly exclude speech (Barrass and Kramer 1999; Hermann 2008), in his understanding of auditory displays Kramer (1994a) includes any form of designed acoustic information (i.e. also verbal information). According to Johannsen (2004), functional objectives of monitoring applications can be either classified as "alarms and warnings" or as information representing the "state and intent" of a system.[6] While the latter gives evidence about the "function of system components," aspects of functionality for alarms and warnings include the perception of their "urgency," their "distinctiveness" and their "arousal," leading to the different sonification design concepts of continuous and intermittent sonification, which will be explained later in this chapter.

Typical monitoring applications can be found in surveillance environments, such as control rooms for power plants (Barrass 1997; Hearst 1997; Viraldo and Caldwell 2013), electrical control rooms (Dadashi et al. 2009), air traffic control towers (Begault 2012; Cabrera et al. 2005), intensive care units (Bourgeon et al. 2006; Meredith et al. 1999; Paterson et al. 2016; Seagull et al. 2001; Welch 1999), cockpits of airplanes (Burt et al. 1999) or (self-driving) vehicles (Chamberland et al. 2017). Using auditory displays for monitoring purposes, users have predefined expectations of the appearing sounds and will execute some kind of template matching comparing momentary sounds to their predefinitions. Csapó and Wersényi (2013) speak of a "conceptual paradigm," where questions that require an answer are a priori predefined and confined. Barrass (1997) denotes auditory monitoring as "a 'listening search' for familiar patterns in a limited and unambiguous set of sounds."

While there are several in situ implementations of auditory monitoring, applications of analytical tasks for process control are still considered to be rather experimental, including data exploration in, for example, network traffic surveillance (Ballora et al. 2011; Hildebrandt and Rinderle-Ma 2015; Vickers et al. 2014; Worrall 2015), analysis of stock market fluctuations (Janata and Childs 2004; Nesbitt and Barrass 2004, 2002; Worrall 2015), seismic data analysis (Barrass and Kramer 1999) or bottleneck analysis in production planning and control (Hildebrandt et al. 2014; Iber and Windt 2012). Performing an auditory analysis task "the listener cannot anticipate what will be heard and is listening for 'pop-out' effects, patterns, similarities and anomalies which indicate structural features and interesting relationships in the data" (Barrass 1997). Analytic approaches based on auditory display tend to be more explorative and hypothetical. Csapó and Wersényi (2013) denote this rarely predetermined approach as an "interactive paradigm." Due to missing reference sounds, "users won't know what sounds to expect precisely" (Barrass 1997) and often need to recondition the examined data or to readjust sonification parameters by recursive procedures. The sonification design for process analysis therefore has to be much more open and adjustable.

While process monitoring is a typically real-time, sometimes even time-critical, task involving mostly peripheral and only occasionally direct monitoring (see above), sonification for analysis purposes can be applied to real-time as well as historical data, generally requiring direct monitoring including high cognitive workload.

## 6.4 Workplace Design for Auditory and Multimodal Displays

Few auditory displays have been designed as stand-alone applications.[7] In general, they are implemented in multimodal interfaces addressing both visual and auditory channels, at times also including haptic feedback devices (Baldwin et al. 2012; Hogan et al. 2017).

For the design and implementation of auditory displays it is essential to consider the acoustic environment of the locations at which the generated acoustic signals will be listened to. Questions concerning our perception of sound are usually answered by the scientific field of psychoacoustics. Walker and Kramer (2004) pointed out that this existing knowledge about the "basic perception of simple sounds" might not be sufficient to "relate to the often complex sounds used in real-world auditory displays including besides aspects of the acoustic environments also the influence of the knowledge, experience, and expectations of the listener." In other words, our understanding of psychoacoustics, how it is usually applied for the perceptual evaluation of auditory displays, is too limited for the complex auditory ecology found in real-world scenarios outside laboratory-like situations. For an appropriate design, the "perspective of the totality of auditory input to the listener" (Begault 2012) needs to be analyzed. This includes acoustic analyses of background noise as well as well as other signal sources, such as "intra-office 'face-face,'" telephone or radio communications. In most cases there will be existing prerequisites that cannot be adjusted at all or only to a limited degree.

The general noise level of a shop floor in the manufacturing industry easily reaches levels requiring workers to protect their ears by wearing ear protection devices. The design of appropriate auditory interfaces may become a challenging task under those circumstances (Robinson and Casali 1999). In order to be resistant against being masked by background noises, alerting sounds should be compounded of at least four spectral components (Edworthy et al. 2017). Spectral richness also supports the localizability of the alerts (Blauert 1983). To be reliably perceived, the loudness of the spectral components should be at least 15 dB above the auditory threshold given by the correspondent frequencies of the noise of the acoustic environment (Patterson 1990). Hellier and Edworthy (1999) suggest to impose warning sounds even 15 to 25 dB above the stated masking threshold. If, for instance, one of the spectral components used in an auditory warning sound already reaches a loudness of 70 dB in the background noise, the minimum loudness has to be 85 dB in order be reliably audible. In many cases this necessary minimum loudness level is exceeded to ensure that the signal is noticed in any case, possibly causing negative side effects by disturbing the cognitive activities of operators and also constraining communication with their coworkers (Guillaume 2011).

For the implementation of new auditory alerting schemes for air traffic control rooms, Cabrera et al. (2005) mention restrictions given by existing hardware that needed to be considered for the design of a new auditory alerting scheme for air traffic control. Loudspeakers were mounted "beneath the console desktop" a small hole in the desktop used for the keyboard cable providing "the only possible direct sound route between the loudspeaker and the operator's ears." Although the general acoustic environment in the regarded multi-terminal (up to 40) operations rooms were absorptive and acoustically optimized and the measured background noise level only occasionally exceeded 50 dBA for 90 percent of the time and 60 dBA for 10 percent of the time, the conditions of displaying acoustic information were rather poor in terms of the localization of the alerts (assignment to a specific terminal) as well as to spectral perception. Sirkka et al. (2014) implemented an auditory alarm concept in a highly communicative environment and positioned each of three loudspeakers next to the control desks operators had to approach in order to respond to a specific alarm. Additionally, all alarms were delivered to a loudspeaker in the lunch room, since operators were required to react to alarms also during their breaks.

### 6.4.1 Design of Future Multimodal Control Centers

An alternative to conventional loudspeakers and headphones are parametric (Reuben and Woon-Seng 2009) and beamforming (Guldenschuh et al. 2008) loudspeaker arrays that deliver—eventually steered by a motion tracking system—acoustic information exclusively to responsible operators (Fuhrmann et al. 2016; Fuhrmann and Amon 2014). Parametric loudspeaker arrays are based on amplitude modulations of ultrasound signals generating audible difference tones to the listener's ears. In comparison to wearing headsets, which impede operators' mobility and can be rather cumbersome, the use of parametric loudspeaker arrays allows free movement (within a certain sound zone) preventing cooperators from being distracted at the same time. Thus, also the general noise floor in the control center can be kept on a lower level. Lee et al. (2011) showed that that the exposure of operators to parametric speakers was less stressful and resulted in faster responses to given tasks than conventional speaker systems. The obviously high potential of parametric speakers systems for future multimodal control centers was demonstrated by Kaiser and Fuhrmann (2014). The authors developed an interactive demonstrator composite of three layers of visual displays simulating the control desk of a traffic monitoring operator. Traditional desktop screens were complemented by large video walls providing overviews and tablet devices for the display of detailed information. Software control was based on gestural interaction driven by several sensor measurements including motion tracking and the rotation of the chair, which also acted as a vibrating element providing additional tactile feedback. Aside from two motion tracked parametric loudspeaker arrays for each of the two workstations, the acoustic interface included a microphone array for directionally optimized intercommunication (Kaiser and Fuhrmann 2014).

## 6.5 Sonification Design for Process Monitoring and Analysis

General principles of sonification design for auditory interfaces have been described by Campo et al. (2004), Hunt et al. (2011), Kramer (1994a) and Peres et al. (2008).

Depending on the type of data or information to be displayed, sonifications can be either designed as intermittent events or as continuous streams (Watson 2006). Most traditional alarms and warnings are implemented as earcons (symbolic mapping) that represent events indicating a specific urgency level. Patterson (1990) suggested a series of bursts indicating the kind and urgency level of an alert. Building on this approach, Hellier and Edworthy (1999) designed auditory warning sounds consisting of two sequential sound events. A "short attention getting sound" (denoted as "attenson"), communicating the urgency of the event, is followed by some voice message containing qualitative information. In order to avoid overlapping and signal masking, serialization of otherwise concurrent events is also the strategy of Brock et al. (2005). Often the semantic connection to the event causing the alarm is arbitrary without any ecological relationship to the referent (Walker and Nees 2011). Also, the same sound is sometimes implemented for several alarms (Frimalm et al. 2014), which makes it difficult for operators to identify the source of the underlying event. Apart from such very basic layouts, earcons are capable of mediating rather complex multidimensional information, especially when they are based on music-related parameters, such as rhythmic or melodic patterns, instrumentation (timbre), tempo or dynamics. Auditory icons (iconic, analogic mapping), on the other hand, often relate directly to the represented events, by the use of natural sounds, which are familiar within the environment they are used in (Brazil and Fernström 2011; McGookin and Brewster 2011). Similar to alarms that are based on speech, however, auditory icons face the risk of being overheard since they are too similar to other sounds in their ecological environments, especially in noisy situations. Besides possible interferences with other verbal communication, speech announcements can also have language and comprehension issues (Frimalm et al. 2014). Leung et al. (1997) compared the learning and retention capabilities of speech-based warnings, auditory icons and earcons and concluded that speech and auditory icons were learned and retained with equal ease while the learning process of abstract sounds appeared to be much more difficult. Auditory icons provide a close relationship to the underlying processes and therefore become much more familiar to operators. This also may explain, why Patterson (1990) recommends to limit the number of abstract auditory warnings (earcons) to only six, while Frimalm et al. (2014) successfully tested sets of up to 30 distinguishable sonic events.

While intermittent sonifications in terms of alerting sounds have been omnipresent for decades, industrial environments are rather reluctant about the implementation of continuous sonifications (Csapó and Wersényi 2013) that permanently display the monitored parameters acoustically. However, considering potentially critical situations arising from the use of intermittent sound, such as false positive alarms or inattentional deafness, continuous sonifications provide a permanent awareness of system states and might enable even “an anticipation of critical situations” (Hildebrandt et al. 2016). For it is far from being trivial to define an appropriate threshold for triggering an intermittent event in order to provide enough time for operators to react.

Arguing that operators in smaller sized enterprises usually have to fulfill more than one task at a time, Hildebrandt et al. (2016) compared the performance of participants in a dual task process monitoring experiment under three conditions: (1) the secondary task was displayed only visually, (2) the secondary task was presented visually and acoustically using intermittent sonification, and (3) the secondary task was presented visually and acoustically using continuous sonification that was based on a forest metaphor representing the momentary states of several machines. Results showed significantly higher performances for the version including continuous sonification with participants acting "less often too late, thus avoiding critical states more often, but also less often too early." In an additional questioning one participant mentioned that the sudden appearance of intermittent sonifications led to more intensive observation of the visual display, while another participant found it difficult to differentiate all the sounds. Another promising statement for future in situ implementations of continuous sonifications was the observation of two participants that "performance would most likely increase over time and the intrusiveness of the sounds would decrease" (Hildebrandt et al. 2016).

Intermittent and continuous sonifications should not be considered as exclusive implementations based on the one or the other. The highest potential is accomplished by thoughtful combinations relating to the specific attributes of the parameters to be displayed. To minimize cognitive control and increase team awareness, Watson and Sanderson (2007) developed an auditory display to monitor the states of patients in operating rooms. In order to have the momentary state present at any time, they choose continuous sonification for critical and significantly changing patient information. This included pulse oximetry, representing the heart rate as the speed of a series of beeps while oxygen saturation is displayed by pitch, blood pressure, respiratory parameters and the level of awareness. For slower changing patient information intermittent earcons were used. The presented parameters included noninvasive blood pressure, temperature and level of muscle paralysis. Further parameters were displayed as auditory icons or alarm interrupts.

Roginska et al. (2006) described four categories of auditory display design denoted as monitoring modes.[8] While "alert mode" and "single track mode" refer to the concepts of intermittent and continuous sonification as they were introduced above, "relationship mode" differs from "single track mode" by displaying convergence, divergence or parallel motion of two or more data streams to be compared. The final "global mode" refers to the concept of "beacons" that was originally introduced (and patented) by Kramer (1994b) and with few exceptions has since been rather neglected in literature. Beacons are sounds that represent a system at certain states displaying too many data streams (e.g. in network or industrial process monitoring) to be perceived independently. Their spectral consistency is too complex to be analyzed in detail by our auditory system. However, we can learn how the "normal state" sounds and we notice when its spectral consistence changes (due to changes in the underlying data or of the system state). The auditory complexity of beacons has assumably affected their implementation in real-world auditory displays since they are rather demanding regarding their usually rather shallow learning curves. However, as studies on "perceptual learning," a term coined by Eleanor Gibson (1969), have shown, there lies significant potential in the discrimination of complex auditory streams for auditory displays, which will be discussed in the last section of this chapter.

## 6.6 Ecological Interface Design for Auditory and Multimodal Displays

Several approaches have been taken to develop general rules and principles for the design of auditory displays including the mentioned aspects of auditory perception, workplace and sonification design. Barrass (1998) defined "some golden rules for designing auditory displays" derived from design principles for graphic display emphasizing its directness ("can be understood almost immediately"), appropriateness ("information required by task: neither more nor less"), range ("any undetectable element is useless"), level ("allows to summarize general behavior, and . . . examine details") and organization ("interactive reorganization . . . can uncover information").

In order to develop appropriate designs for multimodal displays in workspace environments, Watson and Sanderson (2007) suggested a processing scheme for the design of auditory displays using ecological interface design (EID). In contrast to technology-centered, user-centered, or control-centered design approaches, EID considers not only the human-machine "system" but foremost the work itself, which needs to be executed by this system. Quoting E. Hollnagel, Flach et al. (1998) asserts that the objectives to be regarded are not just to "provide the right information at the right time and in the right way" but primarily to find answers to the question of what is "right." This approach makes EID particularly suitable for highly complex and dynamic systems, which cannot be completely understood or embraced even by experienced operators.

Based on the abovementioned example of processes in anesthesia work environments, Watson and Sanderson (2007) developed and evaluated a general processing scheme for the implementation of auditory displays into EID, comprising four main sections, which are summarized as follows:

1. Problem identification:
   - Gather evidence from incident reports, operator reports, filed observations, and so on.
2. Needs analysis:
   - Determine functional structure of work with work domain analysis.
   - Extract variables, constraints and temporal properties.
   - Determine whose work is affected by state changes and when.
   - Determine correct levels of cognitive control over control tasks in work domain using SRK-based behavior level identification (see above).
3. Design synthesis:
   - Identify best modalities to achieve required levels of cognitive control (e.g. visual display, intermittent sonification, continuous sonification).
   - Semantic mapping: decide on the kind of auditory display, sound mappings, number of streams (e.g. earcons, auditory icons, parameter mapping).
   - Attentional mapping: preserve appropriate condition for individual attention and team attention.[9]
4. Evaluation:
   - Conduct tests under representative conditions to see if initial problem is solved.

(Modified from Watson and Sanderson 2007)

The application of EID design principles has resulted in functional in situ implementations of overall complex multimodal displays including sonifications of rather basic or intermediate complexity. In industrial process monitoring and analysis applications, sonifications based on highly complex data scenarios have generally not left their experimental and laboratory-scale states yet. At least to some extent this might be caused by a general resistance to the use of sound in workplace environments as it is typically expressed by participants of focus groups or qualitative studies (Hildebrandt et al. 2014; Rottermanner et al. 2017). A survey of auditory and multimodal displays designed for network monitoring (Hildebrandt and Rinderle-Ma 2015) showed that only two (out of twenty) projects were realized as ready-to-use tools, while twelve remained in a prototype state, including the only three projects regarding both real-time and historic data. Six projects never exceeded the concept stage. The fact that only one of these projects mentioned formal user evaluation indicates the size of the gap that needs to be overcome for the acceptance of auditory or multimodal displays in professional scenarios. At least partially this resistance is comprehensible, for instance, when executive officers are averse to changes in their routine handlings in safety-relevant situations in air traffic control because they may initially increase the risk of incorrect reactions and decisions. Considering the competitiveness of the global market (Nyhuis et al. 2009), it is also understandable when CEOs in manufacturing enterprises hesitate to include sonification-based applications as support for production planning and control (PPC) if the economic advantage cannot be sufficiently quantified, since any optimizing measure in operation procedures also poses the risk of failure.

## 6.7 Potential for Future Developments

Since multimodal displays that follow the processing scheme of EID are developed and evaluated in close collaboration with the responsible operators, their implementations are customized, resulting in respectively high acceptance rates. The scope of the application is well defined and risks to overburden or annoy operators are rather limited. However, by focusing on the user perspective, innovative approaches that utilize a major portion of the potential our auditory system is capable of might remain disregarded, since even slightly more ambitious sonification approaches as the one mentioned above (Hildebrandt et al. 2016) are easily confronted with overburdened test participants, not to mention responses to highly complex sonifications as they are represented by the concept of beacons.

On the other hand, exactly this potential of highly complex auditory displays for data analysis has consistently been emphasized by researchers in the field, to cite only Bruce Walker (as quoted in Feder 2012): "The best pattern recognition system that we know of is our auditory system. . . . We know that with music, we can convey melody, tension, expectancy. There is a lot of similarity between trends and melodies. We extract a huge amount of information, so there are lots of reasons to expect sound to be a good medium."

Worrall (2014) points out that aspects of composition, notation and interpretation should not be confused with considerations of sonification design, since "they have different intents and epistemological imperatives." Also, there are only weak to nonexistent correlations between musical experiences with auditory displays (Walker and Nees 2011). However, Walker's reference to music is quite eligible to bring to mind how and to what extent our auditory system is able to extract information from acoustic signals depending on different levels of (trained) expertise and proficiency: To most people it will be sufficient to catch a snippet of music incidentally on the radio to identify a song they are familiar with. Listeners with a background in, say, classical music or jazz might even assign this very snippet to its composer or the improviser without ever having heard the piece before. They are able to identify the music because they are familiar with certain stylistic attributes or other characteristic features. Furthermore, a classical pianist who has memorized and performed Johann Sebastian Bach's one-hour-long Goldberg Variations several times will manage to identify even the tiniest deviation of a colleague's interpretation from his own. The level of listeners' familiarity with the analyzed music, which generally correlates to their level of expertise, is of crucial importance. In the context of auditory display, familiarity refers not only to the sounds, but also to the underlying processes and their interdependencies—as the reference to the contribution of acoustic emissions to working knowledge[10] also suggests.

Considering the general reluctance to highly complex sonifications, the difficulties do not ultimately lie in the sonic complexity itself, but rather in the methods, the ways in which sounds are mapped and composite. Recent publications, such as by Roddy and Furlong (2014), assert that conventional sonification techniques generally refer to the "disembodied[11] and positivistic models" of Western art music relying on the same techniques and technologies. They suggest that the "positivistic cognitivism"[12] underlying common design principles needs to be overcome in favor to models of "embodied cognition"[13] incorporating physical (e.g. microgestural) experiences and auditory perception. Just like Kramer's approach to beacons (Kramer 1994b), the concepts of embodied cognition and abovementioned perceptual learning share a reliance on holistic phenomena other than positivistic structures. Incorporating the potential of perceptual learning is built on long-term learning processes as they are well accepted not only for high-level accomplishments in sports and music performance (Ericsson and Lehmann 1996), but also for accustoming individuals to hearing aids or cochlear implants (Fu and Galvin III 2007; Martin 2007). The auditory achievements of blind people using screen readers (Moos and Trouvain 2007) are also excellent examples of what can be achieved by perceptual learning. Since perceptual learning can only be implemented in situ, one of the challenges for future developments of auditory displays might be to invent novel methods that parallel the established processes of information acquisition by listening to acoustic emissions. These methods should provide information to operators with different levels of experience, contributing to an increase of working knowledge for process analysis and decision-making.

## Recommended further reading

*Auditory Display—Audification, Sonification, Auditory Interfaces* (Kramer 1994c), *Human Factors in Auditory Warnings* (Stanton and Edworthy 1999), *Ecological Psychoacoustics* (Neuhoff 2004), *Auditory Interfaces* (Peres et al. 2008), *Sonification Handbook* (Hunt et al. 2011).

## Notes

1 Flach et al. (1998) favor "ecological" over "environmental," because the latter "tends to be used for things outside the system," while "the term ecology is used to explicitly include the work domain as an intrinsic part of the distributed cognitive system."

2 In literature, the terminology related to "auditory display" and "sonification" exhibits some inconsistencies that Hermann (2008) tried to clarify. According to his specification, an auditory display denotes the apparatus and "encompasses also the technical system used to create sound waves," while sonification is "an integral component within an auditory display system" referring "to the algorithm that is at work between the data, the user and the resulting sound." Hermann further distinguishes between auditory icons, earcons, parameter mapping sonification and model based sonification as sonification techniques (cf. section 6.5, "Sonification Design for Process Monitoring and Analysis," in this chapter), while alternative classifications (Watson 2006) differentiate between (intermittent, event-based) auditory displays and (continuous) sonifications. According to this denotation, auditory displays are based on sonic events such as auditory icons, earcons and their relatives, whereas sonifications provide permanent acoustic representations of the state of the monitored system. Within the scope of this chapter we refer to the taxonomy suggested by Hermann.

3 If not stated otherwise, in the chapter, "operator" will be representatively used for all of these professions.

4 As opposed to "false negative alarms," when no sound is played, although there is evidence (e.g. by some system error).

5 "Stereo" display can be considered as a basic form of VBAP.

6 Referring to the emotional impact of film sound later in his article, Johannsen (2004) also considers "the appeal of products" as an individual category.

7 Exceptions include, for example, devices explicitly developed as subsidiary applications for visually impaired people.

8 Not to be confused with Vickers' "monitoring modes," introduced at the beginning of this chapter.

9 Watson and Sanderson (2007) distinguish several affordances of the auditory in comparison to the visual domain: ubiquitous versus localized properties, obligatory versus optional properties and transitory versus persistent property, together expressing that auditory displays are perceived only at a specific point of time by anyone nearby, independent of a listener's posture.

10 See the beginning of this chapter.

11 "The idea that an aural event could be objectified and studied in its own right, that is independent of the means of its production . . ." (Worrall 2010).

12 That is, the acquisition of knowledge is based on the reasoned and logical interpretation of real, sensually perceivable and verifiable experiences.

13 Antle et al. (2009) describe how bodily experiences of balance (walking, seesawing and carrying a tablet) are embodied in a simple structure serving as (unconscious) metaphors for abstract decision-making.

## References


Aigner, W., Miksch, S., Schumann, H., and Tominski, C., 2015. Visualization techniques for time-oriented data. In: Ward, M. O., Grinstein, G., and Keim, D. (Eds.) *Interactive Data Visualization: Foundations, Techniques, and Applications*. Boca Raton, FL: A. K. Peters/CRC Press, pp. 253–284.

Alexanderson, P., and Tollmar, K., 2006. Being and mixing: Designing interactive soundscapes. *Proceedings of the 4th Nordic Conference on Human-Computer Interaction: Changing Roles, NordiCHI '06*. New York: ACM, pp. 252–261. https://doi.org/10.1145/1182475.1182502.

Antle, A. N., Corness, G., Bakker, S., Droumeva, M., van den Hoven, E., and Bevans, A., 2009. *Proceedings of the 7th ACM Conference on Creativity and Cognition*. ACM, pp. 275–284.

Arrabito, G. R., 2000. An evaluation of three-dimensional audio displays for use in military environments. *Canadian Acoustics* 28, pp. 5–14.

Baldwin, C. L., Eisert, J. L., Garcia, A., Lewis, B., Pratt, S. M., and Gonzalez, C., 2012. Multimodal urgency coding: Auditory, visual, and tactile parameters and their impact on perceived urgency. *Work* 41, pp. 3586–3591. https://doi.org/10.3233/WOR-2012-0669-3586.

Ballora, M., Giacobe, N. A., and Hall, D. L., 2011. Songs of cyberspace: An update on sonifications of network traffic to support situational awareness. *Proceedings of SPIE 8064, Multisensor, Multisource Information Fusion: Architectures, Algorithms, and Applications 2011* 80640P. https://doi.org/10.1117/12.883443.

Barrass, S., 1997. *Auditory Information Design*. Canberra: Australian National University.

Barrass, S., 1998. Some golden rules for designing auditory displays. In: Vercoe, B., and Boulanger, R. (Eds.) *Csound Textbook*. Cambridge, MA: MIT Press.

Barrass, S., and Kramer, G., 1999. Using sonification. *Multimedia Systems* 7, pp. 23–31.

Begault, D. R., 2012. Guidelines for NextGen auditory displays. *Journal of the Audio Engineering Society* 60, pp. 519–530.

Berner, B., 2008. Working knowledge as performance: On the practical understanding of machines. *Work, Employment and Society* 22, pp. 319–336. https://doi.org/10.1177/0950017008089107.

Blauert, J., 1983. *Spatial Hearing*. Cambridge, MA: MIT Press.

Bourgeon, L., Cazalaà, J.-B., Guillaume, A., Jacob, E., Rivenez, M., and Valot, C., 2006. Non Vocal auditory signals in the operating room for each phase of the anaesthesia procedure. In: *Audio Engineering Society Convention 120.*

Brazil, E., and Fernström, M., 2011. Auditory icons. In: Hermann, T., Hunt, A., and Neuhoff, J. G. (Eds.) *The Sonification Handbook*. Berlin: Logos Verlag, pp. 325–338.

Bregman, A. S., 1994. *Auditory Scene Analysis: The Perceptual Organization of Sound.* Cambridge, MA & London: MIT Press.

Brock, D., Ballas, J. A., and McFarlane, D. C., 2005. Encoding urgency in legacy audio alerting systems. *Proceedings of the 5h International Conference on Auditory Display*. Limerick, Ireland.

Burt, J. L., Bartolome-Rull, D. S., Durdette, D. W., and Comstock, J. R., 1999. A psychophysiological evaluation of the perceived urgency of auditory warning signals. In: Stanton, N. A., and Edworthy, J. (Eds.) *Human Factors in Auditory Warnings*. Aldershot: Ashgate.

Cabrera, D., Ferguson, S., and Laing, G., 2005. Development of auditory alerts for air traffic control consoles. *Audio Engineering Society Convention 119*.

Campo, A. de, Frauenberger, C., and Höldrich, R., 2004. Designing a generalized sonification environment. *Proceedings of the 10th International Conference on Auditory Display*. Sydney, Australia.

Carlander, O., Kindstrom, M., and Eriksson, L., 2005. Intelligibility of stereo and 3D-audio call signs for fire and rescue command operators. *Proceedings of the 5th International Conference on Auditory Display*. Limerick, Ireland.

Carlile, S., 2011. Psychoacoustics. In: Hermann, T., Hunt, A., and Neuhoff, J. G. (Eds.) *The Sonification Handbook*. Berlin: Logos Verlag, pp. 41–61.

Chamberland, C., Hodgetts, H. M., Vallières, B. R., Vachon, F., and Tremblay, S., 2017. The benefits and the costs of using auditory warning messages in dynamic decision making settings. *Journal of Cognitive Engineering and Decision Making* 12/2, pp. 112–130. https://doi.org/10.1177/1555343417735398.

Chion, M., Gorbman, C., and Murch, W., 1994. *Audio-Vision*. New York: Columbia University Press.

Csapó, Á., and Wersényi, G., 2013. Overview of auditory representations in human-machine interfaces. *ACM Computing Surveys* (CSUR) 46, p. 19.

Dadashi, N., Sharples, S., and Wilson, J. R., 2009. Alarm handling in rail electrical control. *European Conference on Cognitive Ergonomics: Designing beyond the Product—Understanding Activity and User Experience in Ubiquitous Environments*. VTT Technical Research Centre of Finland.

Edworthy, J., Reid, S., McDougall, S., Edworthy, J., Hall, S., Bennett, D., Khan, J., and Pye, E., 2017. The recognizability and localizability of auditory alarms: Setting global medical device standards. *Human Factors* 59, pp. 1108–1127. https://doi.org/10.1177/0018720817712004.

Ericsson, K. A., and Lehmann, A. C., 1996. Expert and exceptional performance: Evidence of maximal adaptation to task constraints. *Annual Review of Psychology* 47, pp. 273–305. https://doi.org/10.1146/annurev.psych.47.1.273.

Feder, T., 2012. Shhhh. Listen to the data. *Physics Today* 65, pp. 20–22. https://doi.org/10.1063/PT.3.1550.

Flach, J. M., Tanabe, F., Monta, K., Vicente, K. J., and Rasmussen, J., 1998. An ecological approach to interface design. *Proceedings of the Human Factors and Ergonomics Society Annual Meeting*. Los Angeles, CA: SAGE Publications, pp. 295–299.

Frimalm, R., Fagerlönn, J., Lindberg, S., and Sirkka, A., 2014. How many auditory icons in a control room environment can you learn? *Proceedings of the 20th International Conference on Auditory Display*. New York.

Fu, Q.-J., and Galvin III, J. J., 2007. Perceptual learning and auditory training in cochlear implant recipients. *Trends in Amplification* 11, pp. 193–205.

Fuhrmann, F., and Amon, C., 2014. Evaluation of a transaural audio system using parametric loudspeaker arrays. *Proceedings of the 6th Congress of the Alps Adria Acoustics Association.*

Fuhrmann, F., Amon, C., Leitner, C., Maly, A., and Graf, F., 2016. Personalized sound zoning for communication means—user studies and evaluation. *Proceedings of Conference on Human Computer Interaction (HCI) Europe*. Pilsen.

Gaver, W. W., 1994. Using and creating auditory icons. In: Kramer, G. (Ed.) *Auditory Display: Sonification, Audification, and Auditory Interfaces*. Reading, MA: Addison-Wesley Publishing Company, pp. 417–446.

Gaver, W. W., Smith, R. B., Shea, T. O., and Hall, W., 1991. Effective sounds in complex systems: The ARCOLA simulation. *Proceedings of CHI* (New Orleans, LA). New York: ACM, pp. 85–90.

Gibson, E., 1969. *Principles of Perceptual Learning and Development*. Englewood Cliffs, NJ: Prentice Hall College Div.

Guillaume, E., 2011. Intelligent auditory alarms. In: Hermann, T., Hunt, A., and Neuhoff, J. G. (Eds.) *The Sonification Handbook*. Berlin: Logos Verlag, pp. 493–508.

Guldenschuh, M., Sontacchi, A., Zotter, F., and Höldrich, R., 2008. Principles and considerations to controllable focused sound source reproduction. *7th Eurocontrol INO Workshop.*

Hearst, M. A., 1997. Dissonance on audio interfaces. *IEEE Expert* 12, pp. 10–16. https://doi.org/10.1109/64.621221.

Hellier, E., and Edworthy, J., 1999. The design and validation of attensons for high workload environment. In: Stanton, N. A., and Edworthy, J. (Eds.) *Human Factors in Auditory Warnings*. Aldershot: Ashgate, pp. 283–303.

Hermann, T., 2008. Taxonomie and definitions for sonification and auditory display. *Proceedings of the 14th International Conference on Auditory Display*. Paris.

Hildebrandt, T., Hermann, T., and Rinderle-Ma, S., 2016. Continuous sonification enhances adequacy of interactions in peripheral process monitoring. *International Journal of Human-Computer Studies* 95, pp. 54–65. http://dx.doi.org/10.1016/j.ijhcs.2016.06.002.

Hildebrandt, T., Mangler, J., and Rinderle-Ma, S., 2014. Something doesn't sound right: Sonification for monitoring business processes in manufacturing. *16th Conference on Business Informatics*. IEEE, pp. 174–182. https://doi.org/10.1109/CBI.2014.12.

Hildebrandt, T., and Rinderle-Ma, S., 2015. Server sounds and network noises. *6th IEEE International Conference on Cognitive Infocommunications—CogInfoCom 2015*, 6th IEEE International Conference on Cognitive Infocommunications—CogInfoCom 2015.

Höferlin, B., Höferlin, M., Goloubets, B., Heidemann, G., and Weiskopf, D., 2012. Auditory support for situation awareness in video surveillance. *Proceedings of the 18th International Conference on Auditory Display*. Atlanta.

Hogan, T., Hinrichs, U., and Hornecker, E., 2017. The visual and beyond: Characterizing experiences with auditory, haptic and visual data representations. *Proceedings of the 2017 Conference on Designing Interactive Systems*. ACM, pp. 797–809.

Hunt, A., Hermann, T., and Neuhoff, J. G., 2011. *The Sonification Handbook*. Berlin: Logos Verlag.

Iber, M., and Windt, K., 2012. Order-related acoustic characterization of production data. *Logistics Research* 5, pp. 89–98. https://doi.org/10.1007/s12159-012-0084-y.

Janata, P., and Childs, E., 2004. Marketbuzz: Sonification of real-time financial data. *Proceedings of the 10th International Conference on Auditory Display*. Sydney.

Johannsen, G., 2004. Auditory displays in human-machine interfaces. *Proceedings of the IEEE* 92, pp. 742–758. https://doi.org/10.1109/JPROC.2004.825905.

Kaiser, R., and Fuhrmann, F., 2014. Multimodal interaction for future control centers: Interaction concept and implementation. *Proceedings of the 2014 Workshop on Roadmapping the Future of Multimodal Interaction Research Including Business Opportunities and Challenges*. ACM, pp. 47–51.

Kramer, G., 1994a. An introduction to auditory display. In: Kramer, G. (Ed.) *Auditory Display: Sonification, Audification and Auditory Interfaces*. Reading, MA: Addison-Wesley Publishing Company, pp. 1–77.

Kramer, G., 1994b. Some organizing principles for representing data with sound. In: Kramer, G. (Ed.) *Auditory Display: Sonification, Audification and Auditory Interfaces*. Reading, MA: Addison-Wesley Publishing Company, pp. 185–221.

Kramer, G., 1994c. *Auditory Display: Sonification, Audification, and Auditory Interfaces*. Reading, MA: Addison-Wesley Publishing Company.

Kramer, G., Walker, B., Bonebright, T., Cook, P. R., Flowers, J. H., Miner, N., and Neuhoff, J., 1997. Sonification report: Status of the field and research agenda, *Prepared for the National Science Foundation by Members of the International Community for Auditory Display*. Palo Alto.

Lee, S., Katsuura, T., and Shimomura, Y., 2011. Effects of parametric speaker sound on physiological functions during mental task. *Journal of Physiological Anthropology* 30, pp. 9–14.

Leung, Y. K., Smith, S., Parker, S., and Martin, R., 1997. Learning and retention of auditory warnings. *Proceedings of the 4th International Conference on Auditory Display*. Sydney.

Li, X., Powell, M. S., and Horberry, T., 2012. Human factors in control room operations in mineral processing: Elevating control from reactive to proactive. *Journal of Cognitive Engineering and Decision Making* 6, pp. 88–111. https://doi.org/10.1177/1555343411432340.

Martin, M., 2007. Software-based auditory training program found to reduce hearing aid return rate. *The Hearing Journal* 60, pp. 32–34.

Martins, C. H., Aguiar, P. R., Frech, A., and Bianchi, E. C., 2014. Tool condition monitoring of single-point dresser using acoustic emission and neural networks models. *IEEE Transactions on Instrumentation and Measurement* 63, pp. 667–679.

McGookin, D., and Brewster, S., 2011. Earcons. In: Hermann, T., Hunt, A., and Neuhoff, J. G. (Eds.) *The Sonification Handbook*. Bielefeld: Logos Verlag, pp. 339–361.

Meredith, C., Edworthy, J., and Rose, D., 1999. Observational studies of auditory warnings on the intensive care unit. In: Stanton, N. A., and Edworthy, J. (Eds.) *Human Factors in Auditory Warnings*. Aldershot: Ashgate, pp. 305–317.

Moos, A., and Trouvain, J., 2007. Comprehension of ultra-fast speech—blind vs. "normally hearing" persons. *Proceedings of 16th International Congress of Phonetic Sciences*, pp. 677–680.

Mynatt, E. D., Back, M., Want, R., Baer, M., and Ellis, J. B., 1998. Designing audio aura. *Proceedings of the SIGCHI Conference on Human Factors in Computing Systems*. ACM/Addison-Wesley Publishing Company, pp. 566–573.

Nesbitt, K. V., and Barrass, S., 2002. Evaluation of a multimodal sonification and visualisation of depth of market stock data. *Proceedings of the 8th International Conference on Auditory Display*. Kyoto.

Nesbitt, K. V., and Barrass, S., 2004. Finding trading patterns in stock market data. *IEEE Computer Graphics and Applications* 24, pp. 45–55. https://doi.org/10.1109/MCG.2004.28.

Neuhoff, J. G., 2004. *Ecological Psychoacoustics*. Burlington, MA: Brill.

Nyhuis, P., Münzberg, B., and Kennemann, M., 2009. Configuration and regulation of PPC. *Production Engineering* 3, pp. 287–294. https://doi.org/10.1007/s11740-009-0162-4.

Paterson, E., Sanderson, P. M., Paterson, N. A. B., Liu, D., and Loeb, R. G., 2016. The effectiveness of pulse oximetry sonification enhanced with tremolo and brightness for distinguishing clinically important oxygen saturation ranges: A laboratory study. *Anaesthesia* 71, pp. 565–572. https://doi.org/10.1111/anae.13424.

Patterson, R. D., 1990. Auditory warning sounds in the work environment. *Philosophical Transactions of the Royal Society of London. Series B, Biological* 327, pp. 485–492.

Peres, C., Best, V., Brock, D., Frauenberger, C., Hermann, T., Neuhoff, J. G., Nickersen, L. V., Shinn-Cunningham, B., and Stockman, T., 2008. Auditory interfaces. In: Kortum, P. (Ed.) *HCI Beyond the GUI: Design for haptic, speech, olfactory and other nontraditional interfaces. The Morgan Kaufmann Series in Interactive Technologies*. Burlington, MA: Morgan Kaufmann, pp. 147–195. https://doi.org/10.1016/B978-0-12-374017-5.00005-5.

Rasmussen, J., 1987. Mental models and the control of actions in complex environments. In: Ackermann, D., and Tauber, M. J. (Eds.) *Selected Papers of the 6th Interdisciplinary Workshop on Informatics and Psychology: Mental Models and Human-Computer Interaction 1*. Amsterdam: North-Holland Publishing Co, pp. 41–69.

Reeves, B. N., and Shipman, F., 1996. Tacit knowledge: Icebergs in collaborative design. ACM *SIGOIS Bulletin* 17, pp. 24–33. https://doi.org/10.1145/242206.242212

Renwick, J. T., and Babson, P. E., 1985. Vibration analysis—a proven technique as a predictive maintenance tool. *IEEE Transactions on Industry Applications IA-21*, pp. 324–332. https://doi.org/10.1109/TIA.1985.349652.

Reuben, J., and Woon-Seng, G., 2009. 3D sound affects with transaural audio beam projection. *10th Western Pacific Acoustic Conference*. Beijing, pp. 21–23.

Robinson, G. S., and Casali, J. G., 1999. Audibility of reverse alarms under hearing protectors and its prediction for normal and hearing-impaired listeners. In: Stanton, N. A., and Edworthy, J. (Eds.) *Human Factors in Auditory Warnings*. Aldershot: Ashgate.

Roddy, S., and Furlong, D., 2014. Embodied aesthetics in auditory display. *Organised Sound* 19, pp. 70–77. https://doi.org/10.1017/S1355771813000423.

Roginska, A., 2012. Effect of spatial location and presentation rate on the reaction to auditory displays. *Journal of the Audio Engineering Society* 60, pp. 497–504.

Roginska, A., Childs, E., and Johnson, M. K., 2006. Monitoring real-time data: A sonification approach. *Proceedings of the 12th International Conference on Auditory Display*. London.

Rottermanner, G., Wagner, M., Settgast, V., Grantz, V., Iber, M., Kriegshaber, U., Aigner, W., Judmaier, P., and Eggeling, E., 2017. Requirements analysis & concepts for future European air traffic control systems. In: *Workshop Vis in Practice—Visualization Solutions in the Wild, IEEE VIS 2017*. Phoenix, AZ: IEEE.

Sanders, M. S., and McCormick, E. J., 1987. *Human Factors in Engineering and Design*. New York: McGraw-Hill.

Schaeffer, P., 2017. *Treatise on Musical Objects: An Essay Across Disciplines*. Berkeley, CA: University of California Press.

Seagull, F. J., Wickens, C. D., and Loeb, R. G., 2001. When is less more? Attention and workload in auditory, visual, and redundant patient-monitoring conditions. *Proceedings of the Human Factors and Ergonomics Society Annual Meeting*. Los Angeles, CA: SAGE Publications, pp. 1395–1399. https://doi.org/10.1177/154193120104501817.

Shinn-Cunningham, B. G., and Ihlefeld, A., 2004. Selective and divided attention: Extracting information from simultaneous sound sources. *Proceedings of the 10th International Conference on Auditory Display.*

Sirkka, A., Fagerlönn, J., Lindberg, S., and Delsing, K., 2014. The design of an auditory alarm concept for a paper mill control room. *Advances in Ergonomics In Design, Usability & Special Populations* 3, p. 118.

Stanton, N., and Edworthy, J. (Eds.), 1999. *Human Factors in Auditory Warnings*. Aldershot: Ashgate.

Szalma, J. L., Warm, J. S., Matthews, G., Dember, W. N., Weiler, E. M., Meier, A., and Eggemeier, F. T., 2004. Effects of sensory modality and task duration on performance, workload, and stress in sustained attention. *Human Factors* 24, pp. 219–233. https://doi.org/10.1097/01.HJ.0000286505.76344.10.

Truax, B., 2001. *Acoustic Communication*. Westport, CT: Ablex Publishing.

Vickers, P., 2011. Sonification for process monitoring. In: Hermann, T., Hunt, A., and Neuhoff, J. G. (Eds.) *The Sonification Handbook*. Berlin: Logos Verlag, pp. 455–491.

Vickers, P., Laing, C., Debashi, M., and Fairfax, T., 2014. Sonification aesthetics and listening for network situational awareness. *arXiv:1409.5282*.

Viraldo, J., and Caldwell, B., 2013. Sonification as sensemaking in control room applications. *Proceedings of the Human Factors and Ergonomics Society Annual Meeting*. Los Angeles, CA: SAGE Publications, pp. 1423–1426. https://doi.org/10.1177/1541931213571318.

Walker, B. N., and Kramer, G., 2004. Ecological psychoacoustics and auditory displays: Hearing, grouping, and meaning making. In: Neuhoff, J. G. (Ed.) *Ecological Psychoacoustics*. San Diego: Elsevier, pp. 150–175.

Walker, B. N., and Nees, M. A., 2011. Theory of sonification. In: Hermann, T., Hunt, A., and Neuhoff, J. G. (Eds.) *The Sonification Handbook*. Bielefeld: Logos Verlag, pp. 10–39.

Watson, M., 2006. Scalable earcons: Bridging the gap between intermittent and continuous auditory displays. *Proceedings of the 12th International Conference on Auditory Display*. London.

Watson, M., and Sanderson, P., 2004. Sonification supports eyes-free respiratory monitoring and task time-sharing. *Human Factors* 46, pp. 497–517. https://doi.org/10.1518/hfes.46.3.497.50401.

Watson, M., and Sanderson, P., 2007. Designing for attention with sound: Challenges and extensions to ecological interface design. *Human Factors* 49, pp. 331–346. https://doi.org/10.1518/001872007X312531.

Welch, J., 1999. Auditory alarms in intensive care. In: Stanton, N. A., and Edworthy, J. (Eds.) *Human Factors in Auditory Warnings*. Aldershot: Ashgate.

Worrall, D., 2010. Parameter mapping sonic articulation and the perceiving body. *Proceedings of the 16th Conference on Auditory Display*. Washington, DC.

Worrall, D., 2014. Can micro-gestural inflections be used to improve the soniculatory effectiveness of parameter mapping sonifications? *Organised Sound* 19, pp. 52–59. https://doi.org/10.1017/S135577181300040X.

Worrall, D., 2015. Realtime sonification and visualisation of network metadata. *Proceedings of the 21th International Conference on Auditory Display*. Graz.

Xiao, Y., Mackenzie, C. F., Seagull, F. J., and Jaberi, M., 2000. Managing the monitors: An analysis of alarm silencing activities during an anesthetic procedure. *Proceedings of the Human Factors and Ergonomics Society Annual Meeting*. Los Angeles, CA: SAGE Publications, pp. 250–253. https://doi.org/10.1177/154193120004402630.

7

# Incorporating Brand Identity in the Design of Auditory Displays

## The Case of Toyota Motor Europe

**Elif Özcan, René van Egmond, Alexandre Gentner and Carole Favart**

### 7.1 Introduction

Interface sounds (*beeps*) in passenger cars are most commonly used to draw the attention of the user to a graphical information display, but they are underused for the purpose of implicitly informing users about an ongoing system event (e.g. low oil levels, door not well closed, safety belt not buckled). In general, sounds in the automotive industry are considered complementary to the visual displays in the dashboard design but not as a stand-alone information display (i.e. auditory display) the function of which is to communicate dashboard messages to drivers (Bazilinsky and De Winter 2015). For car manufacturers, the brand identity should provide the tonal quality (i.e. timbre) with which sounds should be designed. Currently, design is based on drawing attention without considering the brand values and does not have a natural causal relationship with the event it signals. The design of (functional) sounds for passenger cars comes with great responsibility to ensure passenger and road safety as well as stylistic concerns over conveying brand values through sound. While focusing on safety through functionality enables drivers to take the right action, styling (expressiveness of the sounds) can increase the hedonic value of the sounds, making them easily acceptable and even pleasant.

The combination of style and functionality has been extensively studied by Hassenzahl and colleagues through interactive products (Hassenzahl et al. 2000; Hassenzahl 2001; Hassenzahl and Tractinsky 2006). In short, considering a product's *ergonomic qualities* (how predictable, simple, manageable, identifiable a product is) and *hedonic qualities* (how stylish, novel, creative, premium a product is) in tandem contributes to *product appeal* (how attractive the product is), making the product more user-friendly. Accordingly, product appeal can have behavioral (how easily or how often users interact with a product) as well as emotional consequences (how satisfied users are). In this chapter, we demonstrate how the overall experience of interface sounds can be improved in passenger vehicles. With ergonomics, we focus on the function of the interface sounds and how correctly identified functions guide the vital interactions with the dashboard. With hedonics, we focus on brand values and how they reflect on the expressiveness of the interface sounds.

Furthermore, the tools and methods to study and design for the hedonic and ergonomic values of interface sounds are either limited or not commonly shared for commercial reasons. This chapter also aims to demonstrate how to design auditory displays in which functionality and appeal are equally prominent. In addition, new tools and design methods are presented that complement the traditional design approaches that oblige interface sounds to be perceptually unpleasant in order to attract attention. We propose metaphors as an approach to designing perceptually more pleasant and readily identifiable sounds that also convey brand values. Finally, we discuss and demonstrate how to use metaphors in auditory displays supported by theoretical reflections and a design application.

## 7.2 Auditory Display Design

Auditory displays serve as an interface in order to carry information from a technical system to a user, and they consist of a set of rules that turns technical information into an intelligible auditory message (Blattner et al. 1989; Stanton and Edworthy 1999; Neuhoff 2011). An *auditory message* signals a change in the current status of a system. The signal can have an informational nature such as providing feedback, or refer to more attention-seeking situations such as alerting the user. Therefore, signals need to be designed in such a way that users quickly understand rather complex information streams. Similar to visual messages, auditory messages can take various forms such as text and numbers, symbols and icons (cf., Rensink 2009). Basically, three categories of auditory displays can be distinguished ranging in terms of concreteness of information: speech, abstract sounds and auditory icons (Kramer 1994).

*Speech* is the most intuitive and direct way of communicating a message. The meaning conveyed through speech messages is transparent, leaving little margin for ambiguity. Speech messages are context-dependent (e.g. culture, situation, location) and language specific; thus global applications are discouraged. Recorded spoken text becomes the message itself. Spearcons are another subcategory of auditory messages, in which speech is digitally altered so as to become an abstracted sound with a hint to the phonemes of the original word. For example, Heydra et al. (2014) used the term ANPR (Automatic Number Plate Recognition) as an abstract template to generate a warning signal by sounding out letters. We will disregard speech for the remainder of the chapter as we intend to focus on the contrast between abstract sounds and auditory icons.

Abstract sounds (aka earcons) do not refer to any existing event via their acoustical construction or semantic associations and almost always need to be learned within the context for which they are designed (Edworthy et al. 1991; McGookin and Brewster 2011). Many of the current alarm-like sounds are considered earcons. Earcons are short musical motives that can be recorded or synthesized. The temporal composition of earcons consists of one tone or a short sequence (milliseconds) of tones designed using mostly one type of timbre (i.e. sound quality deriving from an instrument or material such as glass, wood or metal). Although most earcons have a clear distinguishable pitch and a temporal structure, they can also consist of a series of clicks (e.g. Morse code, or Geiger counter clicks). Examples are email notifications on computers, microwave finish beeps, car feedback sounds or intensive care unit alarms. Some earcons result from the sonification of a continuous data stream. Today, digital tools are used to design earcons, and their production depends on what platform they are played on (e.g. speakers or piezo elements). In case of a piezo element, the parameters of design are limited by the resonance properties of the sound production element. For speaker solutions, designers can pay more attention to making the sound harmonically and temporally structured, as in brief musical scores. Such sounds can also be designed by recording musical instruments and changing musical parameters such as pitch, loudness, rhythm and speed. Depending on the context in which they are used, abstract sounds are often designed to evoke an (unnecessary) unpleasant sensation to trigger a sense of urgency.

*Auditory icons* are representational sounds that refer to real events both through their acoustical structure and consequently their semantic associations (Belz et al. 1999; Brazil and Fernström 2011; Edworthy 2017). It is easier to interpret the meaning conveyed through these sounds. Auditory icons are based on everyday sounds, and their spectral-temporal structure will be complex and depend on the object and action causing the sound over a certain period of time. The design of an auditory message can range from being very concrete (i.e. the exact copy of the original sound) to very abstract (i.e. practically an earcon). However, for auditory icons, designers aim at the sound evoking the right semantic association (to a referent event or object) through similarities established in the auditory structure.

These different types of auditory messages have advantages and disadvantages. The conceptual and perceptual relationship between the auditory signal and its cause is meant to be immediate with auditory icons. Thus, the message becomes clear to users. Disadvantages are that because auditory icons simulate real recordings of environmental sounds, they can still be masked by other environmental sounds or these sounds can be confused by other homophonic events. The advantage of abstract sounds is their endless design possibilities, as they can be created from scratch. A designer can map the criticality of the message at any level by playing with mere acoustical and psychoacoustical properties such as timbre, pitch, rhythm, loudness and duration, which change the perception of urgency. For non-critical contexts, abstract sounds can even be designed to sound pleasant. In contrast, the design space is limited for auditory icons. Assigning criticality does not necessarily require manipulating acoustic parameters; it can also entail finding the exact object/event/action as a source for creating immediate associations between the auditory signal and the message.

Auditory displays can consist of a set of auditory messages (i.e. speech, abstract sounds and auditory icons) depending on the function of the message and the desired user response. There is no direct procedure that can be used to design an optimum auditory display, as design teams need to consider the specific needs of the system, the user and the environment. To begin with, auditory displays might be preferred over visual displays if immediate action is required, users are moving around or cannot change their visual focus, the visual system is overloaded or visibility is poor. Furthermore, compliance rates are much higher with auditory warnings. This can be explained by sound's ability to evoke strong affective reactions, from sensory unpleasantness to emotional response. However, repeated occurrences of any visual or auditory messages can cause perceptual fatigue and reduce the compliance rate. Thus, next to designing auditory displays, it is equally important to consider and study how users (listeners) will interact with the auditory messages in certain situations and contexts in order to determine how critical these events are.

## 7.3 Mapping Information to Auditory Messages

There are systematic ways to map information to auditory messages, especially auditory icons. Gaver (1986) distinguished three types of mappings (i.e. symbolic, nomic and metaphorical) to explain how environmental events could represent the desired message to be conveyed. With *symbolic mappings*, designers rely on social convention for meaning (e.g. clapping for appreciation); with *nomic mappings* designers aim at concrete representations with direct physical causation (e.g. the sound of crumpling paper for a trash bin); and *metaphorical mappings* are designed to create associations between two entities that share physical or conceptual similarities (e.g. a sound with a falling pitch denotes a falling object).

These mappings could directly or indirectly refer to the cause of the event (i.e. the purpose of the message) depending on the strength of the associations created by the signal-referent relationship (Stevens and Keller 2004). *Direct mappings* would employ the sound of the causal event as a reference and therefore make an *analogical reference* to the sound as a source based on perceptual qualities. *Indirect mappings* could further be elaborated with ecological or metaphorical references to the source. An *ecological reference* would mean that the reference to the message coexists in the same environment as the event that needs to be communicated and that they are related by association. A *metaphorical reference* would mean that the reference to the message and the event that needs to be communicated have conceptual or functional similarities. Messages with direct relationships to their referent are learned in fewer trials and elicit quicker responses. Within indirect relationships, ecological relations have stronger associations to the reference sound and metaphorical relations are learned slower, through many trials; but once learned, response times do not differ compared to direct or ecological relations.

Let us exemplify this with the design of an alarm clock sound. The alarm clock sound should convey the message "time to wake up." If speech were to be used, a simple voice recording would say "wake up" (direct), "good morning" (indirect and ecological) or perhaps the vocal imitation for a rooster sound could be used (indirect and metaphorical). For auditory icons, the recordings of a rooster crowing (direct) or birds chirping (indirect and ecological) would be effective; alternatively a sound can be synthesized that is inspired by the salient auditory properties (i.e. similar frequency range, loudness levels, or temporal envelope) of a rooster sound (indirect and metaphorical). For abstract sounds (i.e. earcons), mappings most probably occur at a symbolic level, as such sound can be arbitrarily chosen or may remotely resemble the fundamental auditory properties of everyday events. For example, loud and high-pitched sounds such as a rooster crowing or a cat screeching would be experienced as stimulating and unpleasant. Therefore, symbolically synthesizing perceptually similar sounds could create the effect of a wake-up call due to plain psychoacoustical experience.

## 7.4 Auditory Icon Characteristics

The role of auditory displays is considered rather pragmatic, such as to guide and inform users while they are undertaking certain (visual) tasks. Thus, the effectiveness in conveying information with matching representations is of high importance. There are pros and cons to developing representations of different icon classes, which need to be considered in relation to the intended function of the auditory message in the wider sense. Mynatt (1994) developed a design methodology for icons, which included the following perceptual factors that influence design: identifiability, conceptual mapping, physical parameters and user preferences. Edworthy and Adams (1996) pointed out that in the design of warning symbols, legibility, conspicuity, discriminability and urgency mapping is required in order for people to comprehend and learn the symbols. In visual displays, Holmes (2000/2001) discerned various conventions for designing a set of pictograms (e.g. the overall shape, the style of drawings, the subject matter, the context and the color). Similarly, the studies of McDougall and Isherwood (McDougall et al. 2000; Isherwood et al. 2007) give an extensive account of *icon characteristics* and their perceptual and functional (dis)advantages. These icon characteristics (i.e. concrete-to-abstract continuum, complexity, semantic distance and familiarity) could also apply to auditory icons, as both sensory inputs trigger a similar cognitive process of interpreting sensory information.

*Concrete-to-abstract continuum*. The concreteness of an icon is the extent to which it depicts real objects/people/concepts. The depiction of direct and analogical icons can be more concrete than abstract. When visual icons are abstract they are less pictorial and use more lines and shapes. Concrete representations (e.g. icons) are easier to learn initially, but may not possess cross-cultural relevance. Abstract representations (e.g. earcons) can induce ambiguity in meaning and may take a little longer to learn, but may be more effective as a standardized sound design valid across cultures. The design challenge is to find the right level of representation of real-life objects/events/concepts in the spectrum of abstract-concrete depiction. Aesthetically, users may prefer concrete yet abstract icons for interpreting the auditory message in a pleasurable way. The right level of abstraction in the sound evokes a sense of sophisticated design while oozing familiarity.

*Auditory complexity*. Complexity refers to the amount of detail or intricacy in an icon. Structural complexity in spectral-temporal content of the auditory messages is preferable to concretely expressing an idea through an auditory icon, however, it may complicate perceptual processes of search and reaction times. The rules of auditory scene analysis may be a good trade off if structurally complex icons at least have good perceptual organization. Thus, sound designers would rather aim at simplifying an environmental sound when creating auditory icons.

*Semantic distance*. Semantic distance refers to the closeness of the relationship between the icon and its function (i.e. the information that needs to be conveyed). The ability to infer the meaning of an icon depends on semantic distance. Inferring the function of direct icons with analogical references is often immediate due to close semantic distance; for indirect icons with ecological or metaphorical references it may be more delayed or gradual due to a large semantic distance. Because earcons lack any semantic reference to an event or an object, it can be impossible at times to immediately identify their function and they often need to be learned.

*Familiarity*. In the end, a sense of familiarity needs to be prevalent when interacting with icons/earcons for determining their meaning. Familiarity also refers to recognizability and the frequency of exposure. For example, an earcon (e.g. fire siren), although depicted in an abstract way, can be very familiar to people; or an old-fashioned phone bell may not be recognizable by younger generations. Familiarity can be improved by experience (i.e. learning and exposure).

There is a vast amount of research concerning the relationship between display design, perception, learning and understanding, which can be used to predict whether new designs will be successful. Studies demonstrate that for the organization perceptual elements there needs to be optimal balance between complexity and order, which is supposed to increase the sensory pleasantness of the experience (Tractinsky and Hassenzahl 2005). The degree of familiarity is also an important factor to consider; thus, if icons have typical representations it will be easy to recognize their function and if they are novel it will be cognitively costly to derive their meaning and function (Hekkert et al. 2003). However, typical icons may not be experienced as pleasurable sounds and vice versa. Thus, when going for new icons, a certain level of familiarity may be preferable so as to lower cognitive load and increasing sensory pleasure.

Simplicity is an important feature in icon recognizability and learning. Research suggests that while simplicity in design reduces search times, there is little evidence to suggest that there are any particular advantages associated with either simple or complex icons or combinations (Scott 1993; McDougall et al. 2000). To quote E.R. Tufte (1990): "High information displays are not only an appropriate and proper complement to human capabilities, but also such designs are frequently optimal. If the visual task is contrast, comparison, and choice, then the more relevant information within eyespan, the better." Thus, if it is necessary to use a complex icon (because of the amount of information to be displayed) then the key disadvantage might be increased perceptual search times relative to a simpler icon.

It has been demonstrated that concrete icons are the easiest to learn (Stammers and Hoffman 1991; Stotts 1998), but that this advantage soon disappears with increased exposure (Stevens and Keller 2004). Because people are able to learn the meanings of icons, it has been suggested (Isherwood et al. 2007) that the most important underpinning variable is the *semantic distance* between the icon and the function it refers to. Semantic distance can be generally thought of as a continuum running from closely to distantly related. This represents the degree to which the icon and its function are associated, and is not necessarily the same as the level of concreteness. For example, a concrete icon may be easily understood in terms of what it represents (e.g. rooster sound for alarm clock) but its intended meaning (wake up) may not be well understood. This is because the user, though presented with a concrete icon, then has to draw an inference about its meaning within the context presented. The study of Isherwood et al. demonstrated that the degree of concreteness of an icon is important only in the early stages of learning and recognition, and that semantic distance and familiarity are more important factors with repeated exposure to the icons.

Memory plays an important role in deriving the meaning of icons and making sense of an entire label. An earlier study (Özcan and Van Egmond 2007) showed that memory for object sounds was significantly worse when sounds are depicted by pictographic labels in comparison to text-and-image labels. In this study, memory tasks included recall and recognition of the object producing the sound, and matching a label to a product sound. However, this study showed that people were still able to access semantic memory (correct access rate ranged between .3 and .6) and derive the meaning of iconic representations. Furthermore, the same study also showed that memory favors structure; when sounds have a clear structure it is easier to retrieve them from memory.

## 7.5 Hedonic Value of Auditory Displays

In this chapter, we are interested in taking the functional role of auditory displays to the next level in which hedonic values also are emphasized through design. The aim is to develop sound concepts that enrich the user experience in the aspects of functionality and pleasantness as well as to create a stronger impact on the perception of brand value through sound.

### *7.5.1 Function Versus Pleasure*

Recent studies propose design for sensory experiences (e.g. visual design, sound design) as a suitable strategy for creating pleasurable products. A carefully designed sound tackles sensory pleasure and therefore positively influences the overall appreciation of the design (Lageat et al. 2003;

Özcan 2014; Özcan and Schifferstein 2014). For digital products/services it has been demonstrated that visual pleasure has positive effects on the utilitarian function of human-product interactions such as user performance or usability (Hassenzahl 2004; Moshagen et al. 2009; Tractinsky et al. 2000). Pleasure and functionality (i.e. hedonic-utilitarian consumer attitudes) is also considered as a duality in user decision-making processes (Batra and Ahtola 1991; Spangenberg et al. 1997).

### *7.5.2 Sensory (Un)Pleasantness and Urgency Mapping Within Context*

By definition auditory messages are considered alarming; however, they do not need to be. Sensory unpleasantness is often required in order to draw the attention of the user and make them alert. Some contexts require an immediate response from the alarm user with the help of an unpleasant and urging sound (e.g. when a car door is opened while driving). Other non-critical contexts require a more informative sound with less emotional content because users are already vigilant and/or are waiting for information/confirmation or a kind reminder (e.g. incoming call, car needing maintenance). The extent of sensory (un)pleasantness should be appropriate to the users' emotional state and expectations at the moment of interactions with the auditory message. Considering how alert the user is and whether the message is expected will determine how unpleasant the designed sounds may be. Moreover, the criticality of the consequences of events to be signaled should also be considered when designing the pleasantness of the sounds. For example, missing a microwave oven bell may not have hazardous consequences, however, missing a cardiac arrest alarm can be fatal. Auditory messages need to be clear about the criticality of the event they represent.

Affective experiences vary from complex emotions (e.g. pride, desire, frustration, anger) to basic sensations (liking/disliking) and have two essential dimensions: valence and arousal. Valence refers to the subjective feeling of pleasantness or unpleasantness and defines the hedonic value of an experience; arousal refers to a subjective feeling of being activated or deactivated and defines the intensity of an experience (Russell 2003). With alarm-like sounds, these affective experiences are often situated on the negative side of the valence and positive side of arousal dimensions, providing evidence for unpleasant experiences. Sensory (un)pleasantness can be obtained by manipulating the psychoacoustical parameters such as pitch, loudness, rhythm and repetition. The higher the perceived sharpness, loudness, roughness and noisiness of a sound is, the lower the sensory pleasantness (Zwicker and Fastl 1990). Van Egmond (2004) showed that it is possible to map frequency-modulated (FM) sounds on the circumplex of emotions. The FM signals could be presented on the underlying dimensions of pleasure and arousal. If one argues that an alarm signal needs to result in behavioral action, this signal should be in the quadrant of high arousal and low pleasantness. Conversely, a signal that should not draw that much attention and prompts a behavioral action that can be postponed should lie in the quadrant of high pleasantness and low arousal. Designing alarming sounds allows for an immediate connection between the sensorial experience of a sound and behavioral action in such a way that sensory unpleasantness signifies a basic reaction such as flight. Such sounds are apt to evoke collective reactions to fire escape or ambulance alarms.

More granular emotional responses can be attained by the meaning assigned to a sound in relation to the needs and expectations of the users in a particular context. Thus, sound design based solely on adjustments to unpleasant acoustical parameters would be a partial solution in most situations; a qualitative study is required to gain insights into the user and their context before designing an auditory display and its messages for a more effective design. See Sousa et al. (2017) for a good example of user-context research at a pre-sound design phase.

### *7.5.3 Semantic Associations With Brand Values*

We hope to demonstrate that sounds can be developed to form a possible foundation for an advanced system for auditory interfaces, which intends to not only exceed the currently existing audio signals in hedonic qualities, but also implement distinctive semantic meanings. The user should be able to be recognize and link the specific semantic associations to relating information and functions in the vehicle system. Moreover, sound is almost always considered as a property of a product. Even an independent system such as an auditory display would be treated as an additional element in the design of the vehicle. In such cases, the semantic associations of the product will be sought after in the experience of the auditory messages as well. Experiences of sound congruent with experience of the product are more appreciated by users as congruency characterizes sophistication and dedication applied in the design process by the manufacturer (Özcan et al. 2017).

Designing sounds to convey brand values has been widely applied in the automotive industry (e.g. branding the sportiness of a Porsche sound or a reliable thud of a Toyota car door) and its added value has been discussed in the literature (Spence 2012; Argo et al. 2010; Graakjaer and Bonde 2018). However, alarm-like sounds are often discarded as carriers of brand values. We see auditory messages as on the same level as other physical properties of a product such as material of a car seat or power of its engine. Therefore, users expect a more unified experience with regard to different design elements. Manufacturers often distinguish themselves from their competitors with slogans that summarize their current brand values. For example, in 2018, Renault identified their brand with the slogan "Passion for Life" and Audi, with "Advancement through Technology." Both brands aspire to design the features of their vehicles so that the brand values are highlighted in the experience of a vehicle. However, the slogan themselves only serve as a starting point for vision. In order to measure whether the vehicle evokes the right associations, a "semantic construct" needs to be established that specifically depicts the desired verbal qualities of the brand and the vehicle. For example, "Passion for Life" could be underpinned by descriptive words such as "passionate," "exciting," "dedicated," "lively" and/or "enthusiastic." The semantic construct associated with such words needs to be defined internally among designers, marketers and engineers.

### *7.5.4 Poetic Explorations Into Hedonic Values*

Translating brand values, which are high-level abstract notions, into design features is a challenging task for any design team (Özcan and Sonneveld 2009). First, designers need to interpret the underlying meaning of the brand values and agree on a semantic construct that would capture the meaningful associations at its best. Second, designers need to capture the physical qualities in abstract notions such as brand values and devise a plan to embody (i.e. sketch and construct) the brand values with similar physical qualities. Thus, the difficulty lies in cross-cutting the physical and the conceptual design space while aiming at evoking affective experiences (i.e. aesthetic and emotional) through design. This process is rather similar to the processes of artistic creation, as in sculpture or literature.

In order to capture the subtlety of the semantic and affective associations of brand values, designers can avail themselves of artistic tools such as poetry for inspiration as well as communication. Through the lens of poetry, brand values can be expressed in an intertwined but encapsulating manner with emotions, objects, events, people and places. Poetry, in the sense of verse, is one of the oldest forms of art used to facilitate affective experiences. Poetically rich moments induce creativity, facilitate learning, improve critical thinking and develop empathy and insight (Parr and Campbell 2006; Routman 2000; Özcan 2016). In poetic form, objects become imaginable and therefore relatable and engaging. The use of metaphoric language helps imagery and thus anchors specific meanings. Poetic depictions can carry a specific message and trigger conversation. For all these reasons, designers take full advantage of the benefits of poetry and surface the underlying meaning of brand values.

The literature we cite in the previous paragraph indicates that functional properties of interactive products are emphasized by their hedonic quality. Moreover, investing in sensory experiences will have an impact on experiences occurring at the product level. Thus, not considering the pleasure auditory displays can evoke would be a missed opportunity for the overall experience of the car. In addition, auditory messages are often designed to be merely functional and technical, devoid of brand values or even pleasant features. Again, investing in the hedonic qualities of auditory displays will improve their usability, and brand values will be better integrated into the design process. Investigation into hedonic values can be facilitated by using poetry as a tool for conceptual design.

## 7.6 Conceptual Design Process of the Auditory Display for Toyota Motor Europe

The design brief of Toyota Motor Europe (TME) states that "TME intend to explore ways to emphasize the hedonic quality of their in-vehicle interface sounds (i.e. auditory messages) for Toyota and Lexus vehicles while strengthening the functional quality of their sound." TME had three requirements: the auditory messages needed to be (1) identifiable (i.e. the function had to be recognizable), (2) pleasant to hear (on a sensory level) and (3) associated with the brand values. The messages were required to inform, give feedback, alert according to the varying degrees of criticality, prompt the driver to action in a pleasant way without disturbing/irritating the user and be associative on a higher level.

Our approach to the conceptual design of TME sounds proceeds as follows. We first assign function to the auditory messages, then discover the brand values and metaphors using qualitative methods (i.e. poetry and discourse analysis); finally, sounds will be designed digitally using musical and acoustical approaches. As a strategy, we use brand values to create two distinct families of auditory displays, each representing a separate brand with its own timbre; however, we use one set of metaphors for functional sounds to link two brands under the umbrella of TME. That is, each brand uses the same metaphors for signaling certain functions only with a different timbre. Figure 7.1 illustrates the balance between the functionality and brand values in the auditory display design for TME and can be referred to in the following paragraphs.

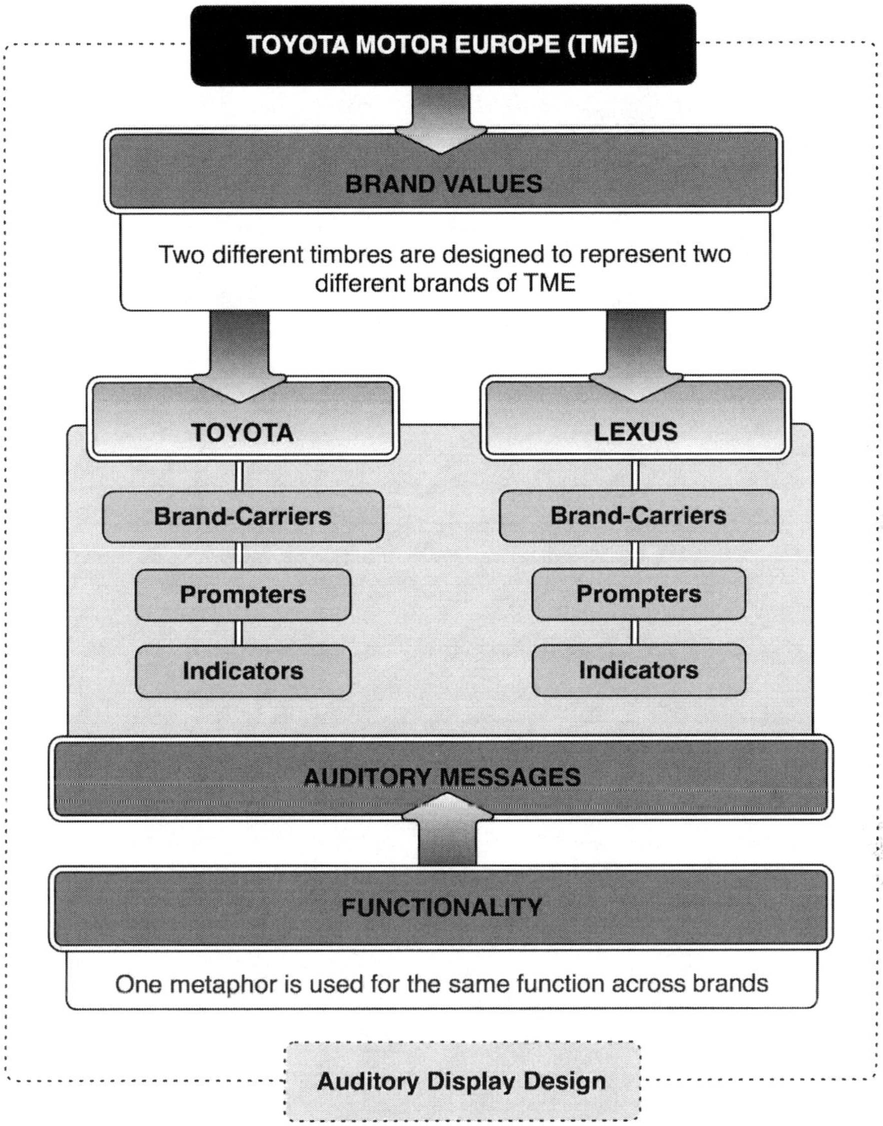


**Figure 7.1** Auditory Display Design for TME Illustrated as the Balance Between the Functionality and Brand Values: Brand-carriers represent welcome and goodbye sounds; Prompters represent lights, seat belts and fuel events; and indicators represent turn indicators and proximity sensors.

## 7.6.1 Assigning Function

Our main approach to designing auditory display is to understand the basic needs of the users in the moments of interactions with the product/system. In the context of TME, considering the driver's needs we distinguished three types of functions for the auditory messages: brand-carriers, prompters and indicators.

Sounds as *brand-carriers* introduce the sound quality (i.e. timbre) that represents the brand values when turning the engine "ON" or "OFF" (i.e. the vehicles are turned on by a button after inserting the key). In this case, we conceived *welcome* and *goodbye* sounds. The role of these sounds is to greet the driver as the voice of the vehicle. These sounds should be inherently pleasant and trigger associations pertaining to the brand value. These sounds are on the musical side of abstract sounds as part of the auditory display and serve a more hedonic role.

Sounds as *prompters* signal a potentially unsafe situation and trigger the user to immediately act upon the event that they failed to perform. Consequently, different levels of urgency can be defined for warning sounds depending on how critical the situation is. In this case, *lights, seat belts* and *fuel* sounds are meant to prompt the driver to switch off the lights, fasten the seat belts and fill the fuel tank. Three levels of urgency have been considered for these sounds: low, medium and high. These sounds are identified as auditory icons as part of the auditory display and serve a functional role.

Sounds as *indicators* provide continuous information (i.e. feedback) on longer-duration interactions and they indicate a certain event taking place during the usage of the product. In this case, *turn indicators* and *proximity sensor* sounds are meant to provide continuous feedback to the driver during the events of turning or parking. These sounds are abstract sounds and serve a functional as well as a hedonic role.

While all these auditory messages need to meet their function, the messages should concurrently carry brand values. Brand values reside more in brand-carriers (welcome and goodbye sounds) and indicators (turn indicators and proximity sensors) but less in prompters (lights, seat belts and fuel sounds). All auditory messages from one brand (either Toyota or Lexus) are meant to have similar acoustic features (i.e. timbre) that are identifiable by people hearing the sounds as one brand.

## 7.6.2 Assigning Brand Values

Both brand concepts are developed parallel to one another although each brand is examined individually and exclusively according to its unique context and character, hence creating one coherent sound concept per brand. The brand values provided by TME were *vibrant clarity* for Toyota and *intriguing simplicity* for Lexus. Thus, these descriptions were used as the starting point for conceptual explorations.

In order to unravel the underlying meaning of these brand values, we used poetic descriptions to gather deeper insights into the semantic construct of these brand values. Accordingly, we conducted a workshop with industrial design students and asked them to think about the word-pairs (*vibrant clarity* and *intriguing simplicity*) for each brand in order to express their first impressions in the form of poems. This assignment put the brand values in context and surfaced immediately relevant semantic associations. The following are two examples of poetic descriptions, one for each brand.

We categorized the main themes and analyzed the metaphors and semantic associations of the poems shown in Table 7.1. These themes were lighthouse, hybrid companion, yellow sunglasses and wooden sword

**Table 7.1** Comparison of Toyota and Lexus Poetic Descriptions.

| *Toyota—Hybrid companion* | *Lexus—Sophisticated ally* |
|---|---|
| you're a good friend who invites to some action.<br>and lightens up my everyday with straightforward satisfaction<br>you can shine full of energy like a dog having fun, and also be loyal on the daily long run, so to me you're a safe bet I can rely on,<br>since you've proven yourself honest and truthful, always freshly inspiring,<br>just like your family and all of its siblings.<br>with you I stay active and can explore unknown routes,<br>and find small adventures, which I'd otherwise had surpassed,<br>all while feeling safe and attended, thanks to, your effort and care. | fresh, sophisticated and focused to surpass<br>you make heads turn and reflect rays of spacious desires and<br>while strong as a giant, you rest like a diamond,<br>calm and serene and proud of your name,<br>you thrive on the challenge of leading the game<br>you show distinction and go beyond all<br>and try to push harder as the young star you are,<br>finesse is your call and you like to answer<br>this in a calm, charming tone,<br>so you will noticed subtly and without overtone.<br>when you're unleashed you make pulses rush,<br>not only then proving precision and touch<br>but also when gliding curves smoothly and shining your lines<br>you keep up your promises far beyond expectations. |

for Toyota and perfume, sophisticated ally, stealth and katana sword for Lexus. These associations reflect the different natures of the brands from a noncommercial perspective.

*Toyota (vibrant clarity)* is universal and playful, and therefore the message (and hence its composition) is friendly, warm and expressive, but through the use of simple sounds it remains accessible. This fits the timbre of the more wooden sounds, which are often created through, for instance, bamboo wind chimes or a wooden xylophone, where the melodies can be complex but the sounds themselves are very simple. Vibrant clarity is underscored with the following descriptive words: *joyful, fresh, friendly, confident, caring, encouraging*.

*Lexus (intriguing simplicity)* is exclusive and serious, and therefore the sound is rich and complex (achieved by applying effects and harmonies). Though because Lexus is not meant for showing off, its message is about capturing and expressing its essence only in a very subtle yet intriguing fashion. In this case, less is clearly more. Intriguing simplicity is underscored with the following descriptive words: *pure, elegant, dynamic, advanced, luxurious, intriguing*.

Toyota is associated with more warm, friendly and playful concepts and Lexus with more sharp, distant and refined concepts. Accordingly, we have come up with a set of descriptive words for the brands (Table 7.2).

**Table 7.2** Examples of Metaphors and Associations Discovered in Poetic Descriptions of Brand Values.

| | |
|---|---|
| **TOYOTA** | |
| **Concept 1—LIGHTHOUSE**: | **metaphor** > Like a beaming lighthouse<br>**associations** > Feeding with energy/Pulsating |
| **Concept 2—HYBRID COMPANION:** | **metaphor** > “like a trusted and reliable companion”<br>**associations** > Playful and fun Human-Robot Interaction |
| **Concept 3—YELLOW SUNGLASSES**: | **metaphor** > “like seeing the world in an uplifting perspective”<br>**associations** > Vibrant Colors/ Euphoria |
| **Concept 4—WOODEN SWORD:** | **metaphor** > “like playfully battling with vivid imagination”<br>**associations** > Leading/ adventurous/dreamy |

| | |
|---|---|
| **LEXUS** | |
| **Concept 5—PERFUME:** | **metaphor** > "like a breeze of expensive perfume"<br>**association** > Overwhelming/Sophisticated |
| **Concept 6—SOPHISTICATED ALLY:** | **metaphor** > "like a superior yet loyal partner"<br>**association** > Cutting-edge/Compassionate |
| **Concept 7—STEALTH:** | **metaphor** > "like a sleek and high tech machine"<br>**association** > feeling shielded and proud when driving |
| **Concept 8—KATANA SWORD:** | **metaphor** > "like a samurai wielding the katana"<br>**association** > aggressive as a weapon/elegant art as it moves |

### 7.6.3 Discovering Metaphors and Analogies for Function and Brand Values

In the design process, we explored functionality and brand values through the use of metaphors; however, for pleasantness and urgency mapping, we used traditional methods of alarm design by altering psychoacoustical parameters (Edworthy et al. 1995; Edworthy et al. 1991). In auditory display design, metaphors serve as a tool for mapping information to auditory icons. However, mapping strategies can also differ for functional sounds and for brand values. On the one hand, messages being conveyed through functional sounds need to be quickly and accurately identified so that drivers are prompted to immediately act upon the message. On the other hand, associations being elicited through brand values do not require immediate attention and only evoke hedonic experiences.

Therefore, two kinds of metaphor mapping strategies are applied: *a direct (literal) mapping for functionality* (i.e. prompters and indicators) using auditory icons and *an indirect mapping for brand values* (i.e. brand-carriers) using abstract sounds (AKA earcons). In discovering the metaphors, it is essential that brand values and the associated meanings are discovered (separately for each brand) first as presented above. This distinction gives rise to different timbre spaces each reflecting specific brand values. For the functional sounds, however, analogies are found and mapped across the brands of Toyota and Lexus.

While discovering metaphors appropriate for direct and indirect mapping, we considered the auditory icon characteristics such as concrete versus abstract representations, auditory complexity, semantic distance and familiarity. For *brand-carriers* (indirect mapping for brand values), we aimed at keeping the auditory messages rather abstract with a large semantic distance from the context to which they belong (i.e. turning on the engine, greeting the driver). Brand-carriers were meant to be novel—thus unfamiliar in the situation—and complex sounds that were musical and needed to be learned in context. Thus, considering the brand values, we kept the metaphors on a high level to evoke an emotive context rather than specific associations. At this stage some analogies were also used for inspiration. For example, starting a Lexus car was described as *"an exclusive perfume, which would be quickly released from the bottle through a gentle push of a button and then the fine particles would slowly float downwards like tiny crystals of a cloud creating a very powerful yet delicate sensation."* Upon analyzing such metaphors, we agreed that "sunny days" could represent the Toyota brand with warm and playful associations. For Lexus, we agreed on "beams of light" with sharp, distant and refined associations.

For *prompters* (direct mapping for functionality), we kept the semantic distance very close and familiarity very high for easy identification of what event the auditory message signals or refers to. These sounds are analogical to the original events as sound sources. For example, we used the seat belt sound with rolling and click mechanism for the seat belt forgotten message, a sputtering engine (due to lack of fuel) for fuel message, and an electric-like sound for lights forgotten message. While the sound events (sources) are familiar to drivers, the auditory icons are situated on the abstract side of the concrete-to-abstract continuum, as the major acoustical patterns in the real events were used as inspiration during sound design. Thus, the sounds from real events were abstracted during design. As a result the acoustic complexity of the original sounds is reduced.

For *indicators* (direct mapping for functionality), the direct mapping derived from using the original sound as analogical reference for turn indicators and imitating impact sounds (mechanical clicks or hitting a bowl) for timbre. As a result, auditory messages imitated impacting events and familiarity was ensured with the temporal structure of the original turn indicators or impacts. These auditory icons are situated on the abstract side of the concrete-to-abstract continuum for two reasons: (1) traditional turn signals are abstract sounds and impact-like sounds alone are short sounds signifying general material interactions rather than specific events and (2) the sound bears little complexity.

### 7.6.4 Designing the Sound Quality of Auditory Messages

Once the metaphors and brand values were decided upon, the conceptual design focused on sketching the sound of the auditory messages. Brand-carriers were treated as musical motives and thus different melodies and timbres were studied for each brand. For Toyota, a playful melody with an upbeat tempo was devised with a wood-like timbre. Structure was built up with leaps between the notes but not steps between the notes. The melodic structure (see Figure 7.2) was chosen such that it would not be easy to remember. This was done to avoid annoyance caused by repeated exposure to the sound on a daily basis. These scores can be trademarked (see for example, NBC sounds). For Lexus, virtually no melody and a minimal rhythmic structure was devised to represent continuity, brightness and sharpness with a metal-like timbre. In Figure 7.3, you can see the visual representation of Toyota and Lexus brand-carriers. Once the timbres were defined and applied to the brand-carriers, the remaining auditory messages were designed to adopt the same timbre but differ in their iconic representation.

### 7.6.5 Making Auditory Messages Pleasant and Urgent

Pleasantness is achieved holistically by proposing auditory icons for *prompters* as opposed to traditional alarming abstract sounds. Brand-carriers should be semantically pleasant as they have associations with desired and positive values. However, for *prompters*, which are inherently functional sounds, a sense of urgency needs to be implemented, which can inversely alter the perceived pleasantness of the auditory messages. Urgency was implemented through temporal structure by repetition and shortening the intervals of the repetition at different levels. Three levels are proposed: low (or no) urgency, medium urgency and high urgency depending on how long the driver takes to act to the message (see Figure 7.4).

**Figure 7.2** Musical Scores for Toyota Welcome and Goodbye Sounds: The goodbye sound reverses the welcome melody. Lexus sounds had virtually no melody.

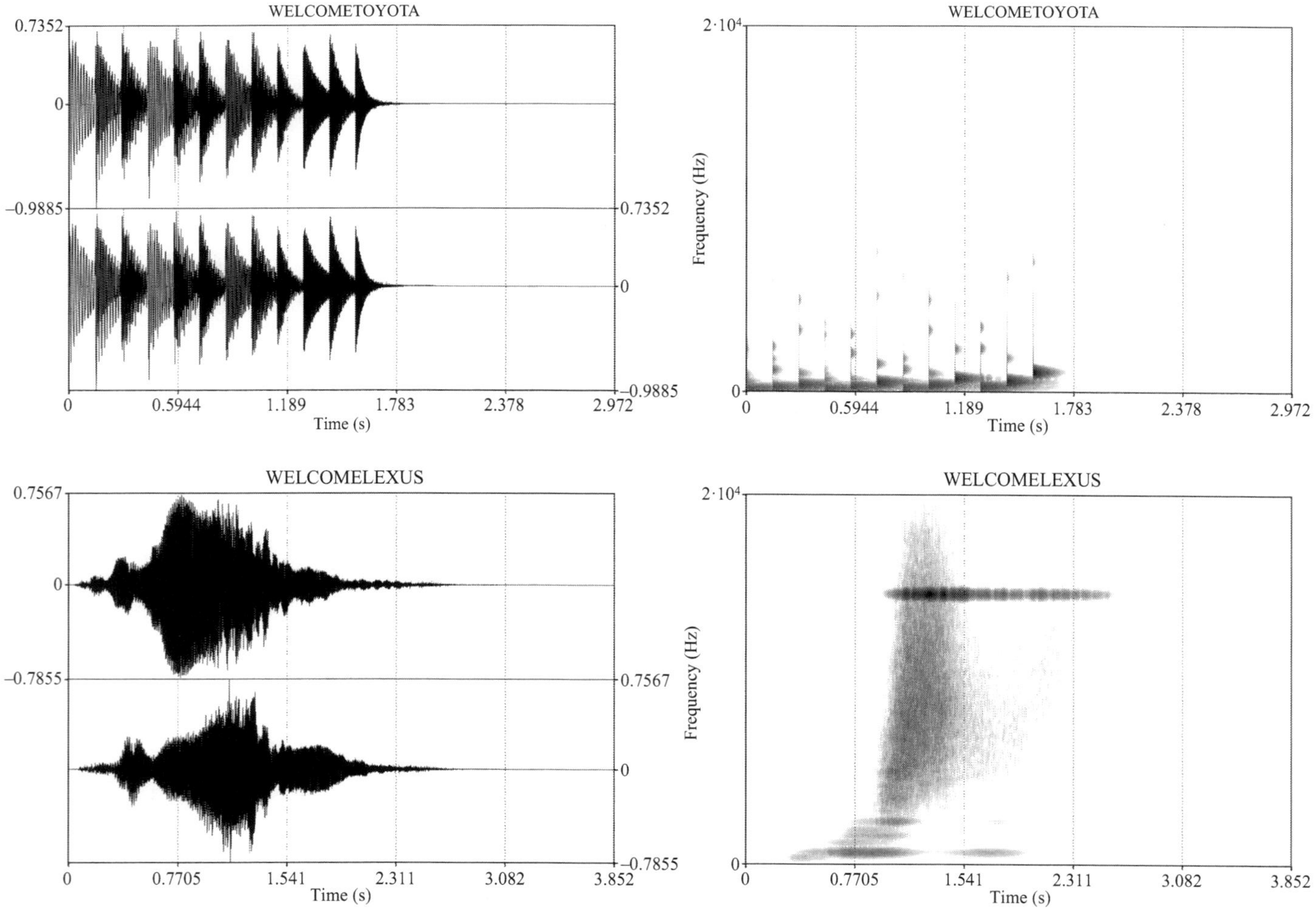


**Figure 7.3** Spectral-Temporal Composition of Toyota and Lexus Welcome Sounds.

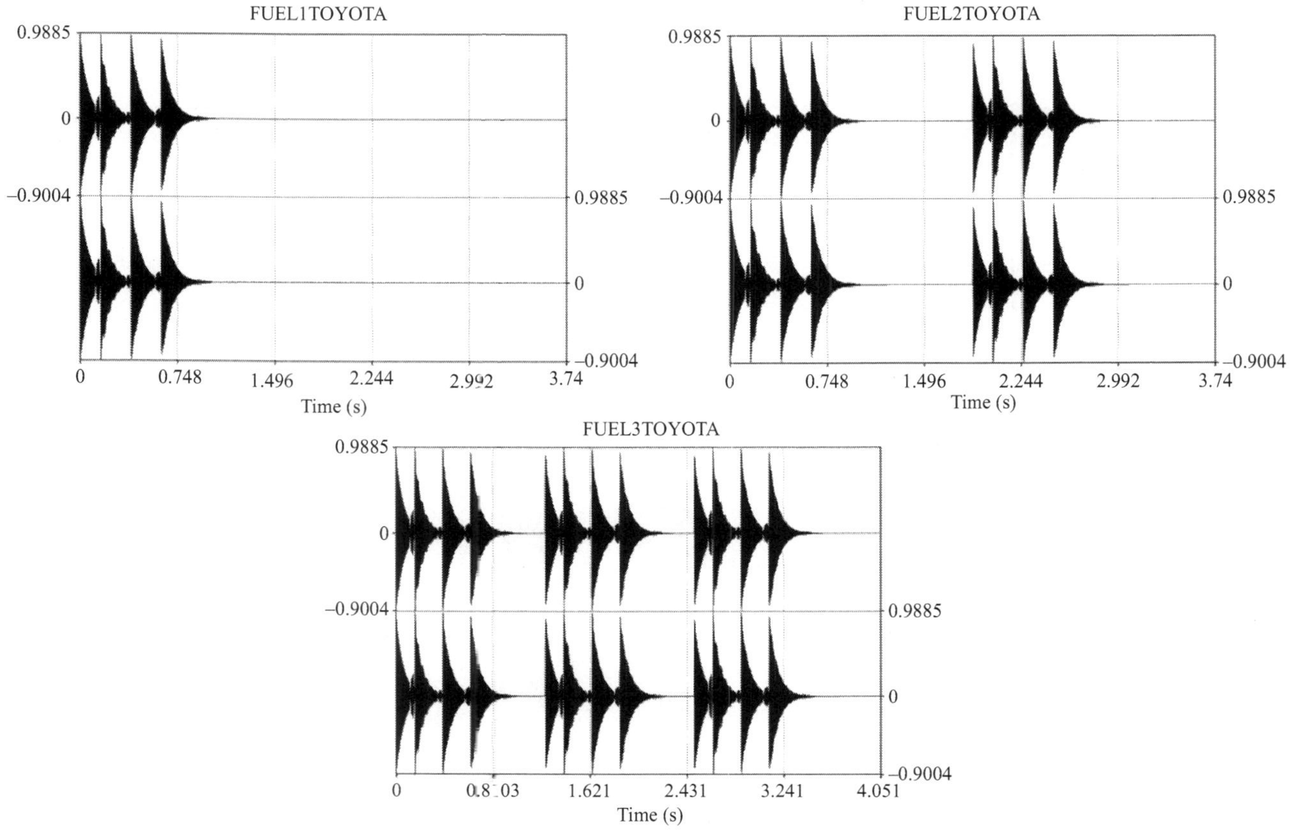


**Figure 7.4** Temporal Progression of the Urgency Applied to the Auditory Messages (Fuel Message for Toyota Auditory Display).

### 7.6.6 Final Sound Design Guidelines and Brand Values Uncovered

Our conceptual design process with metaphors and semantic analysis of brand values has been reflected in the final version of the auditory display design. These final descriptions of sounds can be seen in Table 3 for Toyota and Table 4 for Lexus. Toyota brand values (vibrant clarity) are captured by the *sunny days* metaphor with the following semantic associations: *joyful, fresh, friendly, confident, caring, encouraging*. Lexus brand values (intriguing simplicity) are captured by the *beams of light* metaphor with the following associations: *pure, elegant, dynamic, advanced, luxurious, intriguing*. Tables 7.3 and 7.4 provide further insights into how the auditory messages are conceptualized and which real sound sources would help achieve the desired timbre of the auditory icon. The descriptions even

**Table 7.3** Concept Metaphor, Auditory Icons, Functions and Descriptions of Sound Conceptualized for Toyota Auditory Messages (Brand-Carriers, Prompters And Indicators).

| *TOYOTA* | *Vibrant Clarity* |
|---|---|
| Brand values | *Joyful, Fresh, Friendly, Confident, Caring, Encouraging* |
| **Brand-carrier** | **WELCOME + GOODBYE** |
| Concept Metaphor | **Sunny days** |
| Sound source | *Wooden wind chimes—Water bubbles* |
| Timbre | The timbre of Toyota (wood-like) fits together very well with short notes that are arpeggiated emulating the sound of wind chimes or bubbling water which evoke feelings of energy, pleasure and fun for the user. The goodbye sound is the inverted version of the welcome sound. |
| **Prompter** | **LOW FUEL** |
| Function | Remind the driver to refill the fuel tank in 70 km, 50 km and 10 km. |
| Auditory icon (analogy) | *Sputtering engine* |
| Description of sound | The sound tries to emulate the sound a car engine makes when it is slowly running out of fuel in a very reduced and abstracted manner. The melody decreases in pitch and velocity while it also slightly becomes slower in speed. |

| *TOYOTA* | *Vibrant Clarity* |
|---|---|
| **Prompter** | **SEAT BELT** |
| Function | Remind the driver to buckle the seat belt at intervals of 5 km/h, 15 km/h and >25 km/hr. |
| Auditory icon (analogy) | *Dragging motion and locking in place* |
| Description of sound | The seat belt alarm tries to mimic a dragging motion of pulling the seat belt out through a soft fading-in sound (D4), which ends with two slightly delayed harder impacting tones played on a wooden xylophone (D3+D4). The last impacts resemble the secure locking in of the seat belt clips. |
| **Prompter** | **LIGHTS** |
| Function | Remind the driver to turn off the headlights before leaving and on when driving. |
| Auditory icon (analogy) | *Buzzing electricity* |
| Description of sound | The lights forgotten alarm is sonified through a quick sequence of alternating impact sounds (B3 + C4), which include a short staccato. This should resemble a similar feel to the sound of irregularly buzzing electricity, when strong currents leak or discharge though the air, like when a fuse is blown in slow motion. |
| *Urgency | Only prompters have three levels of increasing urgency (low/no, medium, high) through temporal repetition at decreasing intervals. These messages are regularly deployed depending on the driver behavior, system needs and event criticality. |
| **Indicator** | **TURN INDICATOR** |
| Function | Inform the driver that turn indicators are engaged and (still) on. |
| Auditory icon (metaphor) | *Mechanical clicks* |
| Description of sound | In order to keep the sound basic and simple and therefore accessible, the turn indicators are based upon cultural conventions of the old mechanical clicking combined with a more lively and fun interpretation as if it was played on a wooden xylophone. |

(*Continued*)

Table 7.3 (Continued)

| *TOYOTA* | *Vibrant Clarity* |
|---|---|
| **Indicator** | **PROXIMITY SENSOR** |
| Function | Support the driver while parking to avoid unforeseen impacts |
| Auditory icon (analogy) | *Impacts* |
| Description of sound | The proximity sensor sound is characterized by the reducing of impact sound as the car approaches an obstacle. The sounds go from staccato to legato. This is more in line with the action the user undertakes, slowing down the car while nearing. |

**Table 7.4** Concept Metaphor, Auditory Icons, Functions and Descriptions of Sound Conceptualized for Lexus Auditory Messages (Brand-Carriers, Prompters and Indicators).

| *LEXUS* | *Intriguing Simplicity* |
|---|---|
| Brand values | *Pure, Elegant, Dynamic, Advanced, Luxurious, Intriguing* |
| **Brand-carrier** | **WELCOME + GOODBYE** |
| Concept Metaphor | **Beam of light** |
| Sound Source | *Temple gong—Overwhelming breeze* |
| Timbre | The timbre of Lexus (metal, glass) fits together very well with a long sustained sound, which evokes a feeling of confidence and inner peace. Therefore the welcome sound is inspired by a gong-like sound, which slowly spreads and fills the space before fading away, leaving behind a calm feeling of peace. The goodbye sound differs in pitch as opposed to the welcome sound keeping the timbre the same. |
| **Prompter** | **FUEL** |
| Function | Remind the driver to refill the fuel tank in 70 km, 50 km and 10 km. |

| *LEXUS* | *Intriguing Simplicity* |
|---|---|
| Auditory icon (analogy) | *Sputtering Engine* |
| Description of sound | The sound tries to emulate the sound a car engine makes when it is slowly running out of fuel in a very reduced and abstracted manner. The melody decreases in pitch and velocity while it also slightly becomes slower in speed. |
| **Prompter** | **SEAT BELT** |
| Function | Remind the driver to buckle the seat belt at intervals of 5 km/h, 15 km/h and >25 km/hr. |
| Auditory icon (analogy) | *Dragging motion and bell hit* |
| Description of sound | The seat belt alarm sounds like a metal bell is struck with a bow, creating a dragging motion while the intensity increases until it is completed by two slightly delayed impacting tones on the same object. The hits resemble the belt clips falling into the lock securely, while ending in a zen-like experience, which fades out softly into tranquility. |
| **Prompter** | **LIGHTS** |
| Function | Remind the driver to turn off the headlights before leaving and on when driving. |
| Auditory icon (analogy) | *Buzzing electricity* |
| Description of sound | The lights forgotten alarm is sonified through a quick sequence of alternating impact sounds (B3 + C4), which include a short staccato. This should resemble a similar feel to the sound of irregularly buzzing electricity, when strong currents leak or discharge though the air, like when a fuse is blown in slow motion. |
| *Urgency | Only prompters have three levels of increasing urgency (low/no, medium, high) through temporal repetition at decreasing intervals. These messages are regularly deployed depending on the driver behavior, system needs and event criticality. |

(*Continued*)

Table 7.4 (Continued)

| *LEXUS* | *Intriguing Simplicity* |
|---|---|
| **Indicator** | **TURN INDICATOR** |
| Function | Inform the driver that turn indicators are engaged and (still) on. |
| Auditory icon (metaphor) | *Ringing sound bowl* |
| Description of sound | The turn indicator sounds consist of soft impact sounds on a metal object, which are echoed by a very dampened and wave-like answer, similar to dragging a wet finger over the rim of a wine glass in a rhythmical way that resembles the mechanical working of a turn indicator. |
| **Indicator** | **PROXIMITY SENSOR** |
| Function | Support the driver while parking to avoid unforeseen impacts |
| Auditory icon (analogy) | *Impacts* |
| Description of sound | The proximity sensor sound is characterized by the reducing of impact sound as the car approaches an obstacle. The sounds go from staccato to Legato. This is more in line with the action the user undertakes, slowing down the car while nearing. |

go further in specific details of sounding events and the progression of the sounds. The tables also indicate the type of mapping, that is, whether it is direct (analogical) or indirect (metaphorical).

## 7.7 User Evaluation of the Auditory Display

Following conceptual sound design, the final version of the sounds was tested with potential users according to the agreed parameters: brand values, urgency, functionality and perceived pleasantness. Brand values and pleasantness were measured by rating scales in a questionnaire. Perceived urgency was measured by using a pairwise comparison paradigm. The perceived function of the sounds was determined by using a forced choice paradigm.

*Participants*. Twenty participants (ten male and ten female) volunteered (M = 21.4 years). All participants reported normal hearing.

*Apparatus*. Sounds were played from iTunes playlists. Free field speakers were used that were similar to speakers in a car. Timbre was designed using the Sculpture routine in Logic Pro. Temporal and pitch structure and other effects were applied in Logic Pro, Sound Studio and special filters from MAX.

*Stimuli*. Seven sounds for each brand were specially designed. For three sounds (fuel, light and seat belt) three levels of urgency (low, medium, high) were created. In total, 26 sounds were specially designed based on the following auditory messages and their urgency iterations for each brand: welcome, goodbye, fuel (3x), lights (3x), seat belt (3x), turn indicators and proximity sensor.

*Procedure*. Five sessions with four participants per session were held. All responses were collected on paper. Ratings on brand values and sensory pleasantness were captured using a 7-point Likert scale. The three-level (low-medium-high) perceived urgency of prompters (fuel, light, seat belts) was determined using a pairwise comparison paradigm. The sounds representing different levels of urgency from one function (i.e. prompter) were compared within the same set. For function, a participant had to select the function after hearing that sound. Only one choice could be made and participants were not trained to identify the auditory messages. For each stage of measurement per participants the order of the stimuli was randomized.

For the brand values, *vibrant clarity* for Toyota was represented with a set of six descriptive words (*joyful, fresh, friendly, confident, caring, encouraging*); *intriguing simplicity* for Lexus was represented with another set of six descriptive words (Pure, Elegant, Dynamic, Advanced, Luxurious, Intriguing). These descriptive words were used for the rating of all auditory messages from both brands. That is, Toyota sounds were rated also on Lexus words and vice versa for Lexus sounds and Toyota words.

### *7.7.1 Results*

The rating scale data were analyzed using JMP and SPSS software. The forced choice data on function were analyzed using correspondence analysis. The choices for the paired comparison were analyzed using the logit model for paired comparison of Bradley and Terry (1952). This model yielded three scale values for the urgency variants of each sound. Thus, the judgments on the pairs were reduced to three values (per definition one of the scale values is set to 0).

*Perceived pleasantness*. In Figure 7.5, the perceived pleasantness of the sounds is presented as a function of Brand (Toyota, Lexus) and Sound (seven levels). It can be clearly seen that the pleasantness ratings fluctuate across sounds for the Lexus and Toyota brands. This indicates that pleasantness is not only dependent on the timbre of the sounds but also on the pitch and temporal content of the sounds. Furthermore, the average rating

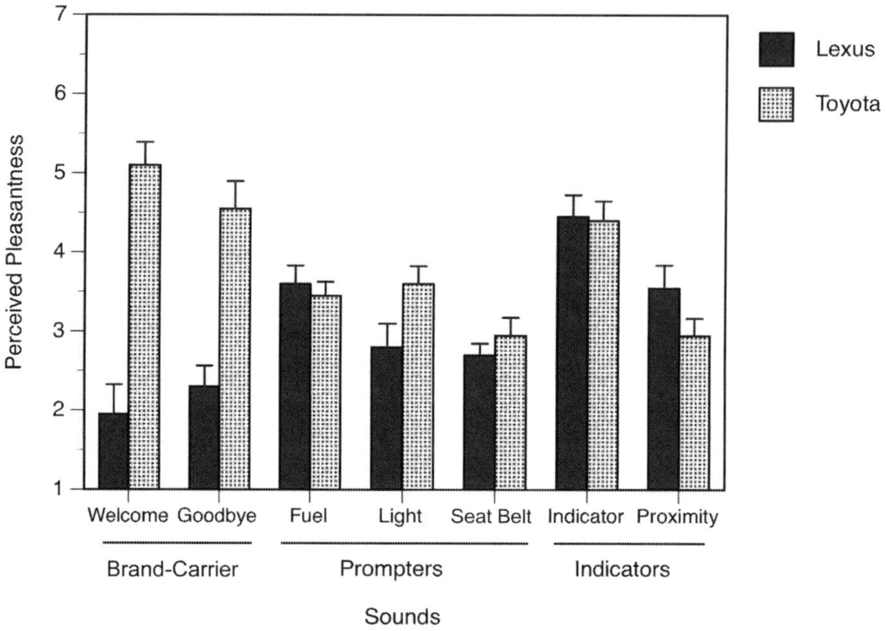


**Figure 7.5** Perceived Pleasantness as a Function of Brand and Auditory Message.

across sounds is higher for Toyota ($M = 3.86$) than for Lexus ($M = 3.05$). The difference between brands was significant, $F(1,19) = 20.82, p < .001$. It can be concluded that the timbre of Toyota is considered more pleasant than that of Lexus because averaging over sounds takes out the pitch and rhythmic structures. The effect of sounds was also significant, $F(6, 114) = 6.06, p < .001$. In particular, the welcome and goodbye sounds of Toyota were judged as being more pleasant than the sounds of the Lexus brand. As can been seen in Figure 7.5, for the fuel, indicator and proximity sounds, the Lexus sounds received a higher rating. A significant interaction effect between sound and brand confirms this, $F(1,114) = 17.62$, $p < .001$.

In addition, the pleasantness ratings were analyzed using "function type" (brand-carrier, prompter, indicators) and "brand" (Toyota, Lexus) as independent variables. Figure 7.6 shows the mean pleasantness rating as a function of brand and function type. It can be readily seen that the prompter and indicator sounds are perceived as being equally pleasant for both brands. However, the brand-carrier sounds for Toyota have a higher pleasantness rating than the brand-carrier sounds for Lexus. The brand-carrier sounds should be higher in pleasantness than the prompter

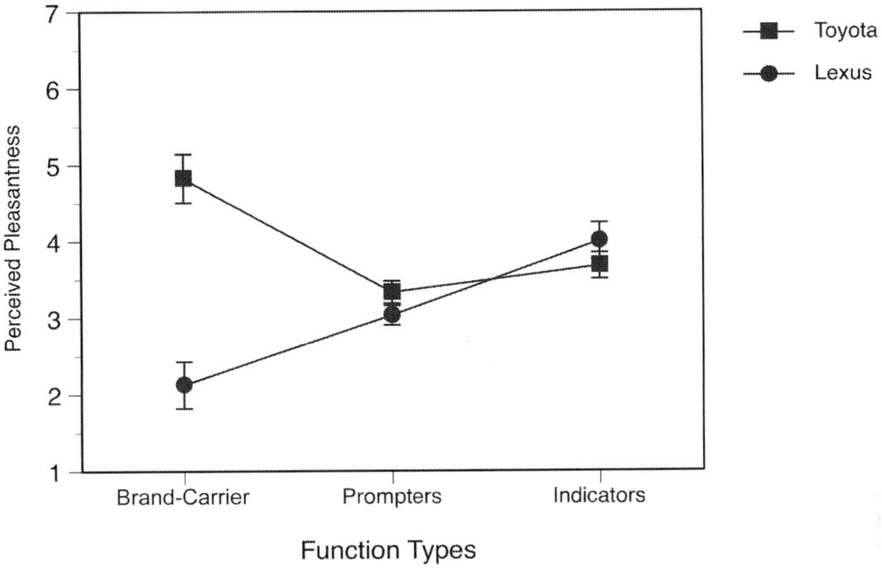


**Figure 7.6** Perceived Pleasantness as a Function of Brand and Auditory Message Type.

and indicator sounds because the latter sounds should draw the attention of the driver to take action. Conversely, the brand-carriers should be perceived as pleasant because the driver wants to feel welcome or likes to return after a "nice" visit.

The aforementioned effects were confirmed by a repeated measures analysis with brand and function as independent variables and perceived pleasantness as a dependent variable. The main effect for brand was significant, $F(1,19) = 24.44$, $p < .001$. The main effect for function type was not significant, whereas the interaction between brand and function was significant, $F(1,19) = 32.07$, $p < .001$.

*Urgency*. The paired comparison data of urgency measurement were analyzed using the Bradley-Terry model. This models yields relative scale values, one of the scale values is zeroed and the other values indicate their relative difference from this point. Because participants judged which sound was more urgent, the scale values reflect the perceived relative urgency. These perceived urgency values are depicted in Figure 7.7 as a function of designed urgency (low, middle, high) and brand.

Designed urgency is indicated beside the scale values. It can be readily seen that perceived urgency is the same as (intended) designed urgency. The range for the Toyota sounds (fuel, light, seat belt) is

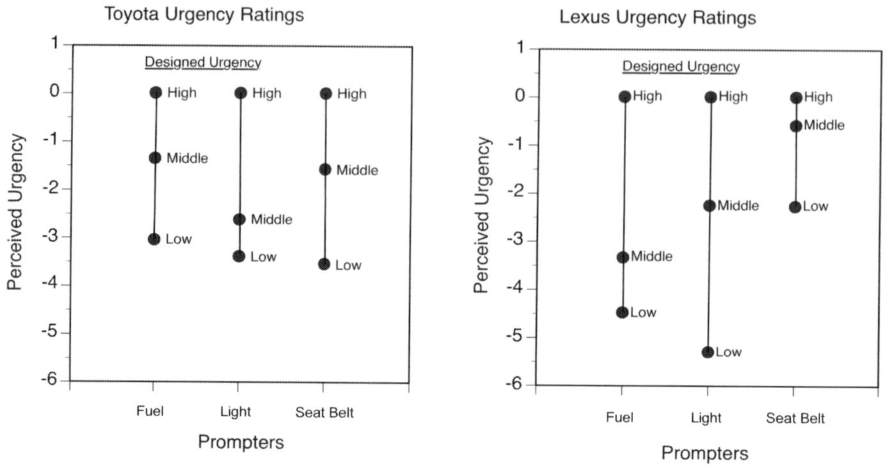


**Figure 7.7** Perceived Urgency as a Function of Auditory Message Type (Fuel, Light, Seat Belt) and Designed Urgency for Toyota and Lexus: The graph on the left depicts the Toyota sounds and the graph on the right depicts the Lexus sounds.

approximately 3, whereas the range for Lexus sounds is 4 for the fuel and light sounds and approximately 2 for the seat belt sounds. It can be seen that for the Toyota fuel and seat belt sounds the middle urgency level is halfway between the high and low urgency, whereas the middle urgency sound for the light lies closer to the low urgency sound. For the Lexus brand, only the middle urgency level for the light sounds lies halfway between the high and low urgency sounds. Note that the urgency was realized by manipulating onset-offset intervals between tones and chunks (melodic fragments that are repeated), thus indicating that the same rhythmic structure with a different timbre may evoke a different perceived relative urgency.

*Function Choice*. Now, we will show whether we were successful in transmitting a certain function through sound design. The participants only heard the sounds and had to assign them to a certain function. They were not told beforehand which sound belonged to which function (this would have been a memory task). Thus, the inherent structure of the sounds should convey their function. Furthermore, this was the third part in the experimental setup. The participants had heard the sounds before in the urgency and pleasantness tasks, but no function was associated with the sounds in these experiments.

In Figure 7.8, two graphs with stacked bar charts are shown. The graph on the left shows the percentage choice of a certain function for

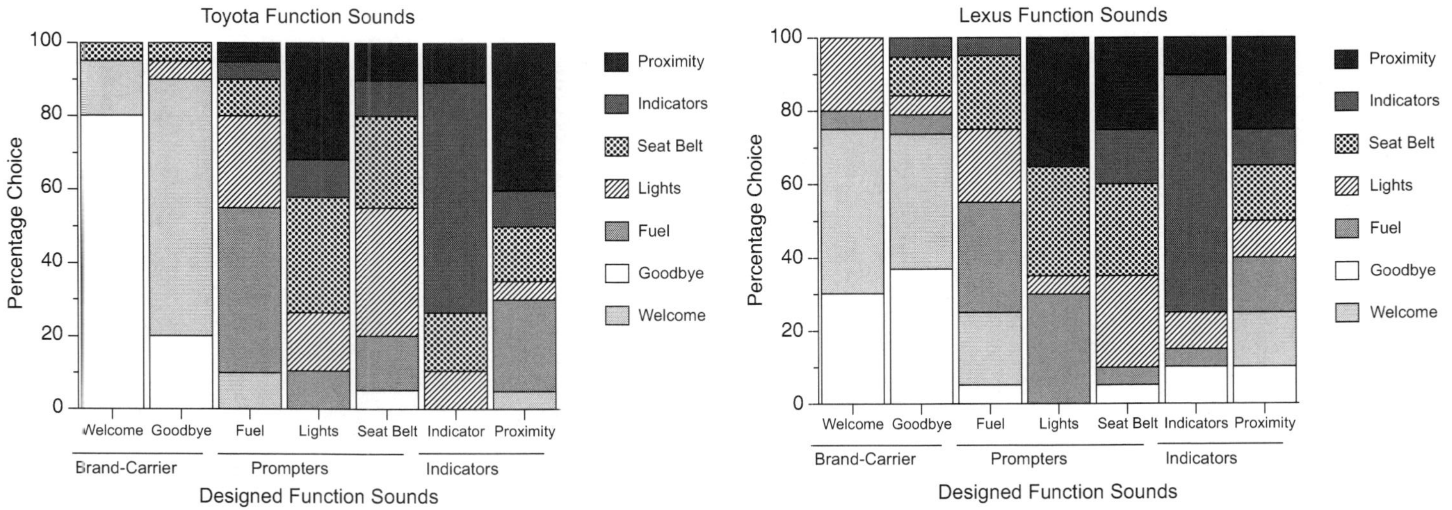


**Figure 7.8** Percentage Choice for a Certain Function in Relation to Designed/Intended Function: The graph on the left depicts the Toyota sounds and the graph on the right depicts the Lexus sounds.

the Toyota sounds in relation to the designed function (x-axis) and chosen sound (legend). The same is shown for the Lexus sounds. It can be readily seen that brand-carriers are better associated with their intended functions. The brand-carrier sounds for Toyota clearly convey their function (i.e. welcome and goodbye) better than those for Lexus. The choice between the welcome and goodbye sounds was more difficult for the Lexus brand. This can be seen where the white and black areas of the bars are almost equal in size. In addition, the intended function of the indicator sounds for both brands are well recognized. This is probably due to the iconic repetition rate that is used in the rhythm of the sounds. For the other sounds the associated function is less clear for the participants, but they experience almost no confusion with the welcome and goodbye sounds. Surprisingly, the proximity sounds were not as well associated with their intended functions. This is probably because the sonification of the distance was used in an uncommon way. In practice, if a car gets closer to an object, the beeping rate increases. In our case the design strategy was to use longer and slower notes to indicate that the speed should not be that high and give a feeling of "you are almost" there; the commonly used beep rate is more annoying and stressful.

*Brand values*. The ratings on the two sets of sounds (Toyota and Lexus) were analyzed with separate principal component factor analyses (PCAs) using Varimax rotation. The PCAs were conducted across subjects ($N = 20$) and types of sounds ($N = 7$). Only two factors for each PCA had values higher than one. Consequently, only two factors were derived for each factor analysis. For the two-factor analyses, Bartlett's tests of sphericity show that the underlying structure (i.e. factors) obtained by the factor analyses are appropriate: Toyota, $\chi 2$ (66) = 836.50, $p < .001$; Lexus, $\chi 2$ (66) = 741.79, $p < .001$. In addition, the Kaiser-Meyer-Olkin measure of sample adequacy shows that the factors represent a good reduction of the brand values for Toyota (.88) and Lexus (.85). Furthermore, the values for Cronbach's alpha show that the factor analyses are internally consistent: Toyota .89; Lexus .87. These tests show that the use of these attributes and the application of factor analysis to reduce the number of attributes is warranted.

In Table 7.5 the results of these factor analyses are shown. The first column contains the attributes sorted on the brand they represent. Thus, the first six attributes were obtained in the metaphor studies for Toyota, and the next six attributes were obtained in the metaphor studies for Lexus. Columns 2 and 3 show the factor loadings for the sounds designed for Toyota, and columns 4 and 5 show the factor loadings for the sounds designed for Lexus. Note that factor loadings above .4 are represented in bold type. The first thing to notice is that the factor 1

**Table 7.5** Factor Loadings as a Function of Brand Attributes and Designed Sounds per Brand

| *Factor Loadings* | | | | | |
|---|---|---|---|---|---|
| *Attributes* | | *Toyota Sounds* | | *Lexus Sounds* | |
| | | *Factor 1* | *Factor 2* | *Factor 1* | *Factor 2* |
| **Toyota** | Caring | **0.76** | 0.06 | 0.14 | **0.77** |
| | Confident | **0.74** | 0.34 | **0.68** | 0.21 |
| | Encouraging | **0.68** | 0.31 | 0.53 | **0.49** |
| | Fresh | **0.56** | **0.46** | 0.47 | **0.55** |
| | Friendly | **0.82** | 0.17 | 0.02 | **0.89** |
| | Joyful | **0.78** | 0.02 | 0.00 | **0.84** |
| **Lexus** | Advanced | 0.17 | **0.83** | **0.83** | 0.02 |
| | Dynamic | **0.60** | **0.45** | **0.79** | 0.16 |
| | Elegant | 0.14 | **0.78** | 0.38 | **0.57** |
| | Intriguing | 0.20 | **0.63** | **0.70** | -0.12 |
| | Luxurious | 0.07 | **0.86** | **0.64** | 0.35 |
| | Pure | 0.36 | **0.68** | **0.52** | 0.35 |
| | **% of variance** | 31.64 | 29.68 | 30.12 | 27.46 |

*Note:* Bold type indicates factor loadings above .4. Explained variance has been indicated below each factor loading per designed sound set (Lexus, Toyota).

loadings for the sounds designed for a specific brand are higher for the brand values. Thus for Toyota sounds, the factor 1 loadings on the Toyota attributes are higher than for the Lexus attributes. The same is true for the Lexus sounds where the factor 1 loadings are the highest for the Lexus attributes. Note that factor 1 is more important (higher variance explained) than Factor 2. Factor 2 loadings show the opposite relation. Thus, the loadings are lower for the non-congruent sounds and attributes. This can readily be seen in the table because columns 4 and 5 (factor loadings Lexus sounds) are the mirror image of columns 2 and 3 (factor loadings Toyota sounds).

### 7.7.2 Discussion

Our main finding is that we can design a family of sounds on the basis of brand values. Even if there are two distinctive brands in one family, like Toyota and Lexus, sounds can differentiate them but can also be seen as a kind of unity through variety. We also showed that these sounds evoke the correct functionality and urgency levels. Although pleasantness scores seem low for functional sounds (prompters and indicators), we speculate that these ratings would likely be higher if they were compared to the original warning sounds. Among the auditory messages, Pleasantness ratings are higher for brand-carriers and lower for prompters and indicators. That is predictable because every functional auditory message needs to evoke attention, which will always lead to less pleasantness. Similarly, designed urgency is well perceived and participants were able to distinguish between low, medium and high urgency sounds.

Comparing brands, the Toyota auditory display is perceived as more pleasant than Lexus. However, in terms of brand values, each auditory display has been rated higher on their respective brand values. This result shows that the method of using poetry and semantic network is worthwhile in defining the meaning structure of brand values. Similarly, the concept metaphors and descriptive words used for inspiring the sound design process helped us create auditory expressions that captured the essence of brand values.

Regarding how well participants identified the function of the sounds, we had mixed responses. Not all the functions were well identified by the participants. In general, participants were able to distinguish between brand-carriers and other auditory messages (prompters and indicators). Beside the brand-carriers, turn indicator messages were easily identified. There may be a couple of explanations for this. In terms of brands, it is possible that Toyota's timbre created an easy-to-identify sound space. When mapping the wood-like timbre, the auditory icon probably stayed intact. Lexus's timbre resided too much in the higher frequency range, which may be contradictory to the frequency range of natural sound events as source of auditory icons. In general, environmental sounds are difficult to accurately identify in the absence of an image or a text/verbal label, making these sounds ambiguous (Özcan and Van Egmond 2009). Although the analogical mapping was well executed, perhaps the participants failed to identify or associate the sound to a natural event. Whereas, the indicator sounds are designed such that they have high resemblance with a typical indicator sound. This finding points to the fact that any auditory message needs to be learned before use. In our study, we excluded the training phase from our user evaluations in order to observe participants' first reactions to the function of the sounds.

## 7.8 General Conclusions

In this section, we will generalize our findings towards improving the domain of auditory display design. TME has provided a unique opportunity for improving the existing theories on auditory display design and applying the outcome of this new theoretical framework in a real design context. With this chapter, we have shown that auditory displays can be designed to be both hedonically and functionally satisfying. Above all, we were able to design sounds for separate brands within one family brand. Brand values fitting the family brand identity (i.e. the umbrella organization Toyota Motor Europe) were related to the specific sub-brands (i.e. Toyota and Lexus).

Through sonic branding, companies determine their brand identity based on the values they want to publicize. With this study, we first practiced conceptual design methods appropriate to capture the essence of "brand umbrellas." Incorporating poetry and metaphors in this design methodology appears to be the right approach based on discussions with business stakeholders (i.e. Toyota Motors Europe) and users. This allowed us to go from an abstract "brand umbrella" to more specific semantic associations of sounds. Then we focused on sonic expressions of brand values with more traditional sound design methods. As a result, we gained insights into on how pleasure and brand values can be mapped on intrinsically functional sounds, also showing that methods used in traditional conceptual design processes can be adapted to sound—and even alarm—design processes.

In general, the sounds designed in this study complied more with the more abstract notion of brand values and evoked the right sense of urgency for specific functions. However, our study also showed the importance of learning and the robustness of the mental representation of the sounds, as our test has been conducted in the absence of any training on learning how to interpret sounds. We consider this an opportunity to further improve the proposed method of designing for hedonic values. Although any auditory message should inherently relate to a specific function, the restricted boundaries for designing feedback and alarm sounds (i.e. total duration, limited amount of pitches) make a learning trajectory eminent. This may suggest sounds that should indicate critical events in a system should be first learned until 100 percent correct identification can take place. Given that each auditory icon in an auditory display will be perceptually dissimilar to other icons, the estimated learning curve for these sounds would be as short as three to six trials.

Moreover, the possibility to train for sound-induced human-system interactions should be facilitated by manufacturers. In general, the correct identification of alarms is a weak point in any alarm design context, ranging from simple beeps in domestic appliances to more complex sounds found in intensive care units or vehicle dashboards. It may require different efforts and strategies to address the specific needs of the users and the system as the communicator of the messages. In dashboard design, it would be worthwhile to consider implementing this on the visual displays in a gamified way. In intensive care units, a virtual reality environment similar to aircraft cockpit simulations can be created to train nurses on system events and their sounds. Both strategies will keep the auditory display users satisfied as well as engaged and confident.

In short, we showed that given a certain brand identity, it is possible to design sounds that prompt users to take designated actions. In future research, we will extend the design methodology presented here to a more longitudinal situation in which users have to learn and interact with these sounds in the long term and in situations in which more systems with different functionalities are present. Furthermore, we also intend to investigate the potential of icon characteristics such a semantic distance and complexity as predictors for the successful design of functionality.

## Acknowledgements

The study presented in this chapter was funded by Toyota Motor Europe (TME) over the period 2012–2014. We would like to acknowledge the contributions of Daniel Esquivel (no longer affiliated with TME) throughout the design and evaluation process. Our gratitude also extends to our lab assistants at the time, Thomas Beernink, Gyán Santokhi and Eric Ringard, for their technical support with co-designing the sounds, running the experiments and documenting the study.

## References

Argo, J. J., Popa, M., and Smith, M. C., 2010. The sound of brands. *Journal of Marketing* 74, no. 4, pp. 97–109.

Batra, R., and Ahtola, O. T., 1991. Measuring the hedonic and utilitarian sources of consumer attitudes. *Marketing Letters* 2, pp. 159–170.

Bazilinskyy, P., and De Winter, J., 2015. Auditory interfaces in automated driving: An international survey. *Peer Journal of Computer Science* 1, no. 13. Viewed September 7, 2018 <https://peerj.com/articles/cs-13/>.

Belz, S. M., Robinson, G. S., and Casali, J. G., 1999. A new class of auditory warning signals for complex systems: Auditory icons. *Human Factors* 41, pp. 608–618.

Blattner, M. M., Sumikawa, D. A., and Greenberg, R. M., 1989. Earcons and icons: Their structure and common design principles. *Human-Computer Interaction* 4, no. 1, pp. 11–44.

Bradley, R. A., and Terry, M. E., 1952. Rank analysis of incomplete block designs: I. The method of paired comparisons. *Biometrika* 39, no. 3/4, pp. 324–345.

Brazil, E., and Fernström, M., 2011. Auditory icons. In: Hermann, T., Hunt, A., and Neuhoff, J. G. (Eds.) *The Sonification Handbook*. Berlin: Logos Verlag, pp. 325–338.

Edworthy, J., 2017. Designing auditory alarms. In: Black, A., Luna, P., Lund, O., and Walker, S. (Eds.) *Information Design: Research and Practice*. New York: Routledge, pp. 371–390.

Edworthy, J., and Adams, A. S., 1996. Symbols. In: *Warning Design: A Research Prospective*. London: Taylor and Francis, pp. 75–100.

Edworthy, J., Hellier, E., and Hards, R., 1995. The semantic associations of acoustic parameter commonly used in the design of auditory information and warning signals. *Ergonomics* 38, no. 11, pp. 2341–2361.

Edworthy, J., Loxley, S., and Dennis, I., 1991. Improving auditory warning design: Relationship between warning sound parameters and perceived urgency. *Human Factors* 33, pp. 205–231.

Gaver, W., 1986. Auditory icons: Using sound in computer interfaces. *Human Computer Interaction* 2, no. 2, pp. 167–177.

Graakjaer, N. J., and Bonde, A., 2018. Non-musical sound branding—a conceptualization and research overview. *European Journal of Marketing* 52, no. 7/8, pp. 1505–1525.

Hassenzahl, M., 2001. The effect of perceived hedonic quality on product appealingness. *International Journal of Human—Computer Interaction* 13, no. 4, pp. 481–499.

Hassenzahl, M., 2004. Beautiful objects as an extension of the self: A reply. *Human Computer Interaction* 19, no. 4, pp. 377–386.

Hassenzahl, M., Platz, A., Burmester, M., and Lehner, K., 2000. Hedonic and ergonomic quality aspects determine a software's appeal. In: Turner, T., and Szwillus, G. (Eds.) *CHI '00, Proceedings of the CHI 2000 Conference on Human Factors in Computing, 01–06 April, 2000, The Hague, The Netherlands*. New York: ACM, pp. 201–208.

Hassenzahl, M., and Tractinsky, N., 2006. User experience—a research agenda. *Behavior & Information Technology* 25, no. 2, pp. 91–97.

Hekkert, P., Snelders, D., and Van Wieringen, P. C. W., 2003. 'Most advanced, yet acceptable': Typicality and novelty as joint predictors of aesthetic preference in industrial design. *British Journal of Psychology* 94, no. 1, pp. 111–124.

Heydra, C. G., Jansen, R. J., and Van Egmond, R., 2014. Auditory signal design for automatic number plate recognition system. In: van Leeuwen, J. P., Stappers, P. J., Lamers, M. H., and Thissen, M. J. M. R. (Eds.) *Creating the Difference: Proceedings of the Chi Sparks 2014 Conference, 03 April, 2014, The Hague, The Netherlands*, pp. 19–23.

Holmes, N., 2000. Pictograms: A view from the drawing board, or what I have learned from Otto Neurath and Gerd Arntz (and Jazz). *Information Design Journal* 10, no. 2, pp. 133–143.

Isherwood, S. J., McDougall, S. J., and Curry, M. B., 2007. Icon identification in context: The changing role of icon characteristics with user experience. *Human Factors: The Journal of the Human Factors and Ergonomics Society* 49, no. 3, pp. 465–476.

Kramer, G., 1994. An introduction to auditory display. In: Kramer, G. (Ed.) *Auditor Display: Sonification, Audification and Auditory Interfaces*. Reading, MA: Addison-Wesley, pp. 1–78.

Lageat, T., Czellar, S., and Laurent, G., 2003. Engineering hedonic attributes to generate perceptions of luxury Consumer perception of an everyday sound. *Marketing Letters* 14, no. 2, pp. 97–109.

McDougall, S. J. P., de Bruijn, O., and Curry, M. B., 2000. Exploring the effects of icon characteristics on user performance: The role of icon concreteness, complexity, and distinctiveness. *Journal of Experimental Psychology: Applied* 6, no. 4, pp. 291–306.

McGookin, D., and Brewster, S., 2011. Earcons. In: Hermann, T., Hunt, A., and Neuhoff, J. G. (Eds.) *The Sonification Handbook*. Berlin: Logos Verlag, pp. 339–362.

Moshagen, M., Musch, J., and Göritz, A., 2009. A blessing, not a curse: Experimental evidence for beneficial effects of visual aesthetics on performance. *Ergonomics* 52, no. 10, pp. 1311–1320.

Mynatt, E. D., 1994. Designing with auditory icons: How well do we identify auditory cues? In: Plaisant, C. (Ed.) *Chi '94 Conference Companion on Human Factors in Computing Systems, 24–28 April, 1994, Boston, USA*. New York: ACM, pp. 269–270.

Neuhoff, J. G., 2011. Perception, cognition and action in auditory displays. In: Hermann, T., Hunt, A., and Neuhoff, J. G. (Eds.) *The Sonification Handbook*. Berlin: Logos Verlag, pp. 63–86.

Özcan, E., 2014. The Harley effect: Internal and external factors facilitating positive experiences with product sounds. *Journal of Sonic Studies* 6, no. 1, p. A07.

Özcan, E., 2016. Poetically rich experiences: The interplay between affect and cognition. *Paper presented at the 10th Design and Emotion Conference, Amsterdam.*

Özcan, E., Cupchik, G. C., and Schifferstein, H. N. J., 2017. Auditory and visual contributions to affective product quality. *International Journal of Design* 11, no. 1, pp. 35–50.

Özcan, E., and Schifferstein, H. N. J., 2014. The effect of (un)pleasant sound on the visual and overall pleasantness of products. In: Salamanca, J., Desmet, P., Burbano, A., Ludden, G. D. S., and Maya, J. (Eds.) *Proceedings of the 9th International Conference on Design & Emotion, 6–10 October, Bogota, Colombia*. Bogota, Colombia: Ediciones Uniandes, pp. 601–606.

Özcan, E., and Sonneveld, M., 2009. Embodied explorations of sound and touch in conceptual design. *Paper presented at the Design and Semantics of Form and Movement (DeSForM) Conference, Taiwan, Taipei.*

Özcan, E., and Van Egmond, R., 2007. Memory for product sounds: The effect of sound and label type. *Acta Psychologica* 126, no. 3, pp. 196–215.

Özcan, E., and Van Egmond, R., 2009. The effect of visual context on the identification of ambiguous environmental sounds. *Acta Psychologica* 131, no. 2, pp. 110–119.

Parr, M., and Campbell, T., 2006. Poets in practice. *The Reading Teacher* 60, no. 1, pp. 36–46.

Rensink, R. A., 2009. Visual displays. In: Goldstein, E. B. (Ed.) *Encyclopedia of Perception*. Vancouver, BC: University of British Columbia.

Routman, R., 2000. *Kids' Poems: Teaching Third and Fourth Graders to Love Writing Poetry*. New York: Scholastic.

Russell, J. A., 2003. Core affect and the psychological construction of emotion. *Psychological Review* 110, no. 1, pp. 145–172.

Scott, D., 1993. Visual search in modern human-computer interfaces. *Behaviour and Information Technology* 12, no. 3, pp. 174–189.

Sousa, B., Donati, A., Özcan, E., Van Egmond, E., Jansen, R., Edworthy, J., Peldszus, R., and Voumard, Y., 2017. Designing and deploying meaningful auditory alarms for control systems. In: Cruzen, C., Schmidhuber, M., Lee, Y. H., and Kim, B. (Eds.) *Space Operations: Contributions from the Global Community*. New York: Springer, pp. 255–270.

Spangenberg, E. R., Voss, K. E., and Crowley, A. E., 1997. Measuring the hedonic and utilitarian dimensions of attitude: A generally applicable scale. *Advances in Consumer Research* 24, pp. 235–241.

Spence, C., 2012. Managing sensory expectations concerning products and brands: Capitalizing on the potential of sound and shape symbolism. *Journal of Consumer Psychology* 22, pp. 37–54.

Stammers, R. B., and Hoffman, J., 1991. Transfer between icon sets and ratings of icon concreteness and appropriateness. *Proceedings of the Human Factors Society 35th Annual Meeting, 2–6 September, 1991, San Francisco, CA, USA*. Santa Monica, CA: Human Factors and Ergonomics Society, pp. 354–358.

Stanton, N. A., and Edworthy, J., 1999. *Human Factors in Auditory Warnings*. Aldershot: Ashgate.

Stevens, K., and Keller, P., 2004. Meaning from environmental sounds: Types of signal-referent relations and their effect on recognizing auditory icons. *Journal of Experimental Psychology: Applied* 10, no. 1, pp. 3–12.

Stotts, D. B., 1998. The usefulness of icons on the computer interface: Effect of graphical abstraction and functional representation on experienced and novice users. *Proceedings of the Human Factors and Ergonomics Society 42nd Annual Meeting, 5–9 October, 1998, Chicago, IL, USA*. Santa Monica, CA: Human Factors and Ergonomics Society, pp. 453–457.

Tractinsky, N., and Hassenzahl, M., 2005. Arguing for aesthetics in human-computer interaction. *i-com—Zeitschrift für interaktive und kooperative Medien* 3, pp. 66–68.

Tractinsky, N., Katz, A. S., and Ikar, D., 2000. What is beautiful is usable. *Interacting with Computers* 13, no. 2, pp. 127–145.

Tufte, E. R., 1990. *Envisioning Information*. Cheshire, CT: Graphics Press, p. 50.

Van Egmond, R., 2004. Emotional experience of frequency modulated sounds: Implications for the design of alarm sounds. In: de Waard, D., Brookhuis, K. A., and Weikert, C. M. (Eds.) *Human Factors in Design*. Maastricht, the Netherlands: Shaker Publishing, pp. 345–357.

Zwicker, E., and Fastl, H., 1990. *Psychoacoustics: Facts and Models*. Berlin: Springer.

8

# Introduction to Sonification

Ivica Ico Bukvic

## 8.1 Introduction

Imagine you are driving a car. As an experienced driver you may be able to identify and monitor a number of aural cues, such as the car engine and wind noise, as well as the dashboard sounds designed to better inform you of the current driving conditions. Whether you are driving a vintage car with a manual gear shift or an automatic, listening to the engine pitch as it increases its rotations per minute (RPMs) may help you pinpoint the optimal time to switch gears. Or, it may help you identify a potential problem with the transmission that suddenly fails to shift into a higher gear, leaving the engine stuck at screeching high RPMs. As you speed up, the wind noise increases, reinforcing the sense of speed. Concurrently, your dashboard plays a symphony of cues or earcons (Oswald 2012), reminding you that you may have forgotten to fasten the seat belt, fully close one of the doors, or that you may be running low on fuel or electrical charge.

It starts to rain. As raindrops begin hitting the windshield, the sudden increase in their rate reminds you of summer storms and the ensuing flash floods that often lead to dangerous driving conditions. You instinctively slow down.

Now snap back out of our imaginary car ride and consider just how many different aural cues were present in the scenario. We had the engine noise, the wind noise, multiple dashboard sounds and then the rain. We could also dig deeper and consider other noises that may have punctuated various actions, like shifting gears. Perhaps the car was old and every time it shifted gears the transmission generated a disturbing clunk. Or, moving the gear shift offered a satisfying thump as it reassuringly landed into a new position.

All of these sounds can be seen as a form of sonification. Some of them are simply byproducts of natural laws that govern our universe, such as the wind noise, engine pitch, or the raindrops hitting the windshield. Others were designed and engineered by humans for humans, such as the dashboard earcons whose purpose is to remind the driver of items that may require their prompt attention.

## 8.2 So, What Is Sonification and Why Should You Care?

The origins of the term *sonification* are not entirely clear. Arguably its first official mention in an academic publication dates back to 1990 (Rabenhorst et al. 1990) when it referred to a complementing aural counterpart to data visualization. Since then, its definition has continually evolved, resulting in a number of similar yet distinct definitions, many of which continue to be used concurrently. Worrall, in his overview of sonification (Worrall 2009), presents a comprehensive review of the term's evolution, while also offering the all-encompassing definition:

> Data sonification is the acoustic representation of data for relational interpretation by listeners, for the purpose of increasing their knowledge of the source from which the data was acquired.

Another, considerably simpler, version offered by Kramer et al. (1999) defines sonification as "the use of non-speech audio to convey information." In this chapter, allow me to propose what may be arguably the simplest and most inclusive version, which defines sonification as "audio that conveys information," thereby leaving all the possibilities for sonification open, including human speech, music and even environmental sounds, as is the case with the rain and wind noises in our imaginary car ride example. Unlike the dashboard sounds that have been engineered by humans, the environmental sounds simply exist due to the laws of nature that govern our universe. As such, they could be seen as unintentional but nonetheless useful byproducts whose usefulness is attained through repeated exposure and growing sensitivity to their nuances.

While this definition stretches the conventional boundaries of sonification that are limited to solely intentional manipulation of sound by humans for humans, it offers unique opportunities for us to observe and learn from how we interact with the environment. Such observations may in turn serve as an inspiration for optimal ways to design an intentional data sonification.

When considering any of the aforesaid definitions, it is important to understand the difference between data, information, and knowledge. Data can be seen as a measurement of change in a phenomenon that commonly exists regardless of our awareness. Consider today's weather monitoring systems that are capable of providing incredible amounts of weather data from all over the globe. This data is simply human-readable representation of changes in the environment that would exist regardless whether we are consciously monitoring them, or whether we are doing so physically (e.g. by being physically present at a particular location and using our sensory mechanisms) or through technology. We could probably spend an entire chapter on the philosophical discussion of whether the data can exist without conscious human involvement. This, however, is not our focus. Instead, I will close this thought by repurposing the old adage: if a tree falls in a forest and no one is there to account for it, does it generate data?

Back to the global weather data example—if we feel compelled to do so, we could go and read all the exciting (or boring) temperature, humidity, pressure and wind speed and direction data from any place on Earth. Most of the time, however, we don't do this. Instead, we focus on the data that most immediately affects us, namely that related to our current location, or possibly a travel destination. When observed, such data becomes information. By reviewing it, we acquire information, such as the temperature or humidity levels, or the percentage of the cloud cover. And, if we have adequate prior experience that we may be able to cross-correlate with the newly acquired information, we may utilize such knowledge to make appropriate observations and decisions, such as it will be warm and that we may need to dress lightly or pack appropriately, or that we may want to bring an umbrella.

So, what does all this mean for someone who has potentially no prior experience with sonification and more importantly, why should one care? For this, we need to take a side trip to the world of perception and cognition.

## 8.3 Human Perception

Consider that your understanding of the surrounding environment is rooted in your ability to perceive it and interact with it. We perceive the world through our senses of sight, hearing, taste, smell and touch. In addition, we can also sense temperature, balance, movement, position, and (with varying aptitude and consequently punctuality) the passage of time. On an existential level our very understanding of reality is inextricably tied to our individualized senses and their inherent and unique idiosyncrasies. Yet, what we all have in common is that we strive towards a state where we can process more information more efficiently, so that we can better anticipate change and therefore make better decisions. Every sense plays an important role in this effort and offers its own *unique affordances*. Studying such unique affordances may help us uncover new ways to perceive data and extract information, as well as *broaden our cognitive bandwidth*, thereby allowing us to consume and interpret more data more efficiently. Doing so will empower us to reinforce existing knowledge and build entirely *new knowledge*. Such knowledge will in turn enable us to spot previously learned patterns or uncover new ones, and apply them in other contexts. Below, we will briefly focus on each of the three aforementioned key elements of human perception and, consequently, sonification, namely unique affordances, the broadening of cognitive bandwidth, and new knowledge.

### 8.3.1 Unique Affordances

When it comes to experiencing data, we most often resort to a visual rendition, a chart or a bar graph. Yet, our vision, as mighty as it may be, is not the end-all solution. After all, we have multiple senses for a reason—to cover aspects the other senses are either unable or not particularly good at addressing. When compared to human vision, one of the unique affordances of human hearing is a much higher data resolution we can consciously process per second. Consider when playing a fast-paced video game, arguably one of the more broadly accessible time-critical vision-centric activities. In most situations we are perfectly satisfied with 60 pictures per second (aka frames per second or FPS). And, if we were ever to experience a second where we only received 59 frames instead of the usual 60, possibly because our computer's video card or graphics processing unit (GPU) was unable to render all 60 frames in time, chances are we would never notice such a drop in the overall frame rate, let alone pinpoint the moment in which such a frame drop may have occurred. When it comes to sound, even the now dated compact disc (CD) quality offers 44,100 points of data per second and while we may not be able to continually detect every subtle data point individually, when combined with neighboring data points, even an untrained ear is capable of identifying changes in pitch, timbre, loudness and location. More importantly, an impulse whose length is 1/44,100th of a second is surprisingly easily identified as a simple "click." Although we are clearly not comparing apples to apples, consider the stark contrast between the aforesaid click that takes place for only a fraction of a second to one dropped frame among 60 that remains by and large unnoticed. This is in good part why in our imaginary car ride the aural cues emanating from the dashboard can be incredibly brief—consider the sound of a turn signal typically going tick-tock that is not unlike the aforesaid impulse, versus a considerably more sluggish rate of the blinking LED that serves as its visual counterpart.

While not entirely unique, another particularly potent affordance of hearing perception is our ability to recognize patterns. Consider a CD that is stuck looping a small chunk of music due to damage to its surface and how quickly you are capable of identifying the ensuing skipping or looping just by listening. In this context, the looping audio signal is a repeating data set consisting of potentially thousands of data points, much like a repetition one may encounter in a graph, such as the "head and shoulders" pattern commonly encountered in stock market data that may signal looming change from bullish to bearish market and vice versa. Our hearing is incredibly sensitive to such temporal patterns offering a unique opportunity to apply this same strength towards other forms of data, including abstract data, charts, bar graphs and beyond.

#### *8.3.1.1 Psychoacoustic Properties*

When perceiving a sound, we are typically capable of identifying pitch (or lack thereof, as is the case with noisy sounds, such as the sound generated by saying "shhh"), loudness, timbre or sound color and duration, as well as location in respect to our position. All of these dimensions of the sound are psychoacoustic interpretations of its acoustic properties. The dominant frequency or the rate of repetition of a periodic waveform per second expressed in hertz (Hz) is an acoustic property we can interpret as pitch. Here we use the term *periodic* to suggest the sound in question has a clearly repeating waveform pattern, which in turn allows us to perceive the dominant frequency. It is interesting to consider that our perception is logarithmic in nature (Varshney and Sun 2013). Consider that the note identified as A4 in Western musical notation (~880 Hz) is double the frequency of A3 (~440 Hz). At the same time, A5 (~1,760 Hz) is double that of A4 (~880 Hz) and therefore effectively double the frequency range between A3 and A4. Yet, in our ears we hear relationships between A3 and A4, and A4 and A5 to be equivalent, both distances being perceived as an octave apart. Pitch is not necessarily always perceivable, as is the case in the aforesaid "shhh" example, or the wind noise in our imaginary car ride scenario. Both examples resemble spectral composition (or timbre) of a white noise. In such cases there is no distinct dominant pitch, a property that in and of itself allows us to identify such sound and separate it from others that may be pitched. We refer to such sounds as non-pitched or aperiodic. Most sounds in nature, however, exist somewhere in between these two extreme cases where they possess pitched qualities while also being noisy. Consider, for instance, different cymbals in a drum set, each of which may have a different implicit pitch.

Loudness is derived from the sound waveform's amplitude. The greater the amplitude, the louder the sound. Just like pitch, it is logarithmic in its nature, although our perception resolution of different loudness levels is arguably less than that of pitch, as is evident in a considerably sparser choice of dynamics in music notation (commonly pianississimo, pianissimo, piano, mezzo-piano, mezzo-forte, forte, and fortissimo, and fortississimo). Timbre is a curious beast. Essentially, any sound you have ever perceived is simply a combination of a large number of individual frequencies with distinctive and continually varying loudnesses and whose cumulative result is a unique color. This is why a note A3 played by a violin and a piano sound different and clearly distinguishable, despite the fact both sounds have the same dominant pitch. Human voice can generate a rapidly changing collection of dominant frequencies in terms of both their pitch and loudness, producing vowels, consonants, syllables and words. On a purely philosophical level, silence can be seen as a form of sound that is devoid of any perceivable frequencies or a sound whose loudness is zero, whereas the white noise is composed of all perceivable frequencies. As such, when comparing these outlier cases with the visual domain, one may regard silence as the color black versus the white noise as the color white.

Another implicit perceived property of sound is its duration, or what is more broadly defined as its temporal component. The reason we do not want to limit ourselves to using time solely for measuring duration is because extreme changes in duration may potentially dramatically alter other sound's properties. Consider a pitched sound, an A4 sustained note played by a violin. If we were to splice a recording of such a sound into a tiny snippet, thereby using solely the temporal dimension, not only would we lose the ability to perceive the sound's pitch, but would also potentially significantly alter the perceived timbre, ostensibly converting it into an impulse-like "click." Consequently, it is important to consider the inextricably intertwined codependence of the different acoustic and, by extension, psychoacoustic properties of sound. Just like the aforesaid example in which the duration affects the perceived pitch and timbre, consider how loudness changes may also alter aspects of the sound perception. Consider the following theoretical example where the sound's loudness has been made zero. Under such circumstances the sound has lost all its identifying properties, becoming effectively synonymous with silence. As we will soon learn, this seemingly dubious example may prove particularly helpful when considering sonification.

While all of these dimensions are unique affordances of sound perception, perhaps the most compelling unique affordance of human hearing is the ability to locate a sound. Unlike vision, which is predominantly anterior in its nature, meaning we only see what is in front of our eyes, human hearing is effectively spherical. We can hear with varying levels of accuracy in all directions around us even though we only have two ears. Pinnae, or the cartilage some of us pierce to hang earrings, is what gives us that third dimension by filtering sound that comes from behind, above and below. From the moment our ears begin to broadcast signals to our brain, the brain begins to learn and refine its ability to differentiate sound source locations. Given that each of our pinnae is unique, our individual brains are also individually calibrated to account for such deviations in our anatomies.

The ability to hear all around us is also a critical survival skill. Whether it is a saber tooth cat that may have been preying on our ancestors, possibly trying to sneak from behind, or a bus whose horn caught our attention at the very last second helping us avoid becoming a hood ornament, our all-encompassing spherical spatial perception is something truly unique to our hearing. In our imaginary car ride, distinguishing the location of the dashboard sounds, windshield raindrops, strange noise that may be emanating from the rear left wheel well or a plethora of other inherently spatial aural cues is what further reinforces our ability to differentiate them. Sound spatialization also appears to be the last frontier in the emancipation of sound properties. For at least the past two millennia, human civilization has spent time refining pitch, loudness and eventually timbre, allowing each of these dimensions to drive musical expression. Consider how in the twentieth century we encounter for the first time a large volume of music whose structure is driven primarily by timbre, particularly the percussion ensemble and computer music. With this in mind, the twenty-first century may very well be defined by the emancipation of the spatial component where musical motives may be rooted in sound's location, rather than its well-established dimensions of pitch, loudness, timbre and the inevitable temporal component, or duration. The similar trend is also apparent within the field of sonification. Even though we have a large number of spatialization algorithms at our disposal that allow us to simulate sound motion across multiple loudspeakers, their use remains relatively limited. Consider once again our imaginary car ride and how even luxury cars continue to sound out earcons from some dubious location on the dashboard, instead of utilizing one or more of its typically rich six-channel (or greater) surround-sound system. Wouldn't it be a lot easier to locate which door may be ajar if we were to simply play a particular earcon from the loudspeaker mounted on that particular door? Instead, more often than not, we find ourselves taking our eyes away from the road, the very thing we are not supposed to do, and staring down at the dashboard where a pretty picture of our car is meant to show us what door may be ajar. Or, worse yet, we struggle to decipher what the earcon is even supposed to mean simply because the car manufacturer never bothered to design different earcons for different events and warnings. Indeed, spatialization is an oft neglected aspect of human hearing, and consequently sonification,that offers ample opportunities for broadening the cognitive bandwidth. So what does that mean?

### *8.3.2 Broadening Cognitive Bandwidth*

Broadening cognitive bandwidth by combining multiple senses is a well-established concept within cognitive load theory (Paul A. 2002), which has seen resurgence in the recent big data movement. There are two ways to observe its implications for cognitive bandwidth. One way is to reinforce a particular cue using multiple sensory inputs. For instance, in our imaginary car ride scenario, the turn signal offers both visual and aural cues to make it easier for the driver to be aware of the fact they have a turn signal on. On the other hand, we could also consider there is only so much we can process at any given moment using any single sense. If we couple two or more senses, it is likely we will have greater capacity to process more information simultaneously where each sense may be assigned to a different data source. This may not necessarily mean the ensuing bandwidth or the new whole will be greater than the sum of its parts. Yet, in time-critical situations, even a fraction of an improvement is still clearly better than perception bandwidth through any single sense. Consider the imaginary car ride we discussed at the beginning of this chapter. Most, if not all of the sonification elements occur concurrently while you are driving a car, an action that requires vision, touch, motor skills, a sense of motion, and potentially even a sense of smell that may help us detect that strange odor coming from the engine. Therefore, we can easily process an aural cue from the dashboard informing us that we may be running low on gas while still keeping an eye on the road and controlling the vehicle. Similarly, we may be listening to the radio, which, unless the sound levels are so loud that they prevent us from being able to perceive environmental aural cues (akin to what a blinding light does to our vision), is a straightforward matter that otherwise has no adverse effect on our ability to drive the car.

It may be useful to further differentiate between cognitive bandwidth achievable by cross-pollinating two or more senses and the two or more aspects within the single sense. For instance, perceiving two data sources, one visually, and one aurally, may yield greater cognitive bandwidth than when trying to pack both data through solely sound or sight. On the other hand, one could also consider projecting two different data sources by using pitch and loudness. Lastly, in either example we could use one and the same data point to drive both dimensions, thereby lowering the ensuing cognitive load, or at least increasing chances of us being able to perceive them in a timely fashion. For instance, if the car we are driving is in a heavy rainstorm, the ensuing raindrop noise may occlude sound generated by the car blinkers. In this situation, having an accompanying visual cue on the dashboard may help us minimize chances of having a blinker stuck in the on mode long after we have already changed a lane or taken that turn.

It may now seem obvious that one of the greatest advantages of sonification is its coupling with visual and other sensory inputs, with the recent revolution in virtual/augmented/mixed reality technologies being an obvious target. However, despite decades of research, sonification is still a nascent field (Bukvic and Earle 2018), and as such stands to benefit from the early-stage exploration and codifying of its unique affordances in isolation from other sensory inputs that may conflate any such assessment. As was the case with visualization, doing so may help us identify new approaches to pattern recognition and knowledge building that in turn may be applicable to other domains.

### *8.3.3 Knowledge Building*

Remember the previous skipping CD example and our ability to near instantaneously spot such an anomaly simply by listening? Now consider how once we are exposed to a particular instrument, for instance, a snare drum or a guitar, how well we may have memorized its timbre, and have become capable of identifying the same instrument in a potentially complex soundscape, such as an orchestra performance or a rock band song. This is where our brain's ability to recall vivid details of previously learned timbres and cross-correlate them in real-time against perceived patterns really comes to focus. How about a few more examples? How well do you still remember a song you grew up with or got married to and how well you can anticipate its moments even after only a few repeat listenings? Or, that album that you listened to multiple times in the same order to the point where when one song ends, isn't it fascinating how precisely you know what is coming up next or just how brief are the pauses between the songs? Unlike the skipping CD example where we may be facing an unfamiliar sound and are still able to quickly spot a repeating pattern, the aforesaid scenarios are examples of our brain's powerful ability to recall complex data patterns stored in the form of music and/or sound and voice color memories, much like our ability to spot and recognize familiar faces.

Human hearing indeed offers ample opportunities for exploring such knowledge building and in turn pattern recognition in a rich, immersive, spatially aware environment. What is particularly exciting is that, much like our recognition of familiar faces in a complex image, this process is inherently subconscious and our brains allow us to detect these patterns quickly and with no added conscious effort. Coming back to our imaginary car ride, consider while driving in a potentially noisy environment (raindrops on the windshield, wind and engine noise, radio, etc.), detecting a familiar dashboard sound (aka an earcon) occurs naturally to us even though we may have made no conscious effort to monitor for its potential occurrence.

While the idea of using sonification to identify an entirely new pattern in a known data set (e.g. a stock market or geospace data) seems like the proverbial Holy Grail that would validate the proposed value and importance of sonification, it is worth considering that sonification may be also seen as "an under-utilised dimension of the 'wow!' factor in science engagement multimedia" (Ballora 2014). As such, regardless of our ability to generate entirely new knowledge, on a purely educational and engagement level sonification may prove an indispensable vehicle for facilitating understanding of known concepts.

To summarize, sonification is a nascent field that offers unique affordances that may help us uncover new knowledge in the existing data. When coupled with other senses, it has a potential to broaden the user's cognitive bandwidth and increase efficiency. It can also serve as a potent way of representing known concepts to promote learning and engagement.

## 8.4 Is It Art or Science?

For sonification to be possible it needs a data source. Consider that the data can be found everywhere around us as well as in anything we do or create. It may be the number of hours we slept last night, or how many times we chewed each bite of our sandwich at lunch. Consequently, it can be extracted from any creative domain, whether it is the sciences, arts, engineering, design or, as is more common, a combination thereof. While generating data from a scientific experiment may seem straightforward, imagining the same in music, for instance, may at first seem strange. Consider for a moment all the different ways we can analyze music using music theory concepts, such as harmonic motion, the recurrence of the theme, or the overall form. The ensuing analysis offers ample data points much like a scientific experiment. Moreover, the very audio recording can be seen as a form of data sonification where the data are the notes in the score. Another way to look at this strange and rather malleable relationship between the arts and sonification is through the lenses of a growing number of art works where the aural and/or visual content is generated by sonifying and/or visualizing some sort of a data stream. The ensuing artifact, whether it is an interactive art installation, sound art, or a musical composition, is effectively a sonification of a data set shaped by an artist's aesthetic choices.

## 8.5 Classifications

Considering that sound perception is time-based, sonification is by and large focused on rendering continuous data stream over time, such as the changes in weather or stock market data, or engine noise. The sonification timeline does not have to match the normal passage of time and, like video, it can be played faster or slower, as well as forwards and backwards. Rather than continually outputting sound, sonification can also focus on punctuating a specific threshold, as is the case in our imaginary car ride where a single sound is played once the fuel level reaches a previously designated low point. As a result, such a cue or an earcon can be seen as a subset of sonification. Unlike the continuous data stream examples, however, earcons are typically interpretations of previously known patterns or conditions (e.g. a previously designated low fuel tank level), whereas continuous streams tend to be raw and unprocessed and left up to our brains to interpret and spot any potential patterns, as is the case with our skipping CD example. Furthermore, the continuous data streams can be generated in two different ways—by taking the data and feeding it directly into an audio output (e.g. a loudspeaker) with minimal processing, or by using it to modify a property of an engineered sound, such as its pitch, timbre, loudness, location, or a combination thereof.

*Audification* is a subset of sonification. It is essentially taking data and with minimal processing converting it into sound. The reason processing may be necessary is because some data ranges may not produce a usable audio signal. For instance, we may want to audify changes in daily temperature. In order to do so, we may need to normalize the data ranges, meaning take the lowest and highest possible temperatures and map them onto the waveform ranges that in a digital domain typically span from -1 to 1. We may further need to adjust its frequency or the speed by which we feed it to the loudspeaker, in order to generate a waveform that is audible to human ears. This, by convention, includes frequencies between 20Hz and 20,000Hz, although in practice there are a few outlier cases where the perception of frequencies below 20Hz and 20,000Hz may be also possible.

While potentially indispensable, this kind of sonification has a fairly specific, if not limited, utility.

Let's assume that our observed temperature is measured only hourly. If we had to generate even a single second of sound, we would require 44,100 such data points, which means we would need at least 1,837.5 days worth of temperature data. The ensuing single second of sound would effectively form an example of audification. In this case, however, given such dense packing of data, despite the high resolution of human hearing, listening to a short snippet is unlikely to allow us to accurately observe what may have occurred within one afternoon, which at this point would amount to a mere six samples of audio or barely 1/10,000th of a second. One way to address this is to slow down the playback of such events and allow the same afternoon to take place across several seconds. This, however, may result in a rather inefficient experience of the ensuing data, requiring minutes, if not hours of listening, needless to mention it may also generate a subsonic waveform, or a waveform that has a resulting frequency lower than 20Hz, that is simply not audible to humans. This is where sonification steps in to save the day. Rather than normalizing or slowing down the data, we may want to use it to drive the frequency of a carrier signal, whether that be a synthesized (e.g. an oscillator) or a sampled sound (e.g. the sound of violin). Depending on the temperature ranges and the context, we may want to transpose it to make it more easily audible, or even multiply its temperature changes to make them more easily perceived. This kind of sonification indeed offers greater flexibility over audification. The temperature can also affect the loudness, timbre, or position. For instance, the resulting sound may get louder and "brighter" as a result of using a low pass filter, as well as move from left to right.

When considering earcons, another way to think about them may be that the monitored data is continually sonified, except that its loudness curve has an extreme exponent and that only when it reaches the peak of its normalized range does it make any perceivable sound. Of course, in practice this is typically not the case. However, as a theoretical construct it may offer a way to reconcile the perceived differences between a momentary cue and a sonification of a continuous data stream.

## 8.6 Sonifying Data

A growing body of sonification research, whether a conscious scientific study or a byproduct of an artistic installation, sound art or a composition, has yielded a broad array of approaches to sonifying data. Considering there are near infinite ways one may sonify a particular data set, and given that this area of research is still in its infancy, it should not come as a surprise that the results tend to be by and large case-specific and their lessons difficult to abstract. Nonetheless, over time several common-sense observations have emerged. They are listed below in no particular order.

### 8.6.1 Ability to Perceive Multiple Concurrent Psychoacoustic Properties

We are capable of concurrently perceiving multiple psychoacoustic properties. This means one could map different data streams to each psychoacoustic property of a sound. For instance, in our weather pattern data example the temperature could be used to drive sound's pitch, while changes in humidity could be perceived as changes in timbre. Naturally, there are limits to what we can concurrently observe. Such limits typically depend on many variables, such as the choice of sound, variance in the data, the mapping strategies, potential hearing perception limits of our target audience and environmental conditions. As a result, this remains by and large a case-specific iterative trial-and-error design process.

### 8.6.2 Single Variable Sonification Using Multiple Psychoacoustic Properties Can Lead to Improved Perception

Humans are generally more capable of perceiving differences in sound if such differences manifest concurrently through different psychoacoustic properties. For instance, in the previous example, rather than mapping both temperature and humidity onto pitch and timbre, respectively, we may want to map temperature both on the pitch and timbre. This typically results in making changes to temperature easier to perceive, but there are once again limitations to what combinations of parameters yield useful results. For instance, having a low pass filter set too low on a high-pitched sound may render such a sound inaudible, thereby making its pitch and pitch-variance imperceptible to the listener. Another obvious downside is that by doing so we limit the number of variables we can concurrently present using one sound.

### 8.6.3 Ability to Perceive Multiple Concurrent Sounds

We have an innate capacity to perceive multiple concurrent sounds. Once again, the limits of our capacity vary depending on the aforesaid variables, namely the choice of sounds, variance in the data, the mapping strategies, potential hearing perception limits of our target audience and environmental conditions. For instance, having two sounds that have similar spectral composition or timbre may make one sound become masked by another, thus making it difficult if not impossible for a listener to identify and separate them. The opposite may be true, as well, whereby one sound may be so rich and spectrally variant that at times our brain may play tricks on us, making us believe we may have heard a sound that simply wasn't there. Has it ever happened to you that while taking a shower you may have heard a faint familiar sound, like your smartphone ringing, and as you scrambled out of the shower to pick up the phone you found out that the phone never rang? Such events are more common than you think, particularly in the aforesaid scenario, where the shower noise closely mimics white noise, or the presence of all perceivable frequencies at continually random and varying rates and intensities. Consequently, as different frequencies randomly vary in their strength, it is not unlikely for the sound to momentarily resemble another familiar sound, like that of our smartphone's ringtone.

### 8.6.4 Capacity to Spot Patterns in Complex Soundscapes

As we continue concurrently mapping different variables onto different psychoacoustic properties of a sound and/or using multiple concurrent sounds, we will eventually cross the threshold beyond which we lose the capacity to monitor all the individual data streams. Yet, all is not lost beyond this threshold. We still have an opportunity to leverage human ability to spot patterns, such as in the skipping CD example. Under such circumstances we may not be able to identify exactly what is going on with each individual variable, but we may be able to spot a cumulative timbre we may have heard before, particularly if it happens to have a distinct feature (e.g. it may be loud, "brighter" than the rest, may have an interesting interaction between two or more sounds due to phasing, and so on).

### 8.6.5 Difficulties in Tracking Steadily Changing Data

Spotting specific milestones in the sonification of steadily changing data is not particularly easy, especially as the rate of change and its range increase. Consider for a moment a violin that glissandos continually up and down. In situations like these, akin to the low fuel earcon in our imaginary car ride, such milestones can be further reinforced by punctuating them with added sounds. Another way to think of this is as if the said earcons were equivalent to the tick marks on a ruler.

### 8.6.6 *The Cocktail Party Effect and the Auditory Scene Analysis*

In designing a sonification, consider the cocktail party effect (Arons 1992) that allows us to carry a conversation even in noisy environments by focusing our hearing to a specific sound timbre or color. Bregman's auditory scene analysis (Bregman 1994) further elaborates on this concept by providing insight into how our mind is capable of partitioning multiple concurrent sounds based on sequential and simultaneous grouping. According to Bregman, sequential grouping pertains to a continuous stream coming from one source based on its acoustic parameters, including the perceived source location. Simultaneous grouping on the other hand focuses primarily on the perceived pitch. This, in combination with the spatial component, may help broaden our capacity to perceive multiple concurrent events. This is in part possible because our brain can bundle sounds we deem undesirable and relegate them to a background noise, like the sound of a computer fan, or a dishwasher. Naturally, there are limits to how much we can discriminate between the wanted and unwanted sounds, and as the noise floor rises, it becomes increasingly harder to spot individual sounds. In other words, as the environmental sounds we choose to relegate to a background noise increase in loudness and as their cumulative spectra approaches white noise, our capacity to perceive desired sounds diminishes.

### 8.6.7 *Listener Fatigue*

One particularly important aspect of sonification is the concern with listener fatigue. The choice between a pleasing and annoying sound may mean the difference between a useful sonification and a pounding headache. Moreover, improper scaling of the data may mean the lack of noticeable difference in the data over the long term, which may lead to a listener simply tuning out such sounds, subconsciously labeling them as background noise, much like the noise of an air vent or a computer fan. To minimize the chance of fatigue, sonification research often resorts to the use of tonal relationships. For instance, mapping data to the diatonic (tonal) pitches of a scale or using data variance to induce previously vetted changes in the rhythm of an engineered sound or a sound layer may be perceived as more appealing than atonal and seemingly random rhythmic sonifications. Similarly, leveraging of sounds we already naturally perceive as soothing even when observed over an extended period of time, such as the sound of a babbling brook, rain, wind chimes, or evening crickets, may be a step towards minimizing the chance of listener fatigue. Other consideration is our greater sensitivity to the higher ranges that we may perceive as being more piercing.

### 8.6.8 Context Is Everything

Not all target audiences have the same hearing ability. As we age, we tend to lose the ability to perceive higher frequencies. Those who work in noisy environments may lose their hearing ability faster. It is important to consider who the sonification is for and what may be the conditions under which such a sonification may be perceived.

### 8.6.9 Minimize the Arbitrary

When considering data sources, it may be useful to focus on those with properties that limit the need for their arbitrary assignment. For instance, consider geographical or geospatial data that lends itself to spatialization given such data is inherently spatial. Spatializing such data to reflect its location in space, in turn, limits the need for the sonification to introduce yet another arbitrarily assigned spatial mapping or a variable that may need to be accounted for in the sonification's assessment and whose potentially varied implementation may adversely affect the human ability to understand the data and its patterns.

### 8.6.10 Ecological Perception

Lastly, when studying sonification from a scientific perspective, we may want to approach it by studying individual acoustic properties in isolation. In doing so, we need to be aware that, depending on the chosen context and conditions, such an approach can potentially lead to misleading or outright inaccurate conclusions. For instance, one may want to consider using a sine tone, the most basic sound as a means to study the human ability to perceive spatial sound. Yet, our ability to spatialize sound depends on the sound being spectrally rich. In other words, a sound needs to consist of more than one concurrent frequency to offer a decent chance of having its location accurately identified by a listener. As a result, a sine tone offering a single frequency is all but impossible to consistently locate in space and as such, any study that leverages sine tone for this purpose may lead to an inaccurate conclusion that humans are not capable of accurately localizing sounds in space. Another way to describe this is the need to ensure that the sonification offers ecological perception and validity (De Gelder and Bertelson 2003) that leverages all the affordances of our environment (Gibson 2014) and thereby sidesteps any potential pitfalls of studying and leveraging acoustic properties in isolation, while exposing and engaging the full perception potential of a perceiver or, more precisely, listener.

## 8.7 Leveraging the Psychoacoustic Meaning of Environmental Sounds

Throughout this chapter we have revisited examples of environmental sonification, or the unintentional byproducts of our physical world that provide us with invaluable information about our immediate surroundings, such as the wind and engine noise in our imaginary car ride. We now return to these examples as an inspiration for a sonification that leverages preexisting conditioning we have acquired over the course of our lifetimes to facilitate the understanding of the underlying data. For instance, consider a sonification that represents weather forecast through sound. Rather than relying on an abstract synthesized sound, we may want to use the sound of rain to signify the rain forecast. Moreover, we could use various kinds of rain sounds to suggest the anticipated amount of rain. A relatively silent and steady rainfall sounds significantly different from a violent summer storm, and these are events we can easily recognize provided we have been exposed to them in the past. If we expect wind gusts, we may want to use the wind noise. Or, if we expect both high winds and rain, we could use both. By doing so, we are effectively leveraging the psychoacoustic or conditioned meaning of an environmental sound and thereby facilitating the connection between the data and the sound. In other words, we are eliminating the intermediary step in interpreting the sonification where a listener needs to learn an entirely new vocabulary to be able to connect a potentially abstract aural event with a specific data stream.

Any environmental sound in this kind of sonification can be further manipulated through its psychoacoustic dimensions. This kind of an approach when compared to a more conventional sonification using abstract sounds requires additional considerations. For instance, we may want to use a particular data stream to affect the pitch of the rain sample. Unless executed slowly and subtly, such a change immediately begins to sound unnatural and therefore potentially distracting, or may lead to a sound that loses its original recognizable association and meaning. On one hand, this may be a powerful way to grab one's attention. On the other, it may also break the sense of seamless integration into our everyday environmental soundscape and therefore increase chances of listener fatigue. While altering pitch or detuning may be tricky, filtering and/or changes in loudness are more natural, as they mimic changes in intensity or a separation between the listener and the source by a wall or some other physical and acoustic obstacle.

It is worth noting not all sounds can be portrayed in this way, so it may be necessary to leverage movie or even cartoon-like approaches to sonifying elements that otherwise do not produce perceivable sound. For instance, consider a sunny day. The sun itself does not generate any sound, or at least none that we can directly perceive. It, however, affects other elements in our environment that do produce perceivable sound, such as the crickets and cicadas that try to cool themselves off during a hot afternoon. Or we can resort to human-centric activities that may take place on such a day and that have an implicit relationship with the sun, such as a sound of a hot, sizzling steak on a grill, as was the case with one of the studies I conducted back in late 2000s (Bukvic, Gracanin, and Quek 2008; Bukvic and Kim 2010)

To date, sonification research has typically shied away from this kind of an approach in favor of more controlled lab-like experiments that focus on synthesized sound, in part because psychoacoustic meaning may have been deemed difficult to quantify and defend within the scientific community. As such, it remains an exciting and woefully underutilized research opportunity.

## 8.8 A Case for Immersive Exocentric Sonification

Our interaction with the world depends on multiple sensory inputs to create a more comprehensive picture of our environment. Moreover, our experiences are often reinforced when perceived concurrently through multiple senses. For instance, reading a text aloud may help us retain more information than silently reading "in our heads" in part because we are engaging multiple sensory mechanisms and therefore reinforcing learning by concurrently activating multiple areas of our brain. Likewise, consider eating an onion with your nose closed (yes, that came out of the blue). Suddenly, this flavorful vegetable seems tasteless and bland. It is the combination of taste and smell that makes an onion seem not only flavorful but also at times so potent our eyes may tear up.

It appears that studying the way we perceive the world needs to be inherently holistic or, as Gibson put it, ecological (Gibson 2014). In other words, the ecological perception suggests that none of our senses can reach their full potential on their own—eyes need eye, head and body muscles to move to be able to attain a full picture of our immediate surrounding. Similarly, to localize sounds more accurately we need to move our head and body, therefore improving both our spatial image. Likewise, by positioning ourselves closer to the desired audio source (remember the cocktail party effect?) we can improve the desired signal-versus-background-noise loudness ratio and consequently the sound's clarity and, in case of human voice, intelligibility. Yet, such a holistic approach is not always possible, particularly considering how complex and multivariate such a study may become when factoring in all the senses. Inevitably, the early efforts in studying human perception have gravitated towards a lab-like experiment approach where scientists isolate a particular sense or focus on a small subset of its affordances. This has yielded important information about individual sensory mechanisms and has paved a way towards more holistic approaches to studying perception that have built upon this foundational knowledge. Today, we are witnessing a growing focus on the holistic study of human perception and its impact on various scenarios (Lindstrom 2005; Bukvic and Earle 2018).

Similarly, much of the research in data sonification conducted so far has focused on applying data on one or a few acoustic sound properties in isolation from other sensory inputs. This may be in part due to the fact the sonification is still a nascent field in need of unpacking its multiple dimensions before it can be cross-pollinated with other senses. It may also be due to the technical limitations of the existing audio infrastructure. For instance, as I write this chapter, there are only a handful of facilities around the world that offer access to a large number of loudspeakers, or what we refer to as high density loudspeaker arrays (HDLAs) capable of providing a high-resolution immersive sound environment (Lyon et al. 2016). As a result, most sonification studies have focused on a subset of psychoacoustic properties. Many have either avoided the use of spatial audio, or have resorted to a simple stereophonic rendition using headphones. Some have tried using the binaural approach that offers greater spatial resolution but currently suffers from an inability to consistently account for variance in the anatomy of individual pinnae, as well as its inability to accurately portray sounds that emanate in front of a listener. A handful of more recent research projects have tried to integrate hearing with other senses, such as head rotation, or body position, both of which are indispensable variables to accurately locating sound sources. A few studies have dabbled with using HDLAs to avoid the challenges and idiosyncrasies inherent in simulating spatial audio (Bukvic and Earle 2018).

It is worth noting that the spatial dimension of the human hearing capacity is inextricably intertwined with our other senses, particularly head rotation and position. For instance, to minimize the so-called cone of confusion for sounds to the left and right of our head where there is no pinna to offer filtering variance, or the front-back confusion, we commonly rotate our head to change our perspective to the sound source and use pinnae and interaural time difference (ITD, or a difference in time it takes for a sound to arrive to each ear due to different physical distance between the source and each ear) to better pinpoint a sound location. The inability to do so vastly limits our capacity to locate and consequently discriminate sounds in a more complex soundscape (again, remember the cocktail party effect) and may yield misleading results. Consequently, studies that do not account for this capacity may conclude that our spatial hearing capacity seems lower than it actually is.

Even when studying human sight in isolation from other senses, we do not prevent users from moving their eyes and in most cases their head. These are, after all, critical elements that enhance our ability to see because our eyes only have a very narrow area where we can clearly perceive details and the rest is generally blurry and only sensitive to motion and change. It is our brain that constructs a holistic image as our eyes scan around and whose range is further enhanced by our head motion. It only makes sense that we should also allow for our hearing to be studied in consort with head movement. Moreover, by moving around in space, we are capable of further refining our ability to locate sound sources. Have you ever misplaced a smartphone in your home? As you go around looking for it, you may opt for calling it from another phone and then listening to where its ring may be coming from. The ability to move throughout the space enables you to use the loudness of the ringtone to your advantage and thereby narrow down the area where it may be located. As you get closer to the smartphone, you have fewer walls and obstacles between you and the phone playing acoustic tricks with sound reflections that may deceive your ability to localize the sound source. This is where you can finally realize the full potential of your spatial aural perception and locate the phone with relative ease.

Coupling head and body movement and position with our hearing perception brings us closer to the way we interact with our environment, something that comes natural to us. We don't have to train for it. We simply interact. Now, consider a walkable HDLA environment with loudspeakers all around you, including above and below. This three-dimensional immersive space populated with numerous aural cues is much like a cocktail party, except instead of being surrounded by people, you are surrounded with multiple sonified data sources. As you move around the space, you are able to get closer to a particular sound, thereby naturally attenuating the perception of other sounds while concurrently using your spatial memory to generate a more comprehensive spatial map. In such an environment, you can interact with the data in the same way you interact with the real world. We can refer to such an environment as the *immersive exocentric environment*. Consequently, we can refer to the sonification that takes place in it as *immersive exocentric sonification* (Bukvic and Earle 2018). Unlike the use of the term *exocentric* within the context of immersive environments that suggests an out-of-body experience (Dede 2009), here we refer to a sonification that emanates from a perimeter of the space and does not require the user to remain in the center of the space where the distance and the ensuing loudness between individual loudspeakers and the listener's ears may be near identical.

In our review of hearing perception's unique affordances we briefly touched upon the spatial component. Here we revisit it as a critical element of the aforesaid immersive exocentric sonification. Spatialization algorithms and approaches are indeed numerous and include:

- The ubiquitous stereophonic spatialization that simply utilizes either a pair of loudspeakers or stereo headphones;
- The binaural spatialization that has seen a growing popularity within the gaming and more recently virtual, augmented and mixed reality contexts, in good part due to its limited hardware needs (Carty and Lazzarini 2009);
- Ambisonics that can also scale to a large number of loudspeakers ideally arranged to form a dome-like icosphere or its subset (Furness 1990);
- Wave field synthesis (WFS) that recreates a waverfront using Huygens's principle and typically requires a large number of tightly packed loudspeakers (Boone et al. 1995);
- Vector base amplitude panning (VBAP) that benefits greatly from a large number of loudspeaker arrangements that ideally form a series of triangles, and cumulatively some subset of an icosphere (Pulkki 1997);
- The manifold-interface amplitude panning (MIAP) algorithm that also targets icosphere-like structures (Seldess 2014); and,
- Most recently, layer based amplitude panning (LBAP) designed in part for the exploration of the immersive exocentric sonification and the spatial mask concept with loudspeakers separated into layers that do not require any particular shape or arrangement (Bukvic 2016).

These are merely a subset of a much larger ecosystem of algorithms and approaches to sound spatialization that may be utilized within the context of the immersive exocentric sonification and beyond, each of which offers its own unique advantages and limitations. Their proliferation in the twenty-first century points to the anticipated musical emancipation of the spatial component of human hearing we mentioned earlier.

When considering the aforesaid immersive exocentric sonification, it is worth noting that the varying distance between the loudspeakers and a listener as they navigate the space should not be seen as a detriment, but rather as an advantage that more closely resembles the way we interact with the world. There may be phasing artifacts and other unforeseen interactions. Yet, all these anomalies are something we commonly experience in the real world and as such our brains are typically accustomed to either ignoring or leveraging them to improve our spatial resolution. This is also true of visual stimuli. Consider the way we perceive movies on cinema screens that, depending on where we sit, may be so large we cannot fit all of them within our peripheral vision. Under such circumstances, some of the areas on the edge of the large screen are clearly much farther from our eyes than the areas closest to where we may be sitting. This may lead to unusual distortions to the image's perspective with respect to our vantage point. Yet, our brain is capable of treating the ensuing projection as a flat image and we happily reconstruct it as such in our heads without any concern for its physical distortion. By extension, in the immersive exocentric environment, particularly when using a layered approach to grouping loudspeakers inherent to the aforesaid LBAP spatialization algorithm, the loudspeakers within the same layer share the same elevation, much like pixels on a screen that share the same row. As we move around such space to explore the immersive spatial aural content, loudspeakers that are closer to us may have a perceived higher elevation versus those that are farther from us, despite all of them sharing the same elevation with respect to the ground. Despite this anomaly, once we have the overall awareness of the visual (e.g. TV) or aural (e.g. loudspeaker perimeter or front) canvas, our brains are capable of compensating for such physical inconsistencies in favor of maintaining consistent relationships between the individual loudspeakers, or, as is the case with its visual counterpart, between the pixels on a screen.

The immersive exocentric sonification focuses solely on the space's perimeter generated by the loudspeakers. While there are ways of simulating sounds within the space, particularly when utilizing WFS and ambisonics, they are prone to idiosyncrasies that limit the human ability to move and study such sources from different vantage points. As such they may be seen as being equivalent to studying our ability to visualize data while making certain elements visual only under certain circumstances. If we were to consider the ensuing loudspeaker front or perimeter as one canvas, we could project and move data across it. The most obvious data choices may be inherently spatial data, such as geographical or geospace data that limit the need for arbitrary assignment of variables to the spatial component (Bukvic and Earle 2018). Much like the aforesaid leveraging of the psychoacoustic meaning of environmental sounds, the immersive exocentric sonification is an entirely new opportunity in the sonification research that begs further exploration.

## 8.9 Where to Go From Here?

In this chapter we have introduced ourselves to the world of sonification, why it exists and how we may benefit from it. We have also reviewed some of the common-sense strategies that have emerged from this nascent field of research. Lastly, we have proposed two new areas of research that may allow us to explore the full potential of sonification, the leveraging of the psychoacoustic meaning of environmental sounds and a more natural immersive exocentric approach to sonification that may help us quantify the limits of the human ability to perceive and interpret spatial audio streams while sidestepping some of the key limitations of the current virtual approaches to spatializing sound. Unsurprisingly, what has been covered in this chapter merely scratches the surface of a much larger body of knowledge. Below is a list of references and suggested follow-on reading that may provide additional insight in the various aspects of sonification and related disciplines.

## References

Arons, B., 1992. A review of the cocktail party effect. *Journal of the American Voice I/O Society* 12, no. 7, pp. 35–50.

Ballora, M., 2014. Sonification, science and popular music: In search of the 'wow'. *Organised Sound* 19, no. 1, pp. 30–40.

Boone, M. M., Verheijen, E. N., and Van Tol, P. F., 1995. Spatial sound-field reproduction by wave-field synthesis. *Journal of the Audio Engineering Society* 43, no. 12, pp. 1003–1012.

Bregman, A. S., 1994. *Auditory Scene Analysis: The Perceptual Organization of Sound* [Online]. MIT Press. Viewed October 25, 2016 <https://books.google.com/books?hl=en&lr=&id=jI8muSpAC5AC&oi=fnd&pg=PR11&dq=Auditory+Scene+Analysis:+The+Perceptual+Organization+of+Sound&ots=SFrZH9ABvD&sig=o5uUgvlJZHz_hdi4c1CQQlfG_qk>.

Bukvic, I., Gracanin, D., and Quek, F., 2008. Investigating artistic potential of the DREAM interface: The aural painting. *Presented at the International Computer Music Conference*, Belfast, UK.

Bukvic, I., and Kim, J., 2010. Perception and Interpretation of concurrent aural shapes using DREAM interface. *Presented at the International Conference on Auditory Display*, Stony Brook, NY.

Bukvic, I. I., 2016. 3d time-based aural data representation using d4 library's layer based amplitude panning algorithm. *Presented at the International Conference on Auditory Display, Canberra, Australia*. Viewed November 9, 2016 <www.icad.org/icad2016/proceedings2/papers/ICAD2016_paper_10.pdf>.

Bukvic, I. I., and Earle, G., 2018. Reimagining human capacity for location-aware aural pattern recognition: A case for immersive exocentric sonification. *Presented at the International Conference on Auditory Display, Houghton, MI*.

Carty, B., and Lazzarini, V., 2009. Binaural HRTF based spatialisation: New approaches and implementation. *DAFx 09 Proceedings of the 12th International Conference on Digital Audio Effects, Politecnico Di Milano, Como Campus, Sept. 1–4, Como, Italy*. Department of Electronic Engineering, Queen Mary University of London, pp. 1–6. Viewed February 21, 2016 <http://eprints.maynoothuniversity.ie/2334>.

Dede, C., 2009. Immersive interfaces for engagement and learning. *Science* 323, no. 5910, pp. 66–69.

De Gelder, B., and Bertelson, P., 2003. Multisensory integration, perception and ecological validity. *Trends in Cognitive Sciences* 7, no. 10, pp. 460–467.

Furness, R. K., 1990. Ambisonics-an overview. In: *Audio Engineering Society Conference: 8th International Conference: The Sound of Audio*. Audio Engineering Society. Viewed November 9, 2016 <www.aes.org/e-lib/browse.cfm?elib=5417>.

Gibson, J. J., 2014. *The Ecological Approach to Visual Perception: Classic Edition*. Psychology Press.

Kramer, G., et al., 1999. *The Sonification Report: Status of the Field and Research Agenda. Report Prepared for the National Science Foundation by Members of the International Community for Auditory Display*. Santa Fe, NM: International Community for Auditory Display (ICAD).

Lindstrom, M., 2005. *Brand Sense: How to Build Powerful Brands Through Touch, Taste, Smell, Sight and Sound*. London: Kogan Page Ltd.

Lyon, E., et al., 2016. Genesis of the cube: The design and deployment of an hdla-based performance and research facility. *Computer Music Journal* 40, no. 4, pp. 62–78.

Oswald, D., 2012. *Non-speech Audio-semiotics: A Review and Revision of Auditory Icon and Earcon Theory*. Georgia Institute of Technology.

Paul, A. K., 2002. Cognitive load theory: Implications of cognitive load theory on the design of learning. *Learning and Instruction* 12, no. 1, pp. 1–10.

Pulkki, V., 1997. Virtual sound source positioning using vector base amplitude panning. *Journal of the Audio Engineering Society* 45, no. 6, pp. 456–466.

Rabenhorst, D. A., et al., 1990. *Complementary Visualization and Sonification of Multi-Dimensional Data*. IBM Thomas J. Watson Research Division

Seldess, Z., 2014. MIAP: Manifold-interface amplitude panning in Max/MSP and pure data. In: *Audio Engineering Society Convention 137*. Audio Engineering Society. Viewed January 31, 2016 <www.aes.org/e-lib/browse.cfm?conv=137&papernum=9112>.

Varshney, L. R., and Sun, J. Z., 2013. Why do we perceive logarithmically? *Significance* 10, no. 1, pp. 28–31.

Worrall, D., 2009. *Chapter 2: An Overview of Sonification. Sonification and Information: Concepts, Instruments and Techniques*. PhD Dissertation, University of Canberra, Canberra, Australia.

9

# Image Sonification—A Practitioner's Account

**David Stout**

## 9.1 Introduction

This introduction to the world of image sonification begins with a brief overview of the pioneers and evolving technical innovations underpinning contemporary approaches to visual music and photo-based sound generation. Readers interested in sound design practices utilizing the sonification of images are encouraged to look further into the works of the engineers, filmmakers, composers and artists whose dedicated focus on art and engineering have helped define this field. The chapter concludes with a detailed look into the contemporary techniques and aesthetic concerns I have explored utilizing real-time, interactive image as a means to generate and perform sound synthesis. A fundamental concept present in most of these examples is the use of transcoding. Applied broadly, transcoding allows for one media type to be translated into another. In reference to digital transcoding, any given object, material, media format, language system or data stream can be represented numerically and can thus be transcoded into another representational form or medium. Transcoding makes image-based sonification possible.

## 9.2 The Evolution of Image Sonification

In 1913, the Italian painter Luigi Russolo, in a letter to the Futurist composer Balilla Pratella, wrote one of the defining documents of twentieth-century musical aesthetics. This passionate manifesto, titled *The Art of Noises*, argued for the embrace of all sound as a potential material for a new art of noises (Russolo 1986). While Russolo hailed a revolt against persisting strains of musical romanticism, his ideas spurred a more far-reaching effect, elevating sound as an aesthetic material and the soundscape as a source of a new compositional dialectic. Russolo's further contribution was the design of a series of novel acoustic instruments called the *Intonarumori*, which utilized hand-activated mechanical means to excite a percussive membrane to emit a variety of noise-laden timbres. These works challenged and transformed classical music conventions, signaling the beginnings of sound art and the discipline of acoustic ecology and the future of media performance art, while simultaneously heralding the distant birth of industrial electronic music (Frieze 2013). In the midst of this well-dramatized Futurist commotion, Russolo's paintings have become a footnote, yet they reveal an artist with a profound capacity to visualize the unseen acoustic forces surrounding us.

Russolo's demonstrated ability to depict sound utilizing the abstract visual vocabulary of early modernist painting provides a speculative foundation for contemporary artists, designers, programmers and musicians who seek an effective means to bridge the visual and sonic mediums (WikiArt n.d.). I am specifically referring to music visualization in its many forms from painting, animation and experimental film to the history of color organs and kinetic light displays. These works are most often associated with the visual representation of music as opposed to the sonification of images and visual data. This creative bias towards music visualization over image sonification may stem from several factors including an observation that visual artists and filmmakers often lead this pioneering research, while also noting that technological advancements have only recently given artists a more robust digital tool set with which to fully explore an array of image sonification possibilities.

This rich history of intellectual curiosity regarding the interrelationship of auditory and visual phenomenon evolved as an art form in its own right in parallel with the development of narrative cinema. The field of visual music is often associated with twentieth-century practitioners (CVM n.d.). These artists, composers and filmmakers are in fact part of a longer history joining Pythagoras and Newton among a celebrated continuum of prominent thinkers who have sought to understand the correlation of pitch and color and, by extension, the more complex interrelationship of music and the kinetic image. Russian painter Wassily Kandinsky was an early proponent of visual abstraction inspired by musical motifs. A German émigré to Los Angeles, Oskar Fischinger is often referred to as the father of visual music, and while he insisted that his work stand on its own without the aid of sound, his animated films demonstrate a keen analytical mind in the visualization of complex musical relationships (Moritz 2004; see also *Oskar Fischinger: Visual Music* 2017). The celebrated works of abstract filmmakers John and James Whitney are supported by John Whitney's seminal text, *Digital Harmony: On the Complementarity of Music and Visual Art* (Whitney 1980). Whitney's overarching desire was to formulate a foundational theory on which to base the practice of nonobjective abstract cinema. He believed this theory to exist as a complement to Western music theory with a strong emphasis on demonstrable harmonic relationships between musical ideas such as pitch, tonal modulation and counterpoint with a direct correspondence to visual elements governing graphic motion design.

For those interested in the interrelationship of music and image, there is much to recommend in Whitney's writing; however, artistic practices change over time and the singular dominance of Western harmonic theory is now augmented by an array of approaches influenced by the disciplines of acoustics, psychoacoustics, computational science and multicultural musical aesthetics in tandem with the increasingly pervasive critical perspectives underlying conceptual artistic strategies born of the postmodern era. In light of this rethinking about the nature of music in the larger context of the phenomenology of sound, it makes sense that visual music practitioners should treat their work in a similarly expansive way.

The Russian composer Arseny Avraamov, working in the early half of the twentieth century, represents just such a visionary figure (Izvolov 2009; see also Izvolov 2014). Avraamov opened the door to new musical possibilities and played a seminal role in the birth of image sonification. He actively sought to overturn the limits of Western 12-tone theory and formulated his own a 48-tone microtonal "Ultrachromatic" music system. His exploration of timbre as a primary compositional concern can now be seen as a speculative leap towards the advent of spectral music composition and the future of physical modeling processes. His creative intellect was far ranging and thus well suited to an imaginative and systematic study of the graphic generation of sound. This came about in October 1929 when Avraamov and his colleagues, engineer and inventor Evgeny Sholpo and painter Mikhail Tsekhanovsky, used a magnifying glass to observe the details in photographic patterns that comprised the optical track of the first Soviet sound animated film. The group came to a radical hypothesis that these patterns, transduced through a microphone to the optical track of a film, could be created—directly bypassing the acoustic recording process by utilizing purely graphic means to generate synthetic sound.

Avraamov subsequently formed the Multzvuk laboratory in Moscow, which in addition to Sholpo, included the acoustic engineer and painter Boris Yankovsky and animator Nikolai Voinov. Working in a collaborative capacity and later as individual researchers, this interdisciplinary arts and engineering team established a fruitful area of research resulting in a variety of techniques, tools and machines for the optical creation of sound. Avraamov demonstrated his self-described ornamental sound works in 1930, which employed a visual vocabulary based on elements of Euclidean geometry to synthesize an array of sonic waveforms with modulations of pitch, rhythmic variation and dynamic intensity.

Avraamov's Multzvuk collective appears to be the first to recognize the potential of a graphically generated system of musical sonification; however, they were not entirely alone. Similar experiments were undertaken in the United States, the United Kingdom and Canada. While the scope of optical or photographic sound is not as well known as it should be within the film community as a whole, the experimental filmmakers and animation artists associated with the field of visual music further developed aesthetic application of these ideas through technical experimentation with both hand-drawn and photographic pattern making. In the US, film director and animator Oskar Fischinger made a public presentation of his own version of drawn sound patterns in 1932 (CVM 2014). Norman McLaren, the Scottish-born Canadian animator is renowned for his prolific work at the Canadian Film Board. Among his pioneering achievements was the focused exploration of hand-drawn sound.

Two excellent examples of McLaren's command of hand-drawn and photographic sound are *Dots* in 1940 and *Synchromy* in 1971 (Canada 2012a; see also Canada 2012b). Norman McLaren was a master of painting directly on celluloid film. *Dots* exploits his commanding technique in two separate ways. The image portion of the film and the optical soundtrack are both painted by hand. Based on available documentation of his technical approach, McLaren would make the visual portion of the film first and then line up a second film in parallel to first where he would then draw the changing sequences of patterned shapes that were to become the soundtrack. An important aspect underlying the precision of McLaren's work is that he devised a series of stencil templates that allowed him to accurately repeat the visual pattern of specific pitches and timbre qualities in the sound. The two separate films were then photographically printed as a composite with the second film printed into the optical soundtrack portion of the new composite print. When viewing *Dots*, it is important to recognize that the very tight coupling of sound and image is not in fact a direct sonification of the visual content we see on the film. Instead we are hearing a secondary set of image patterns McLaren drew as a musical accompaniment to the projected visual images.

The technical advancement of a photo-optical approach to sound was taken further by the Russian audio engineer Evgeny Murzin. In 1938, he began development of the ANS, a system that utilized various graphical patterns on spinning glass disks (Kreichi 1997). This early microtonal optical-music instrument was named for the seminal Russian composer, Alexander Nikolayevich Scriabin (A.N.S.). Scriabin, the early modernist godfather of visual music, was a musician and mystic known for his compositional exploration of the interrelationships between color spectra and musical pitch (Academy of Sciences of the RT n.d.). Murzin completed his sophisticated photo-electronic instrument in 1958. A notable feature of the system is that there are multiple spinning disks that can be activated simultaneously at differing user-defined speeds. This allows for a fine degree of controllable pitch modulation and the creation of more complex sounds. This is striking innovation when compared to the use of projector systems that are designed for set frame rates of 24 frames per second.

While a film can be to some degree sped up or slowed down, it is not really capable of the kind of changes in mechanical speed required to exploit a wide range of frequencies. Film-drawn sound techniques thus required varying lengths and sizes of repetitive shapes to create distinctly different pitches. The evocative sound of the ANS was used by the composer Eduard Artemyev in the realization of a series of remarkable films by Andrei Tarkovsky including *Solaris* (1972), *The Mirror* (1975) and *Stalker* (1979). Tarkovsky's interest in electronic sound stands apart from a majority of filmmakers. The following excerpt from his book, *Sculpting in Time*, goes to the heart of the possibilities that photo-electronic music and sound design can achieve:

> Electronic music seems to me to have enormously rich possibilities for cinema. Artemiev and I used it in some scenes in *Mirror*. We wanted the sound to be close to that of an earthly echo, filled with poetic suggestion—to rustling, to sighing. The notes had to convey the fact that reality is conditional, and at the same time accurately to reproduce precise states of mind, the sounds of a person's interior world. The moment we hear what it is, and realize that it's being constructed, electronic music dies; and Artemiev had to use very complex devices to achieve the sounds we wanted. Electronic music must be purged of its "chemical" origins, so that as we listen we may catch in it the primary notes of the world.
>
> (Tarkovsky 1989)

In 1958 during the same time that Murzin was completing the final implementation of the ANS synthesizer, Daphne Oram successfully convinced the British Broadcasting Corporation (BBC) to create the Radiophonic Workshop (Chambers 2008). Oram, who began her career at the BBC as a sound balancer (mixing engineer) for live orchestras, was a pioneering sound designer, electronic musician and visionary inventor. When she was not mixing sound or seamlessly synchronizing recordings for the BBC, she would stay late into the evening to utilize an array of the Network's tape machines to explore the plastic possibilities of a new kind of music. These experiments simultaneously advanced her skills in commercial sound design production using tape-based *musique concrète* techniques for dramatic and often otherworldly sonic effect. She was a brilliant figure who had to endure the difficulties and limitations of working in a largely male-dominated industry. Shortly after the BBC established the Radiophonic Workshop in a cramped space using cast off recording equipment, Oram left to launch her own studio. This is where she began construction of the Oramics machine, which utilized eight tracks of 35-mm filmstrips powered by electric motors run across an optical scanning system. The Oramics device was a multi-track optical synthesizer that allowed the composer to directly draw her musical gestures into sound (TheUntiedKnot 2012). The instrument is also notable as a hybrid system that converted the drawn image into an electronic image and sound using early television cathode ray technology. Oram's first work on the instrument, titled *Contrasts Essonic*, was not recorded until 1968. This and several other works can be found on *Oramics*, an anthology released by Paradigm Disks in 2007 (Oram 2007).

While celluloid film technologies continued to advance well into the late 1900s, analog electronic systems came into their own right. The first television image was transmitted in 1926 using the "televisor" developed by the Scottish inventor John Logie Baird. As television moved out of the engineering laboratories and into the mainstream of competing corporate broadcast companies a new field of entertainment, information and commerce followed. While this new technology was initially out of reach of the avant-garde artists of the era, a singularly creative figure emerged in the early years of commercial broadcast TV. The comedian Ernie Kovacs began his career in radio and then successfully migrated to live television in 1950 (Maggio 2017). Kovacs came to be recognized as a surreal comic genius renowned for his off-kilter approach blending quick-witted improvisation, playful gags and perceptual tricks exploiting both the sonic and visual elements of the emerging medium. In hindsight radio was probably the ideal training ground for Kovacs in that on-air personalities had to rely on verbal spontaneity in combination with learning new ways to manipulate the audio signal chain in a live broadcast system. Likewise, Kovacs's love of classical music, jazz and the humorous potential of sound effects led him to often foreground the sonic potential that television provided. He is remembered for skits that included the reoccurring use of oscilloscopes, live signal manipulation inverting the video signal into a negative black and white image and casting his actors in mounds of unspooled open reel videotape as a self-referential poke at the medium itself. His keen improvisatory mind was at home with the instantaneous medium of live television and he was also among the first to use video editing as an essential technique for expanding the creative potential of his shows (Barbour 2015)

With the advent of videotape recording the repertoire of film editing techniques became possible. The first videotape recorder developed by Ampex Corporation debuted in 1956. These early machines utilized two-inch magnetic tape reels, were extremely large and prohibitively expensive for independent use beyond the major broadcast networks of the time. Similar to what we have observed with the rapid development and dissemination of personal computing technology, early analog video systems underwent rapid evolution. This history of technical innovation involved many people and is a dramatic story of corporate intrigue, financial failure and success. A pivotal moment came when Sony released the first Portapak video systems in 1967. Portapak, as its name suggests, was a portable video rig consisting of an analog tube camera and a half-inch reel-to-reel video recorder. This breakthrough empowered the first wave of video artists and subsequently new possibilities for the integration of real-time image and sound processes.

The early history of video art begins in the 1960s, developing several primary areas of inquiry that laid the foundations for the practice of contemporary media art. These include video as a performance art medium, video as an extension of sculpture, video as a means to produce activist documentaries and essays, and video as a real-time, instrumental system for image manipulation and abstraction. Among the leading pioneers of video as an art medium are Steina and Woody Vasulka. Internationally renowned as new media visionaries, their early work drew many artists into their studio, leading to their living-space-cum-performance-venue being dubbed "The Kitchen." The original building is long gone but The Kitchen remains as a prominent intermedia-oriented arts center in New York City. Steina credits Woody with a fearless quest to push, probe and ultimately catalog the complexities of real-time video. Woody very directly took the audio signal output and plugged it into the video monitor input. This simple revolutionary act began a lifelong aesthetic interrogation of "the signal" as a means of techno-poetic expression (Vasulka n.d.).

Steina, trained as a classical violinist, and Woody, a graduate of the FAMU film academy in Prague, were both interested in music, sound and image. With video, they had discovered an image-making paradigm with the fluid attributes of sound. Shortly thereafter they purchased an EMS Putney synthesizer and interfaced individual sound oscillators to control both the vertical and horizontal frequencies that govern the physical location and dimensional qualities of the video frame. Thus, they set the image into a free-floating temporal space. Steina and Woody followed up these experiments by purchasing a number of identical television monitors so that the image or the visual artifacts of the signal itself appears to move from one screen to the next. These multi-monitor assemblages were influential at the time, inspiring the color-saturated, multi-monitor sculptural installations of Korean-born video artist Nam June Paik. One can argue that this is not image sonification but rather sonic control of the image. This is partially correct but more importantly, when looking at electronic systems and related digital processes, it is the totality of the system that permits the articulation of both audio and visual elements.

Steina says that she and Woody have three modes of working (Vasulka 2018). The first is sending the recorded video signal out into the audio synthesizer to interact and control the image. The second is to use a pre-recorded sound of varying complexity to control the behavior of the video signal during the video recording process. The third is to combine both of these approaches at the same time, allowing both elements, sound and image, to control and affect each other.

As an intrepid artistic team, Steina and Woody pushed forward into new aesthetic terrain by bringing technicians and engineers into their creative dialogue. This began with making modifications to existing hardware and evolved towards custom design of digital circuitry and robotic devices. In the process, their collaborative and independent works opened up new avenues for interactive media performance, live cinema, video-graphic signal processing, robotics and immersive projection environments. In 1996, Steina was invited to be the artistic director of STEIM, the Studio for Electro-Instrumental Music in Amsterdam, where she met the programmer-engineer, Tom DeMeyer (STEIM n.d.). DeMeyer approached Steina with the idea to develop a software-based video synthesizer and processor that could be run on a personal computer. DeMeyer had already been instrumental in the creation of Big Eye, a video sensor-to-MIDI converter that allowed for mapping discrete regions of a video image, often from a live camera, used to trigger and control any MIDI-capable device. This machine vision application oriented towards artists, musicians and dancers provided a simple platform for image sonification. The result of this collaboration was IMAGE/ine, which arrived at the end of 1997 and was arguably the first video synth instrument to be modeled in software. An important but little used feature of IMAGE/ine was the ability to continuously quantize the values in any given row or column of pixels as sample values in an audio waveform oscillator. This raw and immediate sound source was a direct sonification of the image that held many possibilities for sound artists and performers (Demeyer 2011).

Parallel to artists working with analog audiovisual instruments in the 1960s and 1970s, technical development was rapidly proceeding with digital systems as well. While this technology was largely in the provenance of well-funded research centers, often supported by military and other wealthy corporate patrons, it was nonetheless recognized by a knowing few as the probable future of electronic sound and image making. Among these visionary thinkers was composer and mathematician Iannis Xenakis. Xenakis originally relocated to France from his home in Greece and subsequently took an architectural engineering position with the team lead by the celebrated architect Le Corbusier. Xenakis went on to compose *Concret PH* as a contribution to the entryway of the Philips Pavilion designed by Le Corbusier for the 1958 Brussels World's Fair (Archibald 2011; see also Vas 2015). Xenakis's interest with the interactions of spatial, graphic and sonic constructs is reflected in his compositional use of musical motion, sound mass and velocity often facilitated by multi-channel speaker arrays, and graphic notation systems. A significant outgrowth of his interdisciplinary aesthetic experimentation was the development of UPIC (Unité Polyagogique Informatique CEMAMu) in 1977 at the Centre d'Etudes de Mathématique et Automatique in Paris (Arkeion 1982; see also Xenakis-Collection 2012). UPIC is a digital system that utilizes a digitizing pen and tablet interface to draw musical shapes, waveforms and envelopes on a vector display monitor. These visual waveforms were digitally translated into sound that could then be composed using a drawing technique similar to many contemporary software environments, where the x-axis represents time and the y-axis represents pitch. The UPIC instrument was an early example of the capacity of digital systems to create hybrid instruments. UPIC was simultaneously a system for sound design, music composition and live performance allowing the user to improvise sound by drawing. The immediacy of this system gave the composer the ability to work with musical material directly and challenged, to some degree, earlier strictures of Western music methodology by providing a new tool for the organization of sound.

As mentioned in the introduction to this chapter, digital image sonification is really a subset of data sonification as a whole. A still image and/or a moving image is represented within a computational system in numerical form. As a simple technical concept, there is not a great deal of difference between a numerical table describing arctic climate patterns, ocean current fluctuations or a sequence taken from a Charlie Chaplin film. This is not to say that all data sets or matrices are remotely the same when considering the individual characteristics attributed to any given data source. The importance here is to consider numerical data as a kind of concrete material that can be used for dramatic expression. This idea forms the elementary underpinning of why the term "data dramatization" more aptly describes the combined fields of sonification and visualization: it does not preclude the use of data to drive electromechanical or other higher-level system interactions for the purpose of revealing something intrinsic about the underlying data itself. This approach has obvious potentials in the arenas of science and engineering in addition to stimulating creative applications in music, cinema and performance. In the hands of a conceptually minded artist, data dramatization can be fundamental to making critical observations regarding the socio-historical-political context surrounding the use of the source data being employed. For those interested in sound design applications using image sonification, it is important to examine the contemporary practice of data-driven sound and image works with an understanding that this is a field still in its infancy.

For the beginning sound designer, whether working in the context of film, theater or data sonification, there is an immediate confrontation with the complexity involved in combining two or more powerful mediums. One has to experiment and explore the ways in which sound can clarify the meaning of an image and/or create a new meaning that the image alone does not convey. This is a very subjective process. In the event that one is working for hire on a film or related media or theatrical project, the sound designer and composer must satisfy the demands of the director, producers and possibly a test audience as well. For sound and media artists who are pursuing their own independent work, this is not a primary consideration. Thus, what constitutes the successful wedding of sound and image in terms of what an artist wishes to communicate is solely at their personal artistic discretion. In either case, whether a commercial project or an independent work, the sonic artist will respond from their own imaginative intellect when designing soundscapes, musical themes or expressing data as an auditory experience.

A fundamental consideration when approaching music composition and sound design for data dramatization is that this medium still requires a significant element of human creative interpretation to successfully communicate or reveal the data's content. Specifically, I want to emphasize that a data table derived from any given image component has musical potential but no inherent sound of its own other than fluctuating patterns, peaks and valleys represented by the changes, spikes or stasis of numerical values. At a fundamental level, the sonification sound designers must first assess the nature of the data as defined by the ranges, repetitions and boundaries inherent in the assigned numerical values. Even as the artistic, scientific or engineering goals of any given sonification project can be vastly different, the sonification artist must become an expert at articulating a sonic expression from the data. Similar to the task an instrumental musician undertakes when approaching a musical score, image sonification is an act of interpretation, and similar to the composer, every sonification project requires attention to orchestration and decisions as to what frequency ranges and pitch systems might best serve the project. As a general observation, we can refer to these combined compositional design elements as mapping decisions. Mapping a sonic expression to an image or to any potential data stream may or may not have a musical objective. More succinctly, the process of making such a map requires clarity of aesthetic purpose no matter the end goal.

For both the student and professional already involved in image sonification practices, there will be some understanding of the creative intention and scope of possibilities available through existing technologies. It is helpful to delineate a few of the possible goals an image sonification project might have before focusing more exclusively on some of the specific techniques that can be employed in the design and performance of interactive audiovisual works. As image sonification is a means to transform one sensory modality into another, it has applications as a potential aid for the blind. Projects such as PictureSensation, a mobile app for generating real-time, hapto-acoustic response to live camera imagery, demonstrate the functionality of mapping one sensory modality to another. This example illustrates the need for a thorough analysis of the salient features contained within a photographic image in relationship to the needs and interests of the intended user. The careful determination of which picture elements or qualities should be available for sonification is followed by the equally important task of determining the nature of the sounds to be used as visual/physical signifiers (Banf et al. 2016).

One of the most well-known software approaches utilizing two-dimensional imagery to create and manipulate sound is MetaSynth, which was released for the Macintosh platform in 1999 (U&I Software 2018). The software designer, musician and artist Eric Wenger has also developed a number of generative animation and modeling applications released by his company, U&I Software. A central feature of MetaSynth is the ability to import and/or directly draw images into a timeline. The toolset is similar to many two-dimensional paint and drawing programs that afford the user various virtual brush and pen shapes in addition to common cut and paste editing features. Working with MetaSynth one can see direct user interface analogies to the hand-drawn ornament sound experiments of Fischinger and the Oramics of Oram. These earlier methodologies used lengths of celluloid film as both a medium and interface for composing graphical gestures and patterns to structure linear time. In a similar way, MetaSynth also imitates the use of graphic notation represented by scores such as *Treatise* by Cornelius Cardew or *Artikulation* by György Ligeti (Craig 2007; see also Shubin 2013). Composers most often use graphic musical scores as a method to articulate a new approach to musical materials that require interpretation by skilled instrumentalists and singers, whereas software applications like MetaSynth allow for a more direct visual sculpting of sound. Various optical sonification instruments will have their own unique approaches and sound. MetaSynth utilizes a subtractive synthesis method whereby the resulting drawn or imported photographic imagery is used to both filter and shape the amplification envelope of the various available sound waveforms. This is accomplished by mapping the image's horizontal axis to time, vertical axis to audio frequency and pixel brightness to amplitude.

MetaSynth owes much to the concept of graphical synthesizers, particularly the ANS synthesizer and the UPIC system. A more direct interpretation of the ANS machine is a contemporary software recreation, Virtual ANS, developed by Alexander Zolotov as a multi-platform sound generation application (Zolotov 2016). The Virtual ANS software enables the user to draw atmospheric textures for sound design and musical works and provides a means to produce sonograms. A sonogram can be thought of as an inverse to image sonification as it is instead the visual representation of sound as an image. Just as the work of Steina and Woody Vasulka moves effortlessly between sound as image and image as sound, digital transcoding systems are inherently bidirectional. This aspect of digital transcoding allows sound artists the possibility to transform and manipulate elemental sonic properties as a sequential series of transforms. The transcoding process begins with the initial sonification of an image, renders this resulting sonogram and then visually augments or edits the sonogram to produce an altered sound. This additive/subtractive process can be repeated *ad infinitum* to arrive at subtle variations or completely different kinds of transformed sonic material.

Our overview has thus far covered a variety of technical approaches to image sonification with differing goals in mind. In some cases, graphic images are drawn in sequence to generate a dynamic sound or music work with a distinct focus on the resulting sound produced. In other examples taken from the lineage of abstract visual cinema, there is a desire to create a multisensory experience with direct aesthetic linkage of sound to image. A tangentially related approach is the use of programs such as Big Eye, Isadora and a host of others, including surveillance systems, which allows one to map regions within a live video frame and use the resulting data as a method to generate control and/or trigger signals (STEIM 2012; see also Troikatronix 2017). These data streams can be used to generate sound in a variety of ways but can just as easily be employed to activate some other function. The important point being that this form of image sonification is not specifically limited to the shape, color or contrast of a graphic form but rather encompasses a far more detailed set of possibilities available for image analysis used principally for control purposes.

The early history of optical synthesis was essentially two-dimensional in nature, including hand-drawn shapes and patterned forms cut into paper. In contemporary practice, we can include the use of live imagery from video camera sources including infrared depth cameras such as the Microsoft Kinect as well as the modeling of virtual three-dimensional images. Three-dimensional image sonification presents additional complexities and exciting aesthetic possibilities. An important fundamental distinction is the consideration of whether an image is static or in motion. Moving images by their nature produce dynamic sound, unless, of course, one is exploring the most extreme form of minimalism where absolutely nothing changes within the frame. Andy Warhol's film, *Empire* (1964) comes to mind; however, on closer look this would be an ideal film to illustrate how an image-sonification-based score might enhance the filmmaker's objective as the extremely slow visual development is all about the experience of "real time" in contrast to the compressed or elliptical nature of time represented in most narrative films.

Minimal aesthetic approaches often draw attention to the subtle changes that occur over time—changes that we often ignore. In contrast, the direct sonification of an absolutely still, unchanging image will emit an unchanging tone. To create sonic dynamism utilizing a still image, the sono-visual artist will likely create a system for the sequential scanning of designated regions of the still image to produce variety in the sonic output. This idea has relevance in the use of eye-tracking systems to generate and trigger sound. Eye tracking can be seen as a subset of machine vision applications. This brings up an additional aesthetic parameter when considering interactive image sonification—that being the totality of a cybernetic system inclusive of the human element. This includes a consideration of where in the signal chain or network the artist, maker, designer, performer is placed and also where in this systemic totality the viewer-listener-participant is placed.

## 9.3 NoiseFold and Contemporary Practices

In my own work, all of the aesthetic concerns detailed above are explored to varying degrees in the creation and exhibition of cinematic installations and performances utilizing image sonification as a central method of sound design and music composition. This creative inquiry began with my initial study of visual art and dance, which ultimately led to a focus in music and video as an integral means to reimagine the combination of disciplines. I was fortunate to begin these studies during that moment when analog systems were giving way to digital instruments. After some initial experimentation with hardware-based sonification processes using hybrid analog video and sound synthesis systems in the 1980s, I turned my attention to designing sound and scoring music cues for films, installations and theater works using a blend of synthetic and acoustic source material. The fact that I am a practicing video artist skilled as both an editor and director gave me a unique perspective on the ways sound and image can be conceived of as an interdependent whole. This background in making music for film, theater and dance has proved invaluable for the sound design aspect of my advanced work in live cinema and data bending, specifically as it relates to generative sonification and visualization techniques.

After nearly two decades of creating linear performances and fixed video, installation and music works, I began to explore new possibilities in interactive media. This was a natural development coming out of my combined interests in live video and improvisational performance. This early interactive exploration was largely facilitated by the use of hybrid MIDI systems, where musical instruments could be interfaced to control aspects of the digital video image. A simple but effective example comes from a moment within a feature length performance, titled *House of Shadows*, (2003) where I mapped individual keys on a MIDI keyboard to the frames in a video sequence that depicted a close-up of a dead cottonwood leaf skittering and clattering loudly down the street. Using the keyboard, I could reanimate the visual continuity within the video sequence, playing the synchronous diegetic sound to intensify the naturally occurring phenomenon into a ratcheting industrial tumult, while simultaneously triggering a softer contrapuntal layer of sampled instrumental sounds. I do not consider this image sonification in the strictest sense but the immediacy and variability of the system led the way towards ever more fluid integrations of music, sound and image.

The *House of Shadows* project included a number of different approaches to combining contemporary music and video art utilizing visual abstraction to augment and reconfigure a largely narrative and photographically figurative work. In short, this was a culmination of a more traditional approach to experimental film and music. It was also the first time I began to work with my now longtime collaborator, Cory Metcalf, who came on the tour to mix and process the multi-channel sound. I made many technical advances in this particularly demanding project. By the end of a fraught, albeit liberating, European tour that included fire shooting from the outputs of our video mixer, I resolutely committed to strip everything down to the bare essentials. No prerecorded video performers, no live performers other than myself, no video editing or painstaking animation, no video images or music in a conventional sense, no MIDI controllers, no text in any form, no heavy road cases, no extraneous drama—just a laptop, a single multi-effects guitar pedal, a USB volume-control knob and a lavalier microphone with the tip of the output mini-jack filed away to allow the built-in laptop speakers to remain active while the mic was in use. A strict set of limitations is often an effective way to distill a concept and break new ground.

Utilizing the IMAGE/ine software environment, which I had steeped myself in while co-teaching an early interactive media course with Steina, I began to explore the sonification of grayscale (black and white) video noise (Stout 2013). This technique allowed the user to select a single scan line from the video image and use the resulting data as the waveform of a digital sound oscillator. The oscillator could be tuned to specific pitches over a wide frequency range. The potential complexity of the resulting waveform could very easily result in wild harmonic fluctuations, dependent on the nature of the image comprising the video signal. Obviously, such a direct method of the sonification of generative video noise will produce an analogous audio noise. To make this system interactive with a human performer, I activated the built-in laptop mic and taped the second lavalier microphone to one of the built-in laptop speakers to produce a feedback system. I then fed the output of this video-generated dynamic wavetable oscillator to the laptop speakers and then fed the microphone signals back into the visual processing component of the software to selectively filter, rescale, rearrange and displace the pixels of video noise. This audiovisual feedback system could then be controlled by directly cupping the built-in laptop speakers and microphones with the performer's hands to form a primitive yet effective volume and filtering mechanism. What began as an exercise in basic noise-induced chaos produced a rich body of generative audiovisual behavior that successfully integrated image sonification and sonic visualization as an interdependent whole. From this work, I produced several installations and a two-part performance work, titled *SignalFire* (2004).

The primary sonic technique described above is referred to as scanned synthesis. The term comes from the fact that the waveform can be scanned at a user definable frequency to determine and modulate pitch with the added effect that this also influences the perceptual experience of the timbral motion of the waveform. The implementation of scanned synthesis is credited to Max Matthews, Rob Shaw and Bill Verplank, who developed the technique in 1998–99 while working at Interval Research Inc. Interestingly, this feature was included in IMAGINE/ine, which was developed in 1996–97, preceding the work of Matthews's team. A simple technical overview of scanned synthesis or scan-line synthesis, as it is also called, involves the digital representation of visual pixels as a sequence of numerical values (Lucas and Stout 2018, Figure 9.1).

In a hypothetical example, using a relatively low-resolution quantization range of 1–128, the system is designed to measures differing intensities of the grayscale content of individual pixels in a user selectable area of a video image. This selectable region could be a contiguous row or column of pixels or a larger region consisting of both rows and columns. If the image is, for example, a brightly lit white bed sheet, most of the pixel values will be 128. If the image is instead a dimly lit black asphalt street the image values will hover closer to 1. In many cases the

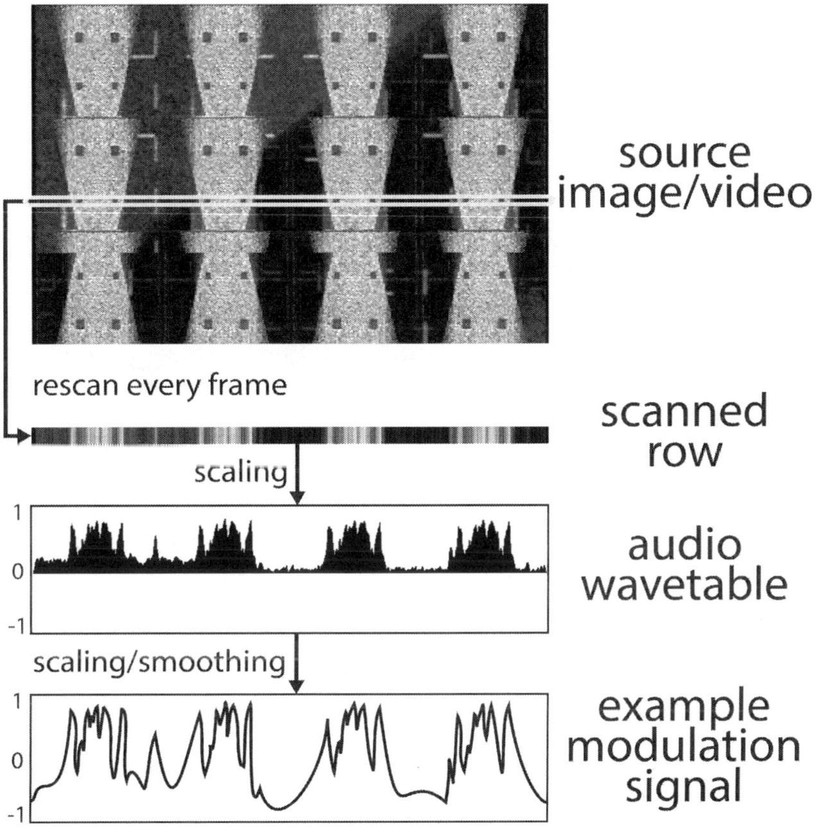


**Figure 9.1** Diagram Illustrating the Process of Selecting Pixel Region of a Single Video Frame and Quantizing Values to be Interpolated as Audio Waveform Data. (Image created 2018 by Lucas and Stout)

user could input a more complex image and the series of adjacent pixel intensities will fluctuate accordingly giving a variable numerical stream as the basis of constructing an audio waveform. Once the user determines the region for sonification the selected pixels are translated into numerical data points, which are reproduced at audio frequency rates. The result is a digital oscillator based on an analogous visual image. The oscillator allows the user to manipulate the frequency or cyclical recurrence of the waveform data to play back the sound at various pitches. In general, scan synthesis of this nature often produces a very edgy or harsh sound. The wider the numerical difference between adjacent pixels the noisier and biting the audio waveform. The example above quantizes the image to limit the number of pixel gradations. If the dimension of the selected visual area is increased and a higher sampling rate is selected the system will produce a smoother sounding oscillation. For this example, I refer to the selection of a fixed region of grayscale pixels. The selected region does not necessarily need to be fixed, as any coordinate position can be dynamically scaled and relocated over time. The sonification system can, likewise, utilize a full color matrix that is designed to measure individual amounts for red, green or blue available at each coordinate position. In addition, the software can feature additional processes that include averaging adjacent pixel values to smooth the data to avoid unwanted audible spikes that might be injurious to speaker systems and eardrums. Taking this kind of digital processing further, the system can be programmed to shape the resulting audio waveforms in a variety of meaningful aesthetic ways.

The next phase of my image sonification research took shape as a generative audiovisual installation titled *The 100 Monkey Garden* (2005). The project employed real-time three-dimensional computer graphics to populate a speculative synthetic ecosystem using a combination of genetic and ecological models. I worked closely with artist-composer-programmer Luke DuBois, who programmed virtual three-dimensional visual forms (NURBS) in MAX/MSP/Jitter software that could emit their own respective sounds. This required a new software oscillator based on three-dimensional geometry that could selectively mix and generate three numeric data streams, one each for the x, y and z planes. This dynamic three-dimensional wavetable oscillator had an additional FM synthesis capability. Hypothetically this FM input could allow for the modulation of one plane against the other or input a modulation source from another independent three-dimensional visual NURBS object.

When looking at the history of graphic sound synthesis, one sees that the digital evolution of real-time three-dimensional image generation has brought with it an important new feature set: the ability to bend, fold, scale

**Figure 9.2** Still Image Taken From the Artist's Early Live Animation Studies That Explored the Interplay Between Generative Digital Image Synthesis and Sonification Techniques. (Image created 2005 by Stout)

and reshape visual forms on the fly, giving the artist-composer the ability to sculpt image and sound simultaneously (Stout 2005, Figure 9.2).

This effectively means that a highly plastic image processing system can become a versatile interface for the design and performance of sound and music. I want that idea to sink in for a moment. Consider that any change to the image not only modulates the scanned oscillator waveform but can also be mapped to dynamically sculpt harmonic content, pitch, amplitude envelopes and any other conceivable digital signal processing that is aesthetically relevant to the work being produced. The *100 Monkey Garden* installation was an automated generative system that utilized a bank of 16 digitally modeled low frequency oscillators and a variety of rule-based triggers to create a wide range of audiovisual behaviors. What became immediately obvious was that we had also prototyped a live performance system.

In 2005, Cory Metcalf and I created NoiseFold, an interactive media duo with a focus on the creation, performance and exhibition of generative sound and image works (NoiseFold 2013). The results of our combined research, music and art-making activities include the development of a hybrid audiovisual synthesis system modeled in MAX/MSP/Jitter, a well-known graphical programming environment. The first iteration of our nFolder software evolved into a sprawling multi-screen GUI that was primarily focused on live performance applications utilizing combined two- and three-dimensional image-to-sound transcoding techniques. As NoiseFold progressed, so did our software, and we eventually reached a mutual decision to scrap version 1.0 of our monolithic sensor activated performance instrument and redesign the software from the ground up as modular environment similar in many respects to hardware modular systems (Stout 2014). This facilitated several new directions as NoiseFold is equally involved in audiovisual performance, music composition, installation, virtual reality and sculptural design (NoiseFold 2010, Figure 9.3).

What essentially began as a duet between two monophonic single voice laptop instruments evolved into a sonographic sound design environment that facilitates a synthesis of skills including visual design, virtual architecture, procedural animation, music composition, instrumental and live cinema performance (Stout 2015b).

A modular system can facilitate all of these activities without the need to reprogram a given digital process from scratch every time we begin a new project. From a sound design perspective, this also means that we can cultivate a growing list of sound modules that can be recombined in various processing configurations. MAX/MSP/Jitter treats visual data as a

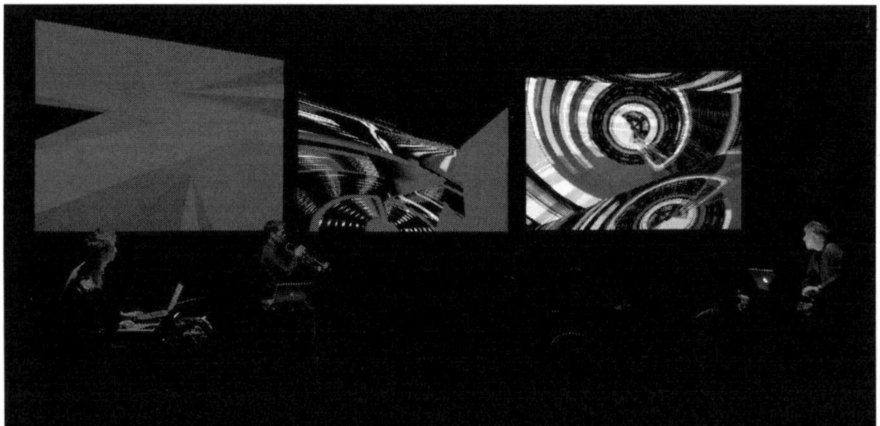

**Figure 9.3** 2010 NoiseFold performance exploring real-time generative image sonification at the Merrill Ellis Intermedia Theatre—NoiseFold, from left, Cory Metcalf, David Bithell (guest musician, trumpet) and David Stout.

series of multi-planar grids. Each plane represents a different scalar value such as luminance, color or three-dimensional axis position. Each coordinate is thus accessible as a graphical pixel, three-dimensional vertex or control point. This makes graphical data interchangeable and well suited to modularization of both media matrices and control signal content. The two previously described, two- and three-dimensional scanned synthesis oscillators, can be interconnected or can form the basis of a larger processing chain that includes a variety of filters, granulators, pitch shifters, resonators, pitch quantizers, delays, and so forth. Using a flexible system means that any given image element can be simultaneously sonified using a multiplicity of techniques. This allows for polyphonic multi-timbral possibilities that can be selectively synchronous or asynchronous to the graphic element generating the sound.

Cory's growing skill as a virtual instrument builder and programmer coupled with our mutually shared restless imaginations has pushed NoiseFold forward into new methods of image sonification. These techniques include modules for dynamic video image-based spectral filters, particle trackers, virtual attractor modeling and an image analysis-based triggering system for activating MIDI events and complex envelope generation (Stout 2015a). While a dominant approach preferred by many media artists and composers is to develop a dedicated tool that will be used in a specific conceptual way within any given art, sound or music work, NoiseFold has sought to create a software environment drawing on multiple techniques as a means of exploration and discovery of hybrid potentials. This approach not only gives rise to tightly integrated, audiovisual relationships but also allows us to orchestrate and blend multiple layers of sonification from the same visual data sources. The result takes various forms including richly layered, multi-voice textures that can be dynamically mixed to fluidly transform sound worlds on the fly using manual mixing or automation data drawn from the visual process itself (Stout 2016, Figure 9.4).

The MAX/MSP/Jitter based nFolder instrument presents a wide field of possibilities for sound making and composition that is intrinsic to the facilitation of poetically engaged audiovisual mapping decisions. The system has been designed as an integral audiovisual synthesis tool, providing a hardware and software interface for the dialogue between image making and sound making. Experienced as an exploratory vehicle, these mapping decisions can happen quickly and be described often as an intuitive process. As a term, "intuition" can at times appear vague and imprecise, but in fact a knowing intuition is experienced by the artist-thinker-performer as quite the opposite. Furthermore, for an intuitive compositional approach to function well, such as the remarkable musical interchange of a seasoned jazz ensemble, the audiovisual composer must have developed a personal

**Figure 9.4** A Video Still Image Taken From The Janus Switch (2016), a Solo Performance Work That Utilizes 3D Image Feedback Techniques to Generate Audio Waveforms and Rhythmic Structures to Create a Wide Range of Tonal, Atonal and Noise Textures.

vocabulary of sonic to visual and visual to sonic relationships. This can be illustrated by considering somewhat improbable questions such as What is the sound of fleeting shadows cast upon an opaquely luminous veil? What is the sound of two jittering agitated visual forms being repelled by each other? What is the sound of a soft blue distance that is slowly enveloping the foreground of the visual frame? When working on a NoiseFold project, these questions come fast and furious. The aesthetic answers or creative strategies required for any given work often rely on the dialectic between an accumulated knowledge of musical styles, skills with sound design and synthesis techniques, and a seasoned understanding of drama for stage and screen to serve the visceral phenomenological experience of sound itself. In other words, I may draw on a stylistic approach that carries an underlying narrative concern, including symbolic, historical or dramatic references, or I may wish to create a sound world that is a solely distinct to the physicality of the object being sonified without any direct reference to a previous sonic meaning. While these ideas can be viewed as decidedly narrative, at its core my aesthetic orientation is a significantly different approach than the sonic strategies often taken in a Hollywood film, where orchestral cues and sound effects are based almost entirely on the repetition of previous references that are largely designed to be familiar signifiers to the viewer-listener. With that said, collectively most of us have been steeped in pop culture, which includes a daily diet of illustrative sound and musical cues as part of radio, TV, internet media and game content. We are so accustomed to consuming the narrative cues imbedded in the soundtrack that we hardly notice the subtle and not so subtle manipulations of meaning at play. Seminal thinker and composer John Cage decidedly stepped outside this kind of narratively based, illustrative use of music and sound to explore a deeper aspect of being a listener discovering the mutable variations within the vibratory nature of the sonic experience itself. His credo, written in 1937, proves remarkably prescient and meaningful to this day (Kostelanetz 1991). At its creative core, NoiseFold seeks similar aesthetic goals, to place the viewer-listener-participant in a compelling state of audiovisual vibratory immersion where conventional descriptive language is at a loss to fully describe the experience even as the audience may come away with its own story and a sense of having traveled far.

The NoiseFold dialogue between sound and image is also a quest for technical advancement facilitated by the design of machine systems that continually push against the limits of computational power. We are interested in future work where one can additively sonify the individual behavior of thousands of particles to create highly detailed sound clouds on a massive scale. The behavior of such large particle fields can be choreographed using algorithms that model a wide range of physical phenomenon from herd migration, flocks, schools and swarms to celestial phenomenon and subatomic interactions. There are obvious technical and aesthetic overlaps between the sonification concerns of a scientific research project and more poetic musical applications. As computational power advances and we move forward with new software and hardware solutions, it will be easier to imagine portable systems that can effectively sonify and visualize complex simulation scenarios in real time.

At this historic juncture, we can apply the visionary experiments of the past with the potentials of contemporary sound design and compositional aesthetics. This inclusive openness feeds NoiseFold's speculative vision well. Early on we discovered the fascinating sonic detail generated by a highly complex dynamically morphing visual form. While we could choose to quantize the resulting frequencies into a specific harmonic scale, this limited notion of what music is or can be would have trivialized the power inherent in the images complexity even as it would lend itself to a certain Disneyesque whimsy. By listening to the ever-changing noise frequencies, we could detect unique moments in the virtual image with our ears that that we could not actually see visually. Thus, we find that noise, microtonality, tonality and atonality can exist simultaneously or sequentially in our work. This effective use of signal processing allows us to tune and polish the output of our visually derived oscillators to create familiar timbres mimicking acoustic instruments and environmental phenomena alongside moments that reveal the primal raw electronics of the unadorned oscillators.

One can only imagine how Luigi Russolo would respond to being handed such a system, where one can perform a live painting that generates sound from the harshest noise to the softest atmospheric murmurs laced with fragmentary melodic clarity that dissolves back into the ephemeral silence of the image alone. Such a system erases the disciplinary boundaries between composer, visual artist and performer.

## Acknowledgements

Special thanks to Stephen Lucas for his technical expertise and invaluable assistance with source material research, to Cory Metcalf for his insight and devotion to this work, to Steina for her candor and lucid descriptions of technical and poetic processes and to Nila Velchoff for her assistance with this text.

## References

Academy of Sciences of the RT, n.d. *Prometheus Institute*. [Online] Viewed May 23, 2018 <http://prometheus.kai.ru/archibald>, 2011. *Phillips Pavilion*. [Online] Viewed May 23, 2018 <http://architectuul.com/architecture/philips-pavilion>.

Arkeion, G., 1982. *Xenakis Aujourd hui en France N° 095*. [Online] Viewed May 23, 2018 <www.gaumontpathearchives.com/index.php?urlaction=doc&id_doc=272896&rang=5&langue=EN>.

Banf, M., Mikalay, R., Watzke, B., and Blanz, V., 2016. PictureSensation—a mobile application to help the blind explore the visual world through touch and sound. *Journal of Rehabilitation and Assistive Technologies Engineering* 27 October, no. 3, pp. 1–10.

Barbour, J., 2015. *John Barbour's Documentary of the Legendary Ernie Kovacs*. [Online] Viewed May 23, 2018 <www.youtube.com/watch?v=HsGyL8y83G8&index=8&list=PL72H2XFA5y3oZgOOLJWuBgw84e2Q8lnyi>.

Canada, N. F. B. o., 2012a. *Dots*. [Online] Viewed May 23, 2018 <https://vimeo.com/32645760>.

Canada, N. F. B. o., 2012b. *Synchromy*. Viewed May 23, 2018 <https://vimeo.com/29399459>.

Chambers, I., 2008. *Wee Have Also Sound-Houses*. [Online] Viewed May 23, 2018 <www.youtube.com/watch?v=NNaqvAH7R34>.

Craig, D., 2007. *Ligeti—Artikulation*. [Online] Viewed May 23, 2018 <www.youtube.com/watch?v=71hNl_skTZQ>.

CVM, 2014. *Ornament Sound Experiments by Oskar Fischinger (c. 1932)*. [Online] Viewed May 23, 2018 <https://vimeo.com/ondemand/26951>.

CVM, n.d. *Center for Visual Music*. [Online] Viewed May 23, 2018 <www.centerforvisualmusic.org/>.

Demeyer, T., 2011. *ImX (Image/ine for OsX)*. [Online] Viewed May 23, 2018 <https://image-ine.org/>.

*Empire*, 1964. [Film] *Directed by Andy Warhol*. New York: Factory Films, Inc.

Frieze, 2013. *The Orchestra of Futurist Noise Intoners*. [Online] Viewed May 23, 2018 <https://vimeo.com/46083321>.

Izvolov, N., 2009. Designed sound in the USSR. *KinoKultura*, no. 24.

Izvolov, N., 2014. From the history of graphic sound in the soviet union; Or, media without a medium. In: Salazkina, L. K. a. M. (Ed.) *Sound, Speech, Music in Soviet and Post-Soviet Cinema*. Bloomington, IN: Indiana University Press, pp. 21–37.

Kostelanetz, R., 1991. *John Cage: An anthology*, New York: Da Capo Press.

Kreichi, S., 1997. *The ANS Synthesizer: Composing on a Photoelectronic Instrument*. [Online] Viewed May 23, 2018 <www.theremin.ru/archive/ans.htm>.

Maggio, G. C., 2017. *Best of Ernie Kovacs*. [Online] Viewed May 23, 2018 <www.youtube.com/watch?v=lWdEr1XaLrw&list=PLURt_7PLeB6GdKQBoDxk10sNHdNntHpps>.

Moritz, W., 2004. *Optical Poetry : The Life and Work of Oskar Fischinger*. Bloomington, IN: Indiana University Press.

NoiseFold, 2013. *NoiseFold*. [Online] Viewed May 23, 2018 <http://noisefold.com/>.

Oram, D., 2007. *Oramics*. [Sound Recording] London: Paradigm Discs.

*Oskar Fischinger: Visual Music*, 2017. [Film] United States: Center for Visual Music.

Russolo, L., 1986. *The Art of Noises*. New York: Pendragon Press.

Shubin, D., 2013. *Cornelius Cardew: Treatise*. [Online] Viewed May 23, 2018 <www.youtube.com/watch?v=JMzIXxlwuCs>.

*Solaris*, 1972. [Film] Directed by Andrei Tarkovsky. Soviet Union: Mosfilm.

*Stalker*, 1979. [Film] Directed by Andrei Tarkovsky. Soviet Union: Mosfilm.

STEIM, 2012. *BigEye*. [Online] Viewed May 23, 2018 <http://steim.org/2012/01/bigeye-1-1-4/>.

STEIM, n.d. *STEIM | Studio for Electro-Instrumental Music*. [Online] Viewed May 23, 2018 <http://steim.org/>.

Stout, D., 2013. *Genesis in Noise*. [Online] Viewed May 23, 2018 <https://vimeo.com/52901327>.

Stout, D., 2014. *Alchimia*. [Online] Viewed May 23, 2018 <https://vimeo.com/73533076>.

Stout, D., 2015a. *Aludel of the Dawn Albedo*. [Online] Viewed May 23, 2018 <https://vimeo.com/119020472>.

Stout, D., 2015b. *David Stout and Cory Metcalf : The Art of NoiseFold*. [Online] Viewed May 23, 2018 <www.academia.edu/10037070/David_Stout_and_Cory_Metcalf_The_Art_of_NoiseFold>.

Tarkovsky, A., 1989. *Sculpting in Time : Reflections on the Cinema*. London: Faber.

*The Mirror*, 1975. [Film] Directed by Andrei Tarkovsky. Soviet Union: Mosfilm.

TheUntiedKnot, 2012. *Daphne Oram Oramics Machine Feature on BBC Click 8/1/12*. [Online] Viewed May 23, 2018 <www.youtube.com/watch?v=7cyHFT2abXE>.

Troikatronix, 2017. *Isadora*. [Online] Viewed May 23, 2018 <https://troikatronix.com/>.

U&I Software, 2018. *MetaSynth 6 for Mac OS*. [Online] Viewed May 23, 2018 <www.uisoftware.com/MetaSynth/>.

Vas, V., 2015. *Iannis Xenakis-Concret PH (1958) HD*. [Online] Viewed May 23, 2018 <www.youtube.com/watch?v=PY3cB3E-Ts0>.

Vasulka, S., 2018. *Personal Communication with David Stout* [Interview] (April 2018).

Vasulka, S. a. W., n.d. *vasulka.org*. [Online] Viewed May 23, 2018 <www.vasulka.org/>.

Whitney, J., 1980. *Digital Harmony: On the Complementarity of Music and Visual Art*. Peterborough: Byte Books.

WikiArt, n.d. *Luigi Russolo*. [Online] Viewed May 23, 2018 <www.wikiart.org/en/luigi-russolo>.

XenakisCollection, 2012. *Upic Xenakis*. [Online] Viewed May 23, 2018 <www.youtube.com/watch?v=7_Gu0qDAys0>.

Zolotov, A., 2016. *Virtual ANS*. [Online] Viewed May 23, 2018 <www.warmplace.ru/soft/ans/>.

10

# Sound and Wearables

**Johannes Birringer and Michèle Danjoux**

## 10.1 Introduction

The idea of "wearing sound," and of designing sound to be worn on the body or on moving objects, is not an altogether new phenomenon in the arts. In a popular cultural sense, however, such an idea may pertain to many twenty-first-century wearable technological gadgets in the music, fashion, sports and health markets. The many people we encounter on a daily basis walking the streets, traveling on trains, jogging, shopping or sitting at their laptops in cafés, with their earphone cables plugged into smartphones or iPods, reflect a changing culture, a process of internalization and privatization that is a symptom of media diffusion and electronic miniaturization, but also of the ever advancing digital technologies, platforms and shifts in distribution through streaming. The advance is also a shrinkage, reducing our sensorial relationship to the world, even as a technology and innovation-driven discourse in postindustrial societies suggests otherwise.

Sound is a critical environmental phenomenon of common experience—we are all "ensounded" and perceive material realities through hearing and listening, in the lively and lifelong manner in which we communicate with others and with the environment, moving through, and existing in, the realm of the senses.[1] Music has played a vital cultural role in our civilizations throughout the history of evolution where sound, organized and performed in song and instrumental-percussive modes of accompaniment, became distinguished, patterned and rhythmicized as musical form for spiritual and communal rituals. These were, in the modern age, gradually superseded by entertainment and the consumption of products in a functional design or functional artistic realm where the distance between metaphysical and instrumental values had diminished or disappeared.

The decisive break with the performance of music arrived through the invention of recording technologies that transformed the idea of being ensounded, through a growing industrialization, marketing and distribution of music, sound and voice through records and mass media (radio, film, television) and, in the later decades of the twentieth century, through individualized home audio systems and portable media, including access now to wi-fi networks and live streaming. The notion of embedding sound on the wearer and into the wearer's behavior is therefore owed to sound's portability, accessibility to transmission, replayability and reproducibility.

In this chapter, we examine some conceptual proposals for "sonic bodies," in the sense that we are interested in introjective, generative and projective aspects of wearable sound, thus in the embedding of sound into clothes—costumes more properly speaking—or onto the body. Embedding may not be the precise term, since we are less concerned with sending sound to a wearer and making it inhabit the wearer than we probe the creation, the making of sound through wearing, and thus the becoming of sounding bodies. Through historical example or contemporary case study—some referencing our own experience working with the Design-and-Performance Lab (DAP-Lab)[2]—we focus not on commercial consumer sectors but on diverse creative arts productions revealing intricate relationships between aesthetic design (fashion, scenography), sonic art (sound design for wearables), and performance (theater, dance, installation, film).

As we define them more specifically in these artistic performance contexts, "wearables" are performative costumes or accoutrements, which to an extent are *sound instruments* (encompassing body-worn technologies and wired or wireless sensors). Yet they are also more than that. Developing the notion of "design-in-motion" coupled with audiophonic wearable concepts (Danjoux) points to a collaborative design fashioning process method, where along with the dancer or performer the designer creates a rehearsal series for constructing the "sound characters"[3] through movement and gestural expression. The choreographic and scenographic side (Birringer) of this process implies the creation of kinetic environments for wearable improvisation, for a "scoring" of their narrative and interactive physical-material potentials in expanded theatrical, architectural, fashion and performance contexts. Finally, in a larger political sense, sounding wearables can have a conductive social dimension that (at least subliminally) reconnects ensoundedness to the kind of cultural ritual and healing endeavors mentioned in the beginning.

## 10.2 Sound Instruments as Body Instruments

As Marshall McLuhan argues in *The Medium is the Massage: An Inventory of Effects*, "all media are extensions of some human faculty—psychic or physical" (1996, p. 26). This includes clothes, which he posits as extensions of the skin. In addition, he posits "electric circuitry" as an extension of the "central nervous system" (pp. 38–40).[4] More recently, Donna Haraway, Anna Munster, Susan Kozel, Don Ihde, Mark Hansen and others have written on issues concerning the body's shifting relation to technology and scientific advancement. On the question of how technology transforms our perception, Ihde for example explains in the context of astronomy and the telescope that the latter becomes an amplifier of perception. Thus, "instrumentally mediated observation," as he calls it in this case, enables extended viewing beyond the limits of normal human perceptual range (cf. Ihde 2002). Furthermore, he adds, the computer can then transform image into data and data into image in a form of reversibility. While we are not focusing on optical technologies and visual techniques here, the data from a performer's body movements—being transmitted to and transformed by the computer—can of course be utilized both for sonic and visual output.

It is the motivation for movement that interests us, and movement's relationship to sound: perhaps we can speak of instrumentally mediated and modulated conduits. Sound, in our experience, can not only be an extension of movement but also work as "intension" or intensification of movement, with the body—and what is worn—as a source for sonic material (and breath sound is a of course a fundamental conduit). With today's digital technologies, in a mediatized world, the various media extensions to the human faculties facilitate an expanded reach (optically, sonically, kinetically, haptically). The technologically equipped body can traverse realms, moving between near and far, real and virtual—its reach stretched through its interactions and mediating tools and through the internet. Furthermore, media extensions offer the experience of remote forms of touching via technological instruments such as virtual reality (VR) headsets and haptic devices. The process of embodiment of new media technologies, argues Anna Munster, has the potential to become both sensate and virtual—beyond pure engagement on a material and corporeal level (2006, p. 17). As *intension*, however, the sound generated or processed by the wearer, can also become a highly affective catalyst or stimulus for movement and a range of expressive and interactional gesture.

"The visual and the tactile, distance and proximity, play a part in shaping our aesthetic perception," writes Ingrid Loschek in *When Clothes Become Fashion: Design and Innovation Systems* (2009, p. 57). She is acknowledging the impacts of design on the "aestheticizing of the subconscious," referring particularly to the materiality of one of the dresses from Alexander McQueen's *Voss* collection (Spring/Summer 2001), a dress which utilized glass microscope slides, "blood plasma slides," and ostrich feathers in its construction. For fashion theorist Caroline Evans, the clothes in McQueen's 2001 collection "almost fetishized materials: feathers, brocade, shells, a wooden bodice, an outfit made from a jigsaw puzzle of a castle. . . " (2003, p. 95) which she identified communicated a certain dysfunctional and psychotic look in the models. Yet, what Loschek describes is more relevant here as it shifts the emphasis away from the visual to the sonic dimensions of the dress-in-motion, a form of body-worn instrument that is animated through a dynamic act of wearing. Worn in a one-off performance by the musician Björk, Loschek explains:

> Her dancing movements caused the glass slides to rattle against each other, and this gentle jingling was integrated as a component of Björk's music: The "blood plasma slides" mutated into percussion instruments.
>
> (Loschek 2009, p. 57)

Loschek also discusses briefly the sounding creations—garments embellished with hundreds of tiny brass bells of differing sizes—in Viktor and Rolf's couture collection *Bells* (2000/2001) and the challenges their sonorous effect posed aurally for a fashion audience (seated in the dark) accustomed to focusing on the visual. In this particular instance, the sounding activated by the movement of the models in the garments adorned in bells is the main focus. Subtle aural irritations and shimmering sonic textures (in the absence of the visual) suffuse the air, offering new sensory stimuli and raising questions for those attuned to a certain sensibility and consciousness of the performing body on the fashion catwalk. The fashion designers mentioned here are not interested in sound creation per se, but it is significant to imagine the sound potential related to movement in the fashioned garment and how that can be experienced, as musician Deniz Peters explains with regard to instrument sounding, as "direct result of a bodily act" (Peters 2012, p. 1), the garment in this case becoming extended as an instrument.

The sculpting of new body shapes through costumes or wearable architectures could be traced back to Oskar Schlemmer's Bauhaus figurines and stage experiments in the early 1920s. Working on the *Triadic Ballet*, a key aspect of Schlemmer's construction of spatial dynamics was the function of the costume. While his later Bauhaus dances have been called "gestural" or "spatial" performances (also involving

**Figure 10.1** "Metal Skirt Sound Sculpture," 1980.

*Source:* © Ellen Fullman. Photo: Anne Marsden

a strong emphasis on light projection), the *Triadic Ballet*—its full version comprising three acts, three performers (two male, one female), twelve dances and eighteen costumes, with each act displaying a different color and mood—displays a predominantly sculptural leitmotif, but it is important to realize that materials (e.g. metal) chosen for the design often imply sonic effects. With exaggerated headdresses and masks, bulbous padded torsos and outfits built with wiring and concentric hoops, extended prop-like limbs and conic or spherical appendages, the *Triadic* "figurines" are constructed to impede movement or shape it in very particular ways, drawing attention to the constructedness of the costumes as well their materials. The stylized motion required to move the costume across the stage would impel arrhythmic, animated steps, intercut with stillness, or a spinning motion that allows the performer to show off the entire 360 degrees of the shape.

Schlemmer's abstraction—perhaps similar to Loïe Fuller's vivid *Serpentine Dance* (1896) during which she whirled voluminous expanses of silk cloth, manipulating the enveloping materials through movement and colored light projected onto them—figuralizes spatial organization. One could almost describe Fuller's and Schlemmer's work (and parallels could be found in Russian constructivist and futurist performances, for example in Malevich's and Lissitzky designs for the opera *Victory of the Sun*) as a wearing-into-space, thus inevitably creating an acoulogical, psychoacoustic dimension. These kinesthetic and choreosonic potentials have inspired contemporary choreographers (e.g. William Forsythe) as well as composers, from Stockhausen and Xenakis to Kagel and Goebbels. Forsythe has coined the expression "choreographic object," and *The Fact of Matter, White Bouncy Castle, Scattered Crowd, The Defenders*, and the online research project *Synchronous Objects*,[5] transpose dance from the stage into other manifestations—participatory installations, architectural environments, soundings, cartographies, digital platforms with animated graphic materials, generative data and algorithms.

Our conceptual proposition for understanding *wearable sound*, therefore, is meant to be complex, open to such transpositions and hybridities. There is no single definition or established practice, much as it is now unnecessary to worry about spurious distinctions between sound and noise. The idea of an instrumental or sonic body is of course primarily owed to music and the many ways in which musicians/sound artists have experimented with forms of electroacoustic or electronic improvisation, with wearable technologies, sensors and actuators as ways of controlling sound (and video) through gesture. This enabled them to move away from the more static synthesizer or laptop scenarios of electronic music—turning their entire bodies into performing instruments through the exploration of the sensory aspects of interaction. The sensorial dimension, which is owed to fashion, is taken even more literally (and also scientifically) in cases where sound is probed through *physiological instrumentation*, where physical and physiological properties of the performers' bodies become interlaced with the material and computational qualities of the electronic instruments. Recent works by Atau Tanaka, Heidi Boisvert, Pamela Z or Marco Donnarumma are good examples of such interactions with audio, video and motion capture modalities, used alongside biosignal-based modalities such as muscle-tension (electromyogram or EMG), heart rate (EKG) or even electroencephalography (EEG). These modes can form a complex system for capturing input modalities from the expressive bodily gestures of a performer. Stelarc had used such interfaces in his work over several decades, when he began experimenting with prosthetic augmentation and robotics (third arm, ear on arm, etc.) and what he calls "extended operational architectures" of the body (2016, p. 93). Even though Stelarc (Figure 10.2) may be an extreme case of a body/media artist exploring biosignals that amplify and intensify internal sounds of the human body, he is of course not alone in the project of attaching the cyborgian body to the network, enabling the physical body and its organs to transmit sound elsewhere, "performing beyond the boundaries of its skin and beyond the local space that it occupies."[6]

The gestural dimension of this remains popular, linked to the desire among digital artists to move away from the prevailing disembodied performance models of the new media aesthetic of the 1980s and 1990s. Fashion, theater, dance and sonic theatricality easily intersect in the drive towards more sensory modes of engagement where corporeal activity sits at the heart of the technological system, where wearable and interactive technologies link to the phenomenology of sounding in staged performance.

Experimental sound artists such as Laetitia Sonami[7] have built an entire performance practice around this notion, using interactive sensor-packed gloves—wearable apparatus—as interface for musical composition. The Lady's Glove,[8] as she named it, is for Sonami first and foremost designed as a controller—the sensors and actuators all highly visible on the surface of the glove (Figures 10.3 and 10.4). She reflects on its unfolding as functional instrument relationally to the software and her musical sensibility:

> I think it becomes an instrument when the software starts reflecting and adapting the limitations and possibilities of the controller and your musical thinking ideas are more a symbiosis between the controller, the software and the hardware.
>
> (Sonami 2010, p. 229)

**Figure 10.2** Stelarc, *Third Hand*, Performance at Tokyo, Yokohama 1980.

*Source:* Simon Hunter. © Stelarc.

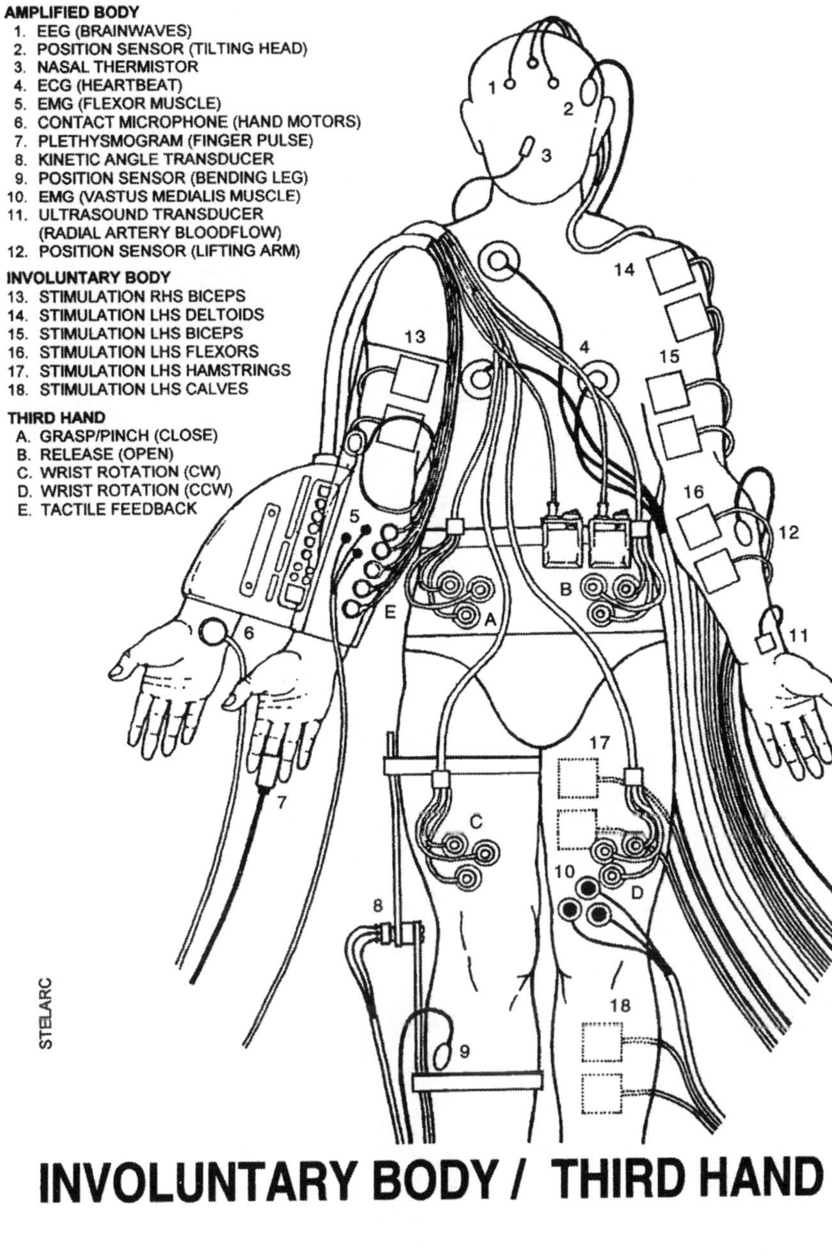


**Figure 10.3** *Involuntary Arm/Third Hand*, Yokohama, Melbourne 1990.

*Source:* Stelarc. © Stelarc.

**Figure 10.4** Laetitia Sonami With Lady's Glove, Stuttgart 2005.

*Source:* Bernd Wendt/falschnehmung [left].

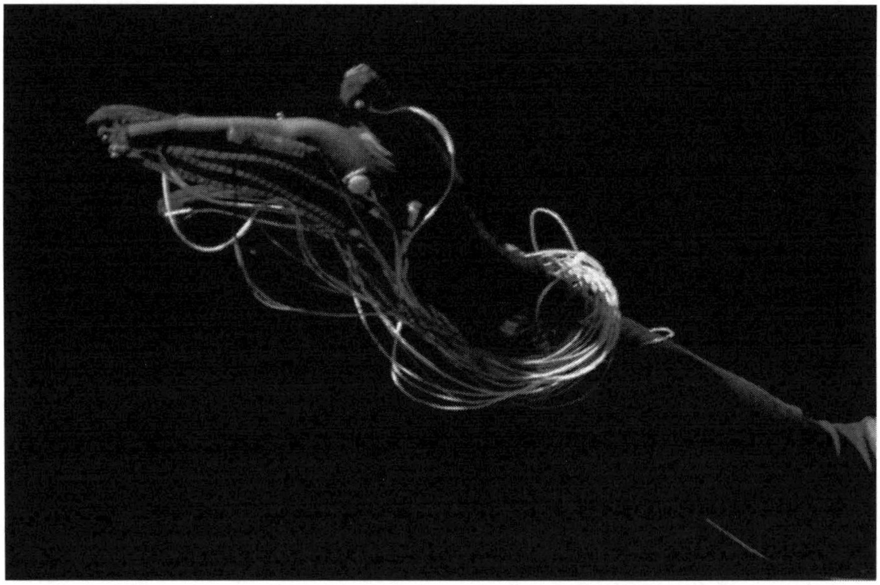

**Figure 10.5** The Lady's Glove, 2005.

Photo: Bernd Wendt/falschnehmung [right].

Sonami's notion of a mutuality of becoming through a process of interaction reflects our experience of sounding through movement in the DAP-Lab performances with dancers and wearable instruments, except that Danjoux does not refer to her prototypes as controllers but positions the costumes and accoutrements as fashion as well as interfacial, aesthetic media that generate a form of distributed agency, as opposed to a form of control. Wearing-into scenographic space, especially when more than one performer is active and when their costumes are inter-referential, necessarily also makes perception and affect of singular gestures (say, an arm or hand movement) more difficult to discern, and yet also clearer, in a call-and-response sense. Such ensemble scenarios are choric and dialogic, and owe more to the lyrical improvisatory blues and jazz of the black avant-garde than to the cool conceptual European tradition evoked by sound theorists obliged to acousmatics (*musique concrète*). Wearing sound, as we understand it, is telling a story.

Composer Tara Rodgers, author of *Pink Noises: Women on Electronic Music and Sound* (a series of interviews with female composers), comments on the physicality and multilayered aspects of Sonami's performances with the glove: "Her compositions have been described as 'performance novels,' because musical form and textual narrative unfold and are transformed through her physical motions" (Rodgers 2010, p. 226). In a similar vein, DAP-Lab stagings of *Suna no Onna* (inspired by Hiroshi Teshigahara's film), *UKIYO* (based on Hokusai woodblock prints and a collage of sources including Russian engineering and a novel by Christian Kracht), and *for the time being*, our version of *Victory Over the Sun*, are designed to link wearables to the sensorium and an unfolding of narrative in performance through multidimensional and intertwined sounding-movement characters (see below).

The notion of activating sound through wearing and simple gestures or everyday motions was explored by Ellen Fullman in her *Metal Skirt Sound Sculpture* (Figure 10.1; 1980). Fullman designed and built this pleated skirt, constructed out of metal as the name suggests, as performance wearable with an integrated system: sound was activated through the simple act of walking, the resultant noise simultaneously generating a soundtrack for her performance. Fullman expands on her motivations and her particular technique of sounding:

> In 1979, during my senior year studying sculpture at the Kansas City Art Institute, I became interested in working with sound in a concrete way using tape-recording techniques. This work functioned as soundtracks for my performance art. I also created a metal skirt sound sculpture, a costume that I wore in which guitar strings attached to the toes and heels of my platform shoes

> and to the edges of the "skirt" automatically produced rising and falling glissandi as they were stretched and released as I walked. A contact microphone on the skirt amplified the sound through a Pignose portable amp I carried over my shoulder like a purse. I was fascinated by the aesthetics of the Judson Dance Theater in their incorporation of everyday movements into performance, and this piece was an expression of that idea; the only thing required for me to do was walk.
>
> (Fullman 2012, p. 3)

Fullman used the wearable sound sculpture skirt in a street performance in downtown Minneapolis during the 1980 New Music America Festival, and a documentary video that exists of the event demonstrates the simple and straightforward execution she had imagined; yet the unexpected sound of the garment created perplexed reactions from the passersby.[9] Her experience demonstrates how body-worn wearables, responding directly to bodily motion, potentially challenge performers and audiences alike when the focus of a work's aesthetic design is directed at the creation of a particular character of sound or sound character that subtly redefines the idea of the "instrument" as well as movement's temporal affects—especially the latter's gestural and narrative characteristics that we find critical for DAP-Lab's theatrical installations.

The conventions of musical theatre and dance position the instrument as both an object (a device created or adapted to produce musical sounds) and a body. The performers engage their instrument and invite the audience to observe, listen to and experience the sonorous body. Just as Fullman arrived at this from a background in sculpture, so did Carolee Schneemann (who began performing around the time of the early 1960s Judson Church Dance Theatre) venture into kinetic body art with an intention of painting-into-space—her earliest choreographic works such as *Glass Environment for Sound and Motion* (1962) and *Newspaper Event*, *Chromelodeon* or *Lateral Splay* (1963) incorporated both performers and audiences as part of the work and led her to develop her conception of a multidimensional, moving-image "kinetic theatre," already at this early point incorporating film as a component of performance. Scheemann's pieces (including her notoriously messy, orgiastic *Meat Joy* happening in 1964) involved scores or task instructions for her fellow performers. With composer James Tenney she performed *Noise Bodies* in 1965 (Figure 10.6), a duet with everyday objects draped around the bodies, reflecting some of the typical Judson Church and Fluxus attitudes towards the mundane and the outrageous (as Schneemann demonstrated in her erotic work, and Charlotte Moorman in her "TV Bra for Living Sculpture" cello

**Figure 10.6** Carolee Schneemann's *Noise Bodies*, with James Tenney, Third Annual Avant-Garde Festival, Judson Hall, NYC, August 28, 1965.

*Source:* Charlotte Victoria © Carolee Schneemann.

performances with Nam June Paik). But it also revealed a keen sense of noisemaking acoustics. Asked about the "sound-producing debris" she wore, Schneemann responded to the interviewer:

> It was a noisy collage. We improvised together regarding what made sound and what gestures would produce varieties of sound. The way my kinetic theatre pieces developed was that parameters were set in terms of certain kinds of duration, position and action and then from studying those we would improvise. So each performance was different. *Meat Joy* has a score and units of specific active improvisation, and then within that motions change and are fluid.
>
> (Qtd. in Enright 2014)

Among the most well-known practitioners working with wearables in sound art and street performance contexts is Benoît Maubrey and his Audio Gruppe. Enacted in public spaces, the costumes Maubrey created for characters such as the *Audio Ballerinas* (1990), *Audio Geishas* (1997) and *Audio Peacock* (2003) were worn by Audio Gruppe members who developed solos with a particular instrument-costume (with built-in amplification). Certain costumes have mutated into highly individualistic and self-contained sound units or "phonic" bodies producing sounds and movements in intimate, close-to-the-spectator performances (Figure 10.7).[10]

Vocalists have also experimented with interactive sensor suits and accessories, for instance, composer/performer Pamela Z with BodySynth®[11] featuring wearable electrode sensors enabling muscle movement to control how her voice is processed, and Julie Wilson-Bokowiec with the Bodycoder System.[12] Rodgers' book features an interview with Pamela Z where she explains her choices for incorporating various technologies into her work stating: "In every piece I do, I incorporate technologies in a certain way. I have kind of a love affair with modern high-tech objects, but I also like the simplicity and directness of mechanical things" (2010, p. 220). Thus, she highlights her interest in both the digital (e.g. cell phone) and the analog (e.g. typewriter) in the compositional processes of her electronic music. Furthermore, works that integrate the glitch as compositional tool, such as Stanley Ruiz' *Barong Analog* wearable synth built into a cheap plastic poncho—a trashy performable noisemaker (exhibited at The Osage Gallery, Hong Kong in 2005 as part of Futura Manila), are pertinent to the exploration of wearable sound we conducted in the DAP-Lab early on, when we compared analog and digital options. Glitch aesthetics, known for the exploitation of dysfunctionalities or accidents in sound and noise music, implied an aesthetic we were keen to explore for its disruptive and affective potentials.

In the remainder of this chapter, we shall focus on such noise aesthetics and the particular subtleties of the poetic dress as sounding instrument in the "expanded choreographic" field, which for us indicates various crossovers between design, theater, art, fashion and music. The wearables described here reflect a historical, critical and reflective sensibility, which makes them less assimilable and commodifiable. They suggest generative performative behavior—each sounding-out affecting a subjective, often quite intimate process of noisemaking that does not comply to any ready-made ideologies of *interactive technology* (the "garments of paradise" Susan Elizabeth Ryan has written about) but seeks to crawl underneath the skin. The wearable instruments we use often tend to be encumbrances, sly inhibitors and misfits, instruments gesturing towards uninstrumentation, thus also questioning the interactive imperative (the contemporary swiping of screens and pressing of buttons) as such.

**Figure 10.7** Benoît Maubrey, *Audioballerinas*, Dancers With Electroacoustic Tutus and Digital Samplers and Motion Sensors Allowing Them to Trigger Their Sounds via Their Choreography 2000.

*Source:* Courtesy of Benoît Maubrey.

## 10.3 Ensounded Wearing

For the types of mediated performance environments in which wearables are most often performed, the short manifesto "After Choreography" proposes that in addition to there being no set choreography, one also cannot speak of free improvisation, but only of the freedom for dancers to move within the technological parameters of the system (Birringer 2008, pp. 119–120). Interactive sensor and capture systems, as we learned during the dance performance *Suna no* Onna (2008), where the dancers' movements "controlled" the digital and auditory space via motion and heat sensor technologies integrated into their garments, tend to be limiting (if accelerometers, for example, actuate simple pitch bends) or disorienting (when delay, feedback, Doppler effects, granular synthesis, etc., are involved). The wearable interfaces enabled the dancers to become embedded in the world they created, but the dancers could not necessarily hear (or see) the kinetic shape-shifting. In encounters with "wearable space" in interactive performance, the dancer will most likely either prefer to learn and internalize what motion sensors do, in order to adopt behavior, or not know or repeat her movements, if the interactive system algorithms are more random, generative and unpredictable.

In *UKIYO* (2010), dancer Anne-Laure Misme, equipped with various clunky sound-generating accoutrements (metal cage/mini crinoline [incorporating curved speaker grills], speakers, contact microphone with transmitter and 12″ vinyl disc), actively explored the technologies that extended her body physically and sonically (Figure 10.8). She was immersed in the long process of making *UKIYO*, and therefore understood the interconnectivity and enfolding of her sounding movement-character within the larger hybrid narrative and sonic landscape, that is, her historic reference, abstract representation of dynamic change—past to future, revolution and mutability. In creating "WorkerWoman" Danjoux had a loose concept for the distorted and dysfunctional sound desired for this character, involving interferences and elements of analog and digital hacker culture, to pull up new sounds and compositional strategies. For her garment instrument design, electronic processes and software coding needed to be known, as well as the basic tools involved in making custom-built interface instruments, which could be small and flexible enough to be worn or integrated into the garments.[13]

The materiality of this prototype connected elements of the old with the new in terms of technologies thus looking back, while simultaneously looking forward in a retro-futuristic fashion. The wireless portable speakers with unstable Bluetooth transmission became motivational worker

**Figure 10.8** Video Still of Anne Laure Misme as WorkerWoman (Act I), Performing in *UKIYO*, Sadler's Wells, London, 2010.

*Source:* © DAP-lab.

tools for Misme, offering unpredictability of performance and flow. The two inverted dysfunctional speakers worn provocatively on the body (as speaker breasts integrated into bra design) intentionally and paradoxically emitted no sound at all—cracked media taken to its most extreme (cf. Kelly 2009). Unexpected sounds were forced by Misme's energetic actions flexing a 12″ vinyl LP—accentuating its materiality—as her motion shifted sound production from standard playback methods of recorded sound on vinyl, through sonic rhymes of air displacement, to detecting and amplifying hidden vibrational sounds. This was made possible using the clip-on radio microphone attached to one of Misme's fingers, with wireless transmitter mounted on her arm. The result was grungy; when she dragged the mike over the vinyl, as one would a stylus across the grooves, the sounds were amplified. Getting down to her knees, pushing the vinyl across her white *hanamichi* strip—generating the sounds of friction of a laborious task—Misme became visibly stimulated by her capabilities to manipulate the sonic landscape. Her movements became more forceful, vigorous and energetic, generating a dark booming crescendo of low frequency sound and hum. She became a noise turntablist—without stylus to delicately traverse the grooves—scratching and applying forceful pressure to the vinyl disc, flexing it in a manner that would eventually cause it to crack.[14]

Helenna Ren's "SpeakerWoman" is another sound character in this installation; she is dressed in an all-white costume that is modeled after early 1960s Cold War fashion (protective spacesuits) but also alludes to workers in rice fields, as she carries a wooden bō across her shoulders from which dangle two spherical speakers, the conical forms swaying gently as she walks across the *hanamichi*, dropping rice grains onto the floor. For a few moments, all we hear are the grains falling, then high frequency sounds begin to sound from her speakers as she moves forward and backward, the wires stretching to the end of the runway and the amp. She begins to swing the speakers, and as they rotate, the sound travels in various directions, growing softer and more intimate, now resembling spectral echoes of bells and percussive music used in Kabuki performances. Her sound travels from her directional speakers outward along the lines she moves, whereas Misme's amplified, distorted noise is diffused from the surround speaker system and subwoofers. Composers Oded Ben-Tal and Sandy Finlayson, who worked with the dancers on these scenes, added a "postdigital" effect at the end of Misme's cacophonous noise performance by letting the volume of her amplified live recording fade to a bare minimum, at which point we hear a locked groove repeating ticks and clippings from an eerie "drum" pattern originally taken from bandoneon tones (an instrument played by another performer, Caroline Wilkins).

Kinesthetically and proprioceptively, gestural interaction with real-time environments (sonic or visual) can deflect both from the physical virtuosity or embodied expressiveness of the performer and from the unpredictable qualities and metaphoric richness of immersive aurality and moving scenographies (films, layered animations, networked video streams). The audience for *UKIYO* was to experience "movability" as a concrete virtuality that was not overdetermined or correlated, in the sense in which software mappings determine, for example, the principal directions and speed of images (forwards, backwards, slow, fast, freeze) or the pitch, amplitude, wave shape and granulation of sound. Our spatial and lighting design aimed at a space both polyphonic and limitless, able to surprise the visitors through unexpected intimacies as the dancers moved with—and through—the audible microsounds they generated.

Our interest in noise and analog/digital sounding characters guided our next production, *for the time being* (2012–14). Danjoux's new prototypes of choreosonic wearables were built to stimulate dialogical partnering between dancers in costumes, affecting both the sound and movement choreography mutually. Rather than solely characterizing the wearables as choreosonic, then, the term now applies to a particular type of *audible improvisation*: costumes and characters in *for the time being* are meant to enter into dialogue, creating a more amalgamated sonic architecture of *relational/transitional* entities. Our re-versioning of the Russian futurist opera *Victory over the Sun* provoked a new dimension of questions about how one garment worn by one dancer can influence the sounding and movement of another in performance. The RedMicro Dress and Futurian ChestPlate prototypes specifically aim to explore such contiguous relations between dancers in wearables, where the dynamics of proximity and distance and the interconnectedness of performers' movements generate sound (Figures 10.9 and 10.11). Designed for both solo and duet performance, the final version of ChestPlate, first tested in rehearsal with flautist Emi Watanabe, then performed by dancer Angeliki Margeti (Figure 10.11) in 2014, had evolved into a fully functional electroacoustic instrument integrating interactive circuitry—incorporating proximity, bend and light sensors to effect sounding of the instrument, and two small amplified wearable speakers into its make up. When partnered with RedMicro Dress, sonic responses from the ChestPlate can be explored synergistically between the two dancers, their relational movements emerging concurrently, as the sensors respond to their movements and proximity.

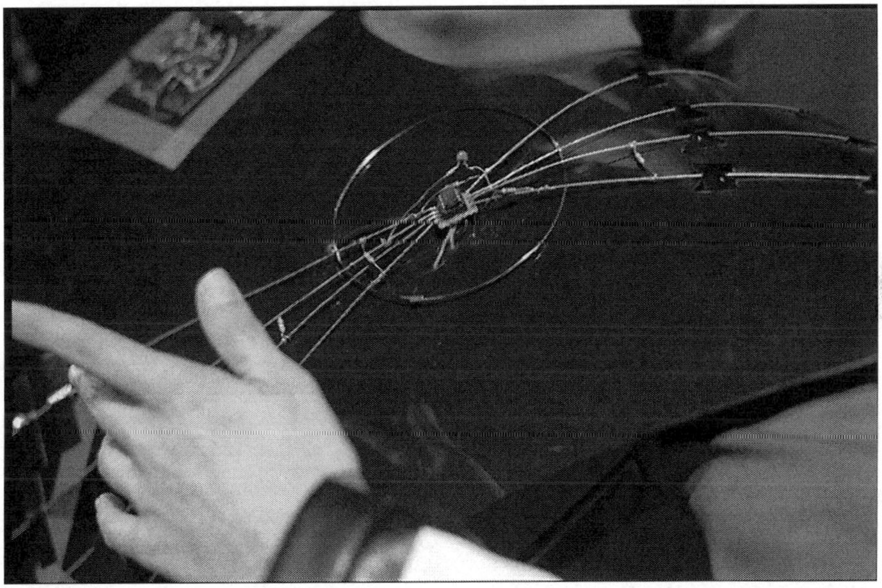

**Figure 10.9** Close-Up of Angeliki Margeti in Futurian ChestPlate Playing her Oscillating Electroacoustic Instrument—Completing the Electronic Circuit Through Touch, 2014.

*Source:* © Hans Staartjes.

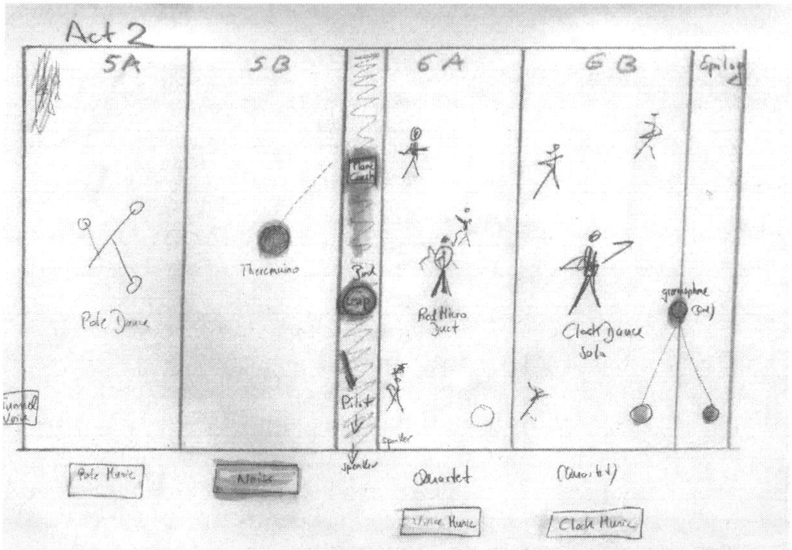


**Figure 10.10** *for the time being (Victory Over the Sun)*—Dramaturgical Sketch Highlighting the Positioning and Choreographic Sequencing of the Four Prototypes: TatlinTower; GraveDigger; RedMicro Dress and Futurian ChestPlate, 2014.

*Source:* © Johannes Birringer.

**Figure 10.11** Video Still of Dancers Vanessa Michielon in RedMicro Dress and Angeliki Margeti in Futurian ChestPlate Performing a Duet in *for the time being (Victory Over the Sun)* Sadler's Wells, London, 2014.

*Source:* © DAP-Lab.

While questions relating to the amplification of analog and electro-acoustic wearable instruments in a digital theater realm had been raised by our earlier research, intensified collaboration with musicians and electronics engineers over the past five years opened up further possibilities for technology-enabled designs and explorations of conductivity.[15] For example, RedMicro Dress, while devoid of its own sounding capabilities, can act as a transceiver—a receiver and transmitter—simultaneously detecting and relaying sounds, picked up in close proximity by its small integrated shoulder microphone, to a larger amplifying system operated by musicians. In the case of this duet, the small speaker system of the Futurian ChestPlate with its limited amplification is enlarged sonically and thus also aurally through improvised partnering.

Act II, Scene VI is in fact a quintet, involving Vanessa Michielon, Angeliki Margeti, Yoko Ishiguro and Rosella Galindo: all partners move relationally but the central duet emerges from the intimate dynamics of Margeti's and Michielon's conjoined improvised performance (Figure 10.11). The Futurian character enters into a proximal relationship with the RedMicro Dress to commence their dance.

Michielon in red executes a repeated series of revolutionary poses, arms held straight and elbows rotating, while Margeti as Futurian approaches in her blue and black garment—a science fiction instrument adorning her chest, its two small speakers attached to her lower back. The light and proximity sensors integrated into the circuitry and construction of the ChestPlate detect the presence of her partner—RedMicro Dress—and mobilize sounding. As bodies draw up close, closer, before retracting again, the sounding emitted from the two integrated speakers is actuated, intensified and distorted by the circuitry interactions (see Figure 10.12). The Futurian's noise is picked up by the dynamic microphone worn by Michielon in the RedMicro Dress, transmitted and amplified—further distorting the sonic textures of noise. Thus, the intimate entwinement of the body instrument is advanced in Act II through dynamic methods of co-creation for compositional purposes. The choreographic here is the choreosonic.[16]

In the opening Prolog and Scene 1 of *for the time being*, Helenna Ren initiates a quartet performing with a central icon of the Russian revolution as wearable sound. The TatlinTower (head)dress prototype was conceived as a wearable electroacoustic instrument to be mounted on the head of the dancer, extending the body through a process of vibrating shivers moving through the body, subtly massaging from the inside-out, in a form of vibrational augmentation. The initial sketch (Figure 10.13) indicates the role of the wearer as a microphone that picks up the sound of the revolution to transmit it to the world at large.

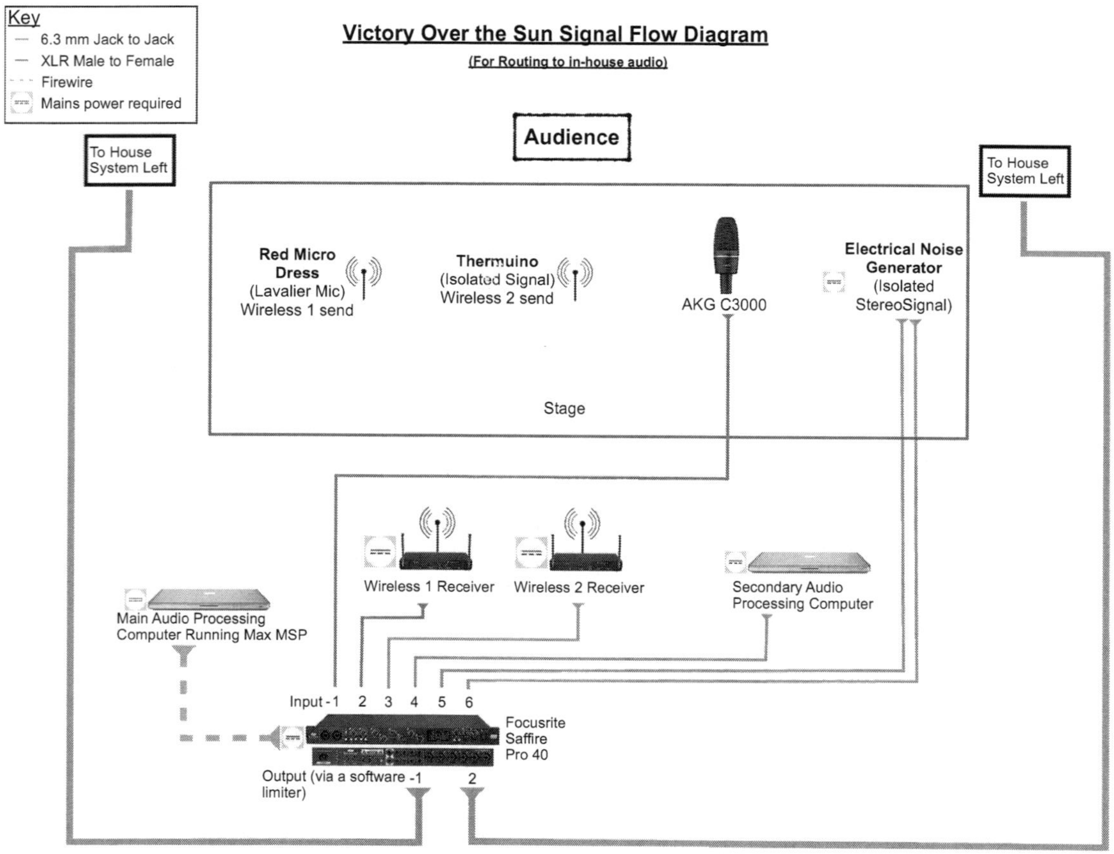


**Figure 10.12** *for the time being (Victory over the Sun)*, Signal Flow Diagram (for Routing to In-House Audio) Produced by Sonic Artist Oliver Doyle, 2014.

*Source:* © DAP-Lab.

**Figure 10.13** Conceptual Design Sketch for the TatlinTower Head(Dress) 2012.

*Source:* © Michèle Danjoux.

The design for the (head)dress follows the double helix formation of Tatlin's famous unrealized tower and is constructed in spring metal. The main body of the instrument design integrates a metal coil attached to a small motor/vibrator at its apex to rotate the coil, a bend sensor for the dancer to control the speed of the motor and subsequent speed of coil rotation, thus altering sonic output, and a black box speaker-amplifier. A piezo contact mic sits within the main construction of the (head)dress; to pick up and amplify the vibrations of the rotating spring which beats the tower, translating the mechanical activity into electrical signals—volts that can then be sent via the small circuit and jack to jack connection mounted at the base of the (head)dress, to the black box speaker, worn on the stomach area of the dancer, for amplification and portability of her minimal sound across the performance space.[17]

DAP-Lab's performance of *for the time being* opens with a double prolog, with Ren stage left wearing the TatlinTower (head)dress, sending out its signals. Across from her tower, downstage right, Khlebnikov's address to the futurians (from the original 1913 *Victory over the Sun)* is recited by actor Ross Jennings who wears a dark blue worker's overall and performs a repetitive transverse movement along a small triangle grid outlined on the floor. Picked up by a condenser mic, the words are processed to disintegrate into noise distortion, in order to demonstrate two points—both the visual relationship on stage between Ren wearing the Tatlin radio tower (and its subtle repetitive signal sounds) and the marching worker on the other side, and secondly their sonic relationship. Their kinaesonic prologs constitute a duet; the TatlinTower dancer operates the radio softly with her white gloved fingers, as the Announcer almost literally performs the function of a "loud speaker" (performing with megaphone) shouting out the bizarre address to the Futurians, obliterating all else.

As Ren transitions from prolog to Scene 1, the complex constraining presence and vibrational touch of the wearable instrument extending her body can be visibly sensed in her restricted movements. After Jennings utters the words "Never/will pass by/like a quiet dream," she rises slowly and skillfully to standing, her center of gravity held low as she explores her body movements in relation to the instrument; head twisting, manipulating the sensor (Figure 10.14); stopping and starting her sound, shifting it into the space, moving the black box away from her body, arms outstretched. Ren wears a white productivist suit with the TatlinTower (head)dress, and is joined by two other women workers in white. In addition to the physical constraining effects placed on her movement habitus by the vibrating apparatus, a psychological dimension to the wearing might also be activated by these clinical if utilitarian suits.

**Figure 10.14** Helenna Ren in TatlinTower, Scene I, *for the time being (Victory over the Sun)*, Sadler's Wells, 2014.

*Source:* © Hans Staartjes.

## 10.4 Needles, Nails and Feathers

Reflecting on the expanded choreosonics elicited through such wearables as we have described them here, it is apparent that we have drawn close links between fashion and art, music and dance-theatre/opera, yet the contexts for the wearables moved from the catwalk to gallery and theater environments. Portable and mobile media have also been used in urban contexts (locative media projects such as the roaming pieces by Blast Theory in England; Susan Kozel's *AffeXity* project with mobile devices in Sweden, or Canadian sound artist Janet Cardiff's audio walks). Our examples from Ellen Fullman and Audio Gruppe were meant to indicate how such works tend towards situationist and sound installation art—rather than involving participant audiences, the wearable sound is enacted by the instrument builders and performers who develop more intimate knowledge of the emergent behaviors of materials and prototypes.

This implies aesthetic criteria for the custom-built design of the wearable, the way it insinuates and/or encumbers movement gestures, the way sounds are placed (in the "score"), and how they unfold relationally and meaningfully. One conclusion to be drawn is that the mutual enfolding of movement and sound generation requires more careful attention to the scope and insistence of the aural—and in the case of DAP-Lab's adaptation of a futurist opera, attention to the music drama and its narrative threads. This is the reason why we speak of "sound characters," and although to some extent they are visual abstractions—and their gestures perceived as an important part of the visual aesthetic and kinetic atmosphere of the performance—they also carry art historical and musical dimensions that undergird the wearable through particular design and noise aesthetics (in their synergies with Japanese ukiyo-e prints, Kabuki, noise art such as KK Null, Otomo Yoshihide, Sachiko M, Toshi Nakamura, and Russian Futurism and Suprematism). The noise aesthetic is nowhere clearer than in the vibrational "radio" coil sonics of TatlinTower (head)dress and the heavy metal guitar associations of Futurian ChestPlate, the latter's visual electronic circuits a subtle allusion to Rodchenko and Stepanova's abstract *Tofts* graphics. In one scene we also hear operatic voices, but they come from an old gramophone record played by Caroline Wilkins' Motley Eye bird-character wearing a cone-shaped beak. She uses her beak as a stylus, and the recorded voice becomes warped as the needle eventually gets stuck in the groove. If noise and crackling distortion are considered a form of interference, it is a pertinent index of a particular design aesthetic elaborated throughout some of our work, and the dancers' somatic and technical experiences of the wearables and their particular encumbrances (Danjoux 2017).

Our more recent work explores large-scale kinetic atmospheres (*kimospheres*) that invite the audience to become immersed in a multisensorial architecture of sonic, visual and tactile elements, "wearable design" here stretched to an overall construction of fabrics and gauzes (a meta-scenic dress) onto which light and film particles are projected. We have also tested the role of wearable VR headsets within the kimospheres, inviting visitors to choose how they negotiate the organic and anorganic through the threshold of augmented virtuality. In such environments, narrative composition is weaker, as emphasis shifts to the experiential and what Danjoux's experiments with conductivity promise to open up to audiophonic wearable performance design—namely heightening the sensory aural-auratic force of garments in the absence of a narrative theatrical frame.

NailFeathersDress (Figure 10.15) opts for a more fashionable design aesthetic than the ones drawn from historical text or cultural era, exploring the notion of sounding-movement design from an essentially abstract point of view. Constructed using a multitude of nails interwoven into the main conductive mesh fabric of the dress, the idea of this garment was to amplify purely the sound of the wearer's movements stimulated by the dress and nothing more. This was achieved through integrating a series of piezo contact mics into the garment, to pick up the vibrational qualities of the nails—animated by the dancer—and then making these audible via a wearable amplifier-speaker carried like a camera on the dancer's body.[18] Elisabeth Sutherland, wearing this garment, confided how inspired she was to use her body literally as an instrument, unencumbered by narrative or any additional factors, to generate sound through her individual steps, crouching and whole body torque. Her body expanded through wearable design to create a fused and intertwined sounding instrument body, generating compositional elements of pure instrumental music kinetically in performance. Wearer sensation and interaction were the only motivations to movement-sounding.

Given the personal associations we all have with the clothes or accessories we wear, it is apparent that performers respond in individual ways to the challenges of the wearable structure and the "felt" presence of body-worn technologies. We can thus state a pertinent outcome of the emergent choreographies of real-time interaction and the amplification of physical presence through costume, namely particularized forms of performance specific to the *character of the wearable*. The dancers adopt or discover movement expressions that are not based on familiar technical vocabularies (ballet, modern dance, tanztheater, etc.) but inspired by the intricacies of the material and sonic design. The designs created by Danjoux for the DAP-Lab productions seek to be both visually highly distinctive yet also distinctively audible when activated through wearing by the dancer in motion.

**Figure 10.15** Video still of NailFeathers Dress Worn by Elisabeth Sutherland in Rehearsal, *metakimosphere no.3*, Artaud Performance Centre, London, 2016.

*Source:* © Michèle Danjoux.

The overarching emphasis of Danjoux's *design-in-motion* research was directed at what she calls the *sonic touch* (2017, p. 217), namely the contained, intimate, proximate movement expression of the dancer articulating her wearable dress as a transceiver instrument. The method used for discovering the sonic touch fundamentally stipulates a practice of designing that attaches electroacoustic instruments onto the dancer's body and costume, testing emergent behaviors of materials, movement and sound in the design process while conceptualizing them as relational, dynamic and active. A significant marker of transfer between this method and artistic movement practices concerned with emergent/improvisatory processes is the importance of initial design form (provided in stages, added on in the process). Each prototype is a kind of machine of its own kinetic and sonic poiesis.

## Notes

1 As anthropologist Tim Ingold reminds us, *being alive* is a matter of realizing how we move and change, and how we are always ensounded moving through the world, which is also a world of sonorities and auditory spaces (Ingold 2011, p. 138; cf. Birringer 2017a).

2 DAP-Lab is a crossmedia lab exploring convergences between performance, telematics, textile/fashion design and movement, visual expression, film/photography, sound and interactive design, founded by Birringer and Danjoux in 2004: http://people.brunel.ac.uk/dap/.

3 DAP-Lab productions have consistently used the notion of wearables as *sound characters* through the specific costumes Danjoux designed for the movement rehearsals out of which the particular choreosonics of a work emerged (for example in *Suna no Onna*, *UKIYO* and *for the time being*). For a comprehensive delineation the design-in-motion method, see Danjoux 2017. Other publications on choreosonic wearables and kinetic atmospheres include Birringer 2013, 2017b; Birringer and Danjoux 2009a, 2009b, 2013; Danjoux 2014.

4 Earlier in the 1960s, and concerning the notion of media as extensions to the communication condition of the present body, sociologist Erving Goffman had discussed (analog) technologies such as microphones and other mechanical devices as "boosting devices"—to amplify and augment the naked senses (Goffman 1963, p. 14).

5 http://synchronousobjects.osu.edu

6 See Stelarc's website for his description of engineering internet organs: http://stelarc.org/?catID=20242. Alongside the growth of net.art and telematic performance (in the current era of internet-based experimentation), collaborations using biosignal data transfers are part of the international new media arts and computer music circuits (which include festivals, conferences and journals). Donnarumma curated biophysical works for the 2015 Computer Music Journal's Sound and Video Anthology; the Brazilian collective Corpos Informáticos has explored the networked body for over two decades (see: https://anthology.rhizome.org/telepresence).

7 http://sonami.net.
8 Documentation of Laetitia Sonami in performance with "Lady Glove" available at: www.youtube.com/watch?v=C8GqbS2w_Lg.
9 Available at: https://vimeo.com/channels/1017437/45207205.
10 www.benoitmaubrey.com/.
11 www.pamelaz.com/.
12 www.bodycoder.com/.
13 Nicolas Collins's *Handmade Electronic Music: The Art of Hardware Hacking* (2006) was helpful for the prototyping process; the book derives from his course (at the Department of Sound, School of the Art Institute, Chicago) for introducing students to some electronic alternatives to the computer, ways to bridge the gap between the sound world of a generation raised in an electronic culture and the "gestural tradition of the hand," as he calls it.
14 See: https://youtu.be/g2yfYrlvOLM.
15 During preproduction and production phases of *UKIYO* (2009–11), all early tests in London and Tokyo focused on designs that explored elements of audiophonic cloth, sounding objects, portable sound, sensor interfaces and wearable speakers. Some of DAP-Lab's tests during *for the time being* (2012–14) were also staged at Interaktionslabor Göttelborn (Germany). Research for the METABODY (www.metabody.eu) project, a large-scale EU Culture Program collaboration between 11 partner organizations, enabled prototype testing during workshops and exhibitions in Madrid, Genova, Dresden, Amsterdam and London. Regarding investigations of conductivity, the Studio for Electro-Instrumental Music (STEIM) invited Danjoux to a three-day research-creation laboratory on e-textiles, movement and sound (October 2014) bringing together experts in the divergent fields with the view to creating synergies through the convergence of these disciplines. Working teams consisted of textile experts, interaction designers, sound artists, choreographers dancers and performers. The short but intensive residency was organized by Marije Baalman, an electronics engineer then based at STEIM, and it was during this residency that a new ConductiveCoat prototype was created, and the seeds for the NailFeathersDress (2016) were planted. For a brief film of latter, created for *metakimosphere no. 3*, see: www.youtube.com/watch?v=Iw4T-uM3n-U.
16 See: www.youtube.com/watch?v=oXBW4oWyK0.
17 Collaborating with Danjoux, electronic artist John Richards built the micro-circuitry, advised on materials for achieving the desired vibrational qualities and designed the quiet disruptive sound for this audiophonic prototype.
18 Neal Spowage assisted in the electronic construction of this prototype design.

## References


Birringer, J., 2008. After choreography. *Performance Research* 13, no. 1, pp. 118–123.

Birringer, J., 2017a. Ensounded: An introduction to special topic: Sound/theatre: Sound in performance. *Critical Stages/Scènes Critiques* (December/Décembre 2017) no. 16. Viewed June 14, 2018 <www.critical-stages.org/16/special-topic/>.

Birringer, J., 2017b. Metakimospheres. In: Broadhurst, S., and Price, S. (Eds.) *Digital Bodies: Creativity and Technology in the Arts and Humanities*. London: Palgrave Macmillan, pp. 27–48.

Birringer, J., and Danjoux, M., 2009a. Wearable performance. *Digital Creativity* 20, no. 1–2, pp. 95–113.

Birringer, J., and Danjoux, M., 2009b. Wearable technology for the performing arts. In: McCann, J., and Bryson, D. (Eds.) *Smart Clothes and Wearable Technology*. Cambridge, MA: Woodhead Publishing Ltd., pp. 308–419.

Birringer, J., and Danjoux, M., 2013. The sound of movement wearables. *Leonardo* 46, no. 3, pp. 232–240.

Collins, N., 2006. *Handmade Electronic Music: The Art of Hardware Hacking*. 2nd ed. London: Routledge.

Danjoux, M., 2014. Choreography and sounding wearables. Special issue on 'critical costume'. *Scene* 2, no. 1–2, pp. 197–220.

Danjoux, M., 2017. *Design in Motion: Choreosonic Wearables in Performance*. Phd Thesis, London College of Fashion, University of the Arts London.

Donnarumma, M., Caramiaux, B., and Tanaka, A., 2013. Body and space: Combining modalities for musical expression. *Paper presented at Conference on Tangible and Embedded Interaction (TEI), UPF: Barcelona*. Viewed June 14, 2018 <<http://marcodonnarumma.com/publications/>.

Enright, R., 2014. Notes on fuseology: Carolee Schneemann remembers James Tenney. *Border Crossings* 132. Viewed January 5, 2018 <http://bordercrossingsmag.com/article/notes-on-fuseology>.

Evans, C., 2003. *Fashion at the Edge: Spectacle, Modernity and Deathliness*. New Haven: Yale University Press.

for the time being [Victory over the Sun] by DAP-Lab (2014) Directed by Johannes Birringer and Michèle Danjoux [Lilian Baylis Studio, Sadler's Wells, London. 3–4 April].

Fullman, E., 2012. A compositional approach derived from material and ephemeral elements. *Leonardo Music Journal* 22, pp. 3–10.

Goffman, E., 1963. *Behavior in Public Places: Notes on the Social Organization of Gatherings*. London: Free Press.

Hansen, M., 2006. *Bodies in Code. Interfaces with Digital Media*. New York: Routledge.

Ihde, D., 2002. *Bodies in Technology*. Minneapolis: University of Minnesota Press.

Ihde, D., 2010. *Embodied Technics*. Copenhagen: Automatic Press/VIP.

Ingold, T., 2011. *Being Alive: Essays on Movement, Knowledge and Description*. London: Routledge.

Kelly, C., 2009. *Cracked Media: The Sound of Malfunction*. Cambridge, MA: MIT Press.

Kozel, S., 2007. *Closer: Performance, Technologies, Phenomenology*. Cambridge, MA: MIT Press.

Loschek, I., 2009. *When Clothes Become Fashion: Design and Innovation Systems*. Oxford: Berg.

McLuhan, M., [1967] 1996. *The Medium is the Massage: An Inventory of Effects*, with illustrations by Quentin Fiore. Berkeley, CA: Gingko Press Inc.

Munster, A., 2006. *Materializing New Media: Embodiment in Information Aesthetics*. Hanover: Dartmouth College Press.

Peters, D., 2012. Introduction. In: Peters, D., Eckel, G., and Dorschel, A. (Eds.) *Bodily Expression in Electronic Music: Perspectives on Reclaiming Performativity*. New York: Routledge, pp. 1–14.

Rodgers, T. (ed.). 2010. *Pink Noises: Women and Electronic Music & Sound*. Durham: Duke University Press.

Ryan, S. E., 2014. *Garments of Paradise*. Cambridge, MA: MIT Press.

Sonami, L., 2010. Laetitia sonami. In: Rodgers, T. (Ed.) *Pink Noises: Women and Electronic Music & Sound*. Durham: Duke University Press, pp. 226–234.

Stelarc, 2016. Excess and indifference. In: Gardiner, H., and Gere, C. (Eds.) *Art Practice in a Digital Culture*. London: Routledge, pp. 93–116.

*Suna no Onna* by DAP-Lab, 2008. Directed by Johannes Birringer and Michèle Danjoux [Laban Studio Theatre, London. 8 December].

*UKIYO [Moveable Worlds]* by DAP-Lab, 2010. Directed by Johannes Birringer and Michèle Danjoux [Lilian Baylis Studio, Sadler's Wells, London. 26 November].

11

# Smart Musical Instruments

## Key Concepts and Do-It-Yourself Tutorial

**Luca Turchet and Mathieu Barthet**

### 11.1 Introduction

Recently, a new strand of research on digital musical instruments (DMIs) has emerged within the field of new interfaces for musical expression (NIME), the so-called "smart musical instruments" (SMIs) or "smart instruments" (Turchet 2019). This type of musical instrument is characterized by embedded computational intelligence and wireless connectivity, and is conceived to enable new forms of interactions between performers, their instruments and audiences. The ubiquitous availability of the internet supports the emergence of SMIs along with advances made in portable embedded systems equipped with wireless interfaces. Such interfaces enable SMIs to be connected on both local area networks and the internet expanding their capabilities.

SMIs belong to a wider category of "Musical Things" which can be defined as computing devices capable of sensing/actuating, acquiring, processing and exchanging data serving a musical purpose. Musical Things that are interconnected through the internet operate within the "Internet of Musical Things" (IoMusT) which refers to the networks, interfaces and music-related information exchanged by interoperable devices dedicated to the production and/or reception of musical content (Tuchet et al. 2018c).

The goal of this chapter is to provide the reader with a basic understanding of the origins and fundamental principles underlying smart musical instruments and to provide guidelines to start designing some of their technological components. We first highlight the main characteristics of smart musical instruments and provide some examples drawn from academia and the industry. We then illustrate how artificial intelligence and connectivity can benefit SMIs through a short tutorial based on affordable commercial technologies and open-source frameworks.

## 11.2 What Is a Smart Musical Instrument?

This section offers an overview of the research areas related to SMIs before presenting their properties and architecture. We provide examples and use cases demonstrating novel performer-audience applications made possible with SMIs.

### 11.2.1 *Background*

SMIs can be seen as an evolution of augmented instruments at the intersection of new interfaces for musical expression, networked music performance systems and the Internet of Things. Although a comprehensive survey of these endeavors is beyond the scope of this chapter, we examine here selected examples of studies in this space.

#### *11.2.1.1 Augmented Instruments and Self-Contained DMIs*

The so-called "hyperinstruments" (Machover and Chung 1989) or "augmented instruments" (Bevilacqua et al. 2006; Miranda and Wanderley 2006; McPherson 2015) have been the subject of much attention in the new interfaces for musical expression community at least since the 1980s. Augmented instruments are conventional musical instruments enhanced with sensor and/or actuator technology and associated digital signal processing. They aim to provide performers with additional degrees of freedom for musical expression to go beyond traditional sonic interaction. This is made possible by combining acoustically generated sounds established by the physics of the instrument with electronically generated sounds or modulations based on sensory information capturing the performer's actions. Augmented instruments are generally controlled with conventional playing techniques but exploit human gesturality to provide new musical affordances in an effort to balance out technical and artistic novelty with rich musical tradition. Some augmentation techniques exploit acoustic mechanisms of sound production by feeding back computer-generated sounds to the instrument body through electromechanical actuation (Overholt et al. 2011), or by using modal active control of the soundboard to modify the natural sound of the instrument (Benacchio et al. 2016).

The sensors added to augmented instruments are generally used to track the gestures of the player, producing musical performance information which is then repurposed via mapping techniques (Hunt et al. 2003). Such techniques typically allow the player to control parameters of electronically generated sounds. Some examples of sensor-based augmentations of musical instruments include the Hyper Flute (Palacio-Quintin 2003), the Augmented Violin (Bevilacqua et al. 2006), the Electrumpet (Leeuw 2009) and the Hyper Mandolin (Turchet 2017).

Actuators can be used to directly control the vibrating structure involved in the instrument sound production (e.g. a guitar soundboard, a drum skin, the piano strings). Examples of use of actuators can be found for different instruments including the Magnetic Resonator Piano (McPherson 2010), the Feedback Resonance Guitar (Overholt et al. 2011), the Overtone Fiddle (Overholt 2011) and the augmentation of acoustic drums using electromagnetic actuation (Gregorio and Kim 2018).

Augmented instruments typically encompass external components including a computing unit (e.g. a laptop), audio interface, loudspeakers, cables and power supply. Another line of research on DMIs has favored compactness and the self-containedness for sound processing and production. DMIs designed following these principles, whether extensions of conventional instruments or entirely novel designs, are typically based on embedded systems like the Bela platform (McPherson and Zappi 2015) or the Satellite CCRMA platform (Berdahl and Ju 2011; Berdahl et al. 2013). Berdahl's embedded acoustic instruments (Berdahl 2014) are examples of self-contained DMIs.

#### *11.2.1.2 Networked Music Performance Systems and Audience Participation*

The field of Networked Music Performance (NMP) originated from the exploration of collaborative musical interaction forms held over local or remote networks. Comprehensive reviews of hardware and software technologies enabling NMPs can be found in (Gabrielli and Squartini 2016) and (Rottondi et al. 2016). NMP is made possible thanks to the transmission of audio or symbolic performance recordings played on traditional acoustic instruments and/or DMIs (Barbosa 2003; Weinberg 2005). NMP systems can leverage various types of network architectures, from wired ones (Local Area Networks, Wide Area Networks) to wireless ones (Wireless Local Area Networks, see e.g. Gabrielli and Squartini 2016; global cellular data networks, see e.g. Essl and Lee 2017). Open Sound Control (OSC) is a flexible protocol that enables the exchange of control information over networked devices in interoperable ways (Wright et al. 2003; Schmeder et al. 2010). The open address space structure of OSC allows developers to configure services matching the requirements of designed DMIs (see the DIY part of section 11.3). For this reason, OSC is often used as a communication protocol in musical applications involving smart devices, for example in (Migicovsky et al. 2014) where the authors use smart watches as musical controllers.

A well-known example of tangible interface enabling NMP is the ReacTable (Jordà et al. 2007). This interface features a tabletop that tracks objects moved on its surface to generate synthesized sounds. The ReacTable allows various players to simultaneously interact with objects placed on a same table or on several networked tables in geographically dispersed locations. Smartphones have also been proposed as ubiquitous technologies enabling networked music performance (see Essl and Lee 2017 for a review).

Some NMP systems aim at engaging audiences in the creative process during a live performance. Examples of technology-mediated audience participation systems include Mood Conductor, where performers follow mood indications expressed by the audience using mobile devices (Fazekas et al. 2013), Open Symphony, which lets audiences shape the musical structure of musical improvisations (Zhang et al. 2016; Wu et al. 2017), and A.bel where interactive content is distributed onto the mobile devices from audience members (Clément et al. 2016). Other participatory systems let audience members produce sounds themselves using mobile devices as musical instruments (see e.g. Essl and Lee 2017). In light of their network capabilities, smart musical instruments enable novel performer-audience interactions.

#### *11.2.1.3 From Augmented to Smart Musical Instruments*

The concept of hyperinstrument introduced by Machover and Chung (1989) entails the integration of computer systems into musical instruments that "monitor data from the input instrument, redefines the controls on that instrument, and acts in accordance with its programmed musical knowledge." Such systems provide "intelligence" to the instrument to various degrees: by enriching the amount of possible musical results produced by performers, by "being aware" of what is being performed and reacting accordingly, up to being computationally creative through expert agents capable of producing accompaniments or sonic enhancements. This seminal vision does not however position hyperinstruments as part of digital networked ecosystems. Network connectivity, the Internet and distributed architectures are examples of network communication features that can be leveraged to design intelligent systems for musical instruments. Smart musical instruments can be seen as self-contained augmented instruments/hyperinstruments enhanced with network capabilities and features from the Internet of Things (Gubbi et al. 2013). We discuss their main characteristics in the next section.

## 11.2.2 Smart Musical Instrument Key Concepts

### 11.2.2.1 Sensing and Actuating

Similarly to augmented instruments, sensors and actuators can be important components of SMIs. While playing, performers manifest a range of intangible expressions, from gestures related to instrumental control to postural movements and physiological responses. In SMIs, the use of sensors can be aimed at capturing some of these intangible expressions. Sensors can also be used to interpret contextual aspects about the performer and the performance. The sensed data that characterize performers' expression and context can be repurposed in various ways using mapping strategies, for instance, to map gestures to control parameters of electronically generated sounds (see e.g. Hunt et al. 2003). For example, the data from an inertial measurement unit (IMU) tracking the spatial orientation of a smart guitar could be used to modulate the feedback parameter of a delay effect (e.g. more delay when the guitarist lifts up the neck of the instrument). Besides sensor-driven control of audio effects, sensors can be used to trigger and generate other types of sounds (e.g. triggering recorded sound samples or synthesized tones from synthesizers and drum machines). Another use of sensors on smart musical instruments is to (de)activate functionalities (e.g. to change sound banks and presets, start/stop a loop station). In contrast to augmented instruments, in SMIs, players' interactions captured by the sensors can also be aimed at delivering control messages to interconnected external devices. For example, an IMU embedded in a smart guitar could be used to modulate the sound from the instrument as well as control live visuals projected on screens during a live performance.

In addition to or instead of sensors, smart instruments may also integrate different types of actuators. For a smart musical instrument to be self-contained, it needs to be capable of diffusing electronically generated sounds. Loudspeakers or vibration speakers can be embedded into the instrument (see the Sensus Smart Guitar by Mind Music Labs[1] and the HyVibe guitar[2] described in section 11.2.4). Vibration motors can also be used to provide haptic feedback to performers for communication purposes.

### 11.2.2.2 Network Connectivity

A characteristic of smart musical instruments is their capability to join local and remote networks wirelessly, and exchange information over them with other connected devices. The exchanged information may relate to real-time content (e.g. audio or symbolic signals produced by the SMI while playing), or asynchronous data (e.g. metadata downloaded from the cloud). Real-time communication imposes strict constraints on communication technologies to enable low-latency wireless communication.

#### 11.2.2.3 Embedded or Distributed Intelligence

Intelligent capabilities can be brought to smart musical instruments using embedded or distributed systems. In SMIs, an embedded computational unit is responsible for all computations such as sensor data processing, sound processing and generation similar to that provided by a digital audio workstation. The computational unit also processes received/transmitted data from/to compatible connected devices. Embedded systems for SMIs must be capable of guaranteeing low-latency audio processing while at the same time handling the connectivity functions. Computationally expensive processes can be distributed for example using cloud computing (Armbrust et al. 2010), which can be used to offload processing for scalability purposes, or edge computing that allows computation to be performed at the edge of the network for real-time requirements (Shi et al. 2016). Semantic audio (see e.g. Slaney 2012; Fazekas et al. 2011) and interactive machine learning (Fiebrink and Caramiaux 2018) are examples of artificial intelligence and knowledge representation frameworks that can be used to design intelligent components for SMIs. We highlight below several interrelated properties conferring intelligent capabilities to SMIs.

- *Personalisation*

This property relates to the ability of the instrument to let performers customize aspects of the musical results such as the identity and sound quality of produced sounds. Personalization is an emergent topic providing users of interactive music systems with more agency and creative control (see e.g. Barthet et al. 2016, for a discussion). Although the timbre of an instrument can vary according to performers' expressive gestures (see Barthet et al. 2010 and 2011), the instrument is confined to its own timbre space dictated by the physics of the instrument (see e.g. Barthet et al. 2010). SMIs' sensor and actuator technology can be used to let performers change the timbral identity of the instrument going beyond the purely acoustical production pathways. The network connectivity feature of SMIs also provides access to online audio content, which can provide sound samples or background tracks that can be played back directly with the instrument rather than through external loopers or digital audio workstations. Examples of web technologies developed for the reuse of Creative Commons audio content are discussed in the chapter "Leveraging Online Audio Commons Content For Media Production" in this series' Volume 1,

*Foundations in Sound Design for Linear Media*. Such web technologies can be aimed at personalizing the audio content available to performers for sample-based synthesis and playback in SMIs, as discussed in Turchet and Barthet 2018.

- *Context awareness*

In pervasive computing, context awareness refers to a system that is "cognizant of its user's state and surroundings and must modify its behavior based on this information" (Satyanarayanan 2001). A performer's context includes attributes such as physical location, physiological state (e.g. body temperature and heart rate), and emotional state (such as calm, excited or angry). This can be extended to the instrument itself being aware of what and how it is being played. The context awareness property can be established based on deterministic rules (some contextual attributes become tied to specific musical outcomes) or intelligent music expert agents that reason and make deductions based on input data. Techniques enabling to recognize a performer's playing actions using real-time music information retrieval (see e.g. Turchet et al. 2018b) can provide data for personalized or proactive musical behaviors. One can mention some speculative usage of context awareness letting SMIs adapt their sound to the acoustic environment and its reverberation condition, or shape their tone according to the emotional state of the performer.

- *Proactivity*

In pervasive computing, proactivity relates to the capability of a system to support the task of the user, self-tuning and adjusting its behavior automatically to fit circumstances (Satyanarayanan 2001). Proactive behavior is closely linked to the context awareness property in that it requires an understanding of the user intent. Several speculative applications exploiting the idea of proactivity for musical instrument can be highlighted: music e-learning—automated recommendation of songs to practice based on the performer level (see e.g. Barthet, Anglade, Fazekas et al. 2011), music performance—generation of automated accompaniment based on musical content and structure (see e.g. Pachet et al. 2013).

#### *11.2.2.4 Interoperability*

Interoperability is another key aspect of smart musical instruments. In contrast to augmented instruments, which are primarily designed to expand the expressive capabilities of bespoke instruments, SMIs are conceived to be able to exchange information with other Musical Things (e.g. other SMIs, smartphones, Augmented/Virtual Reality headsets). The Musical Instrument Digital Interface (MIDI) protocol enables DMIs to exchange musical information but it relies on the serial transmission of data and is limited in resolution. Open Sound Control is a more flexible standard to organize and transmit sound control information over networks. However, OSC requires developers to define a suitable syntax for musical control. Tempo synchronization of networked DMIs can be achieved using the Ableton Link[3] protocol. Further research is needed to extend the interoperable capabilities of SMIs in the context of the Internet of Musical Things (Turchet et al. 2018).

### 11.2.3 Smart Musical Instrument Architecture

The general technological architecture of a smart instrument is illustrated in Figure 11.1. As described below, it comprises three main layers (inputs, processing, outputs) as well as memory and power supply capabilities.

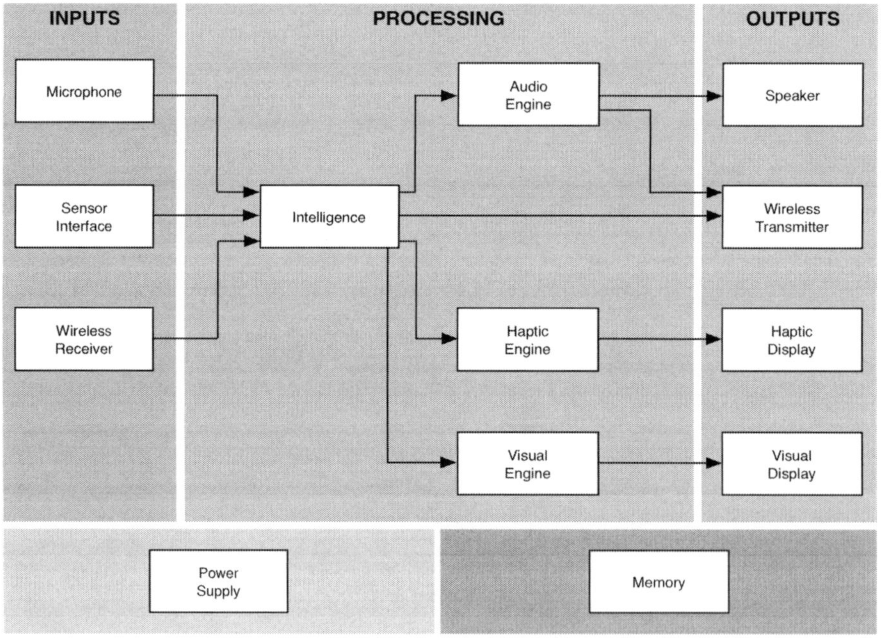


**Figure 11.1** Schematic Representation of the Technological Components of a Smart Musical Instrument.

#### 11.2.3.1 Input Layer

The input layer typically includes:

- A microphone system: to capture the sound generated by the instrument. This can be achieved using a single or a set of acoustic pressure microphones, piezoelectric or magnetic pickups embedded in the instrument.
- A sensor system: to track, for instance, performers' gestures or contextual information. Sensors can be aimed at tracking continuous interactions (e.g. a pressure sensor capturing direct or indirect interactions with a zone of the instrument it is in contact with, a proximity sensor detecting movements from the hands) or discrete interactions (e.g. a push button that is activated or deactivated).
- A wireless receiver: to receive networked data wirelessly from connected devices, including streams of audio and symbolic musical information. The receiver can connect to both local networks and the internet.

#### 11.2.3.2 Processing Layer

The processing layer typically includes:

- An intelligent system: to confer the instrument data-driven or knowledge-driven reasoning and reflective capabilities. This system enables context awareness and proactivity properties discussed in section 11.2.2.1. Different artificial intelligence approaches can be deployed such as real-time music information retrieval, sensor fusion, machine learning or semantic audio.
- An audio engine dedicated to the generation of audio content.
- A haptic engine dedicated to the generation of haptic content.
- A visual engine dedicated to the generation of visual content.

#### 11.2.3.3 Output Layer

The output layer typically includes:

- A loudspeaker: a sound delivery system located in the instrument (for instance via actuation systems or loudspeakers embedded in the instrument).
- A wireless transmitter: a system for wireless transmission of various types of data to networked devices and that can connect both local networks and the internet.

- A visual display to output visual content produced by the visual engine component.
- A haptic display to output the haptic content produced by the haptic engine component.

An SMI also includes a power supply that provides power autonomously without recurring to external power sources (e.g. embedded batteries). Memory is required to store and access digital data (e.g. backing tracks or sound samples, recordings of played songs in symbolic or audio formats).

### 11.2.4 Examples of Smart Musical Instruments

To date, only a small number of smart musical instruments have been developed in academia and industry. To illustrate the concepts presented above, we now describe examples of smart musical instruments and use cases.

Sensus Smart Guitar. This instrument is crafted and developed by the company MIND Music Labs.[4] As can be seen in Figure 11.2, it is a hollow body electric guitar augmented with sensors embedded in various parts of the instrument, onboard processing, a system of multiple actuators attached to the soundboard, and interoperable wireless communication

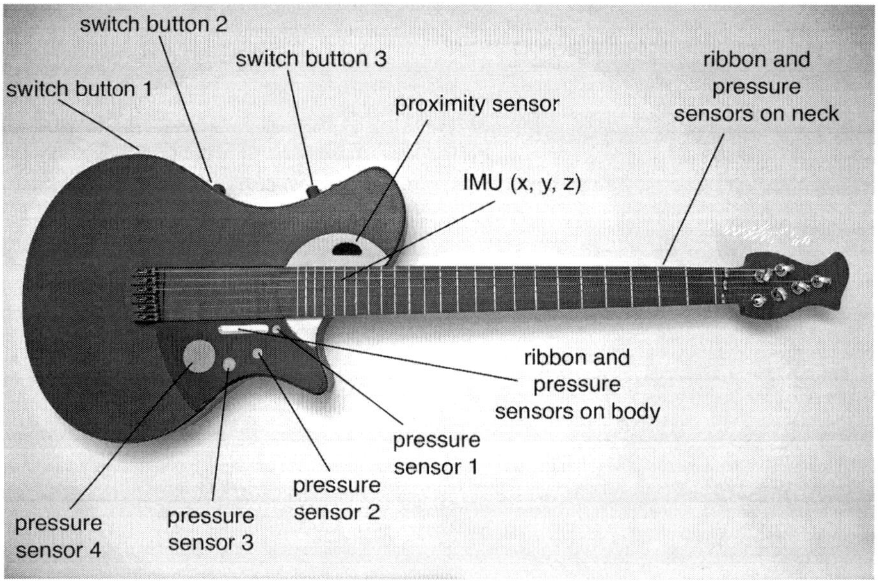


**Figure 11.2** MIND Music Labs' Sensus Smart Guitar, With indications of Sensor Types and Placement.

(Turchet 2019). The utilized embedded system runs ELK,[5] a music operating system based on Linux, which guarantees round-trip latencies of 1 ms, supports music software plugins and provides wireless connectivity via Bluetooth, Wi-Fi and 4G. The internal sound engine that runs over ELK affords a large variety of sound effects and sound generators, and is programmable via dedicated apps for desktop PCs, smartphones and tablets. The low-latency wireless communication implemented in the smart guitar supports the exchange of MIDI and OSC messages as well as audio signals. Some use cases for the Sensus Smart Guitar are reported in (Turchet et al. 2017). The instrument has been used to wirelessly control (1) visuals delivered on screens by VJ programs such as VDMX by Vidvox, (2) audio plugins and functions of digital audio workstations running on laptops and (3) elements of virtual environments provided through virtual reality headsets. The smart guitar has also been used to record audio files and share them directly on social networks.

HyVibe guitar. The HyVibe guitar[6] is an acoustic guitar which enables the production of multiple audio effects such as chorus, reverb, distortion and delay directly from the body of the guitar using a vibration control system (Benacchio et al. 2016). The guitar also lets players record song ideas and guitar licks to create backing tracks or share them on social networks using the cloud. The settings of the guitar can be controlled using a dedicated mobile app.

Smart Cajón. The smart cajón prototype described in (Turchet et al. 2018a, 2018b) is illustrated in Figure 11.3. It is a conventional cajón

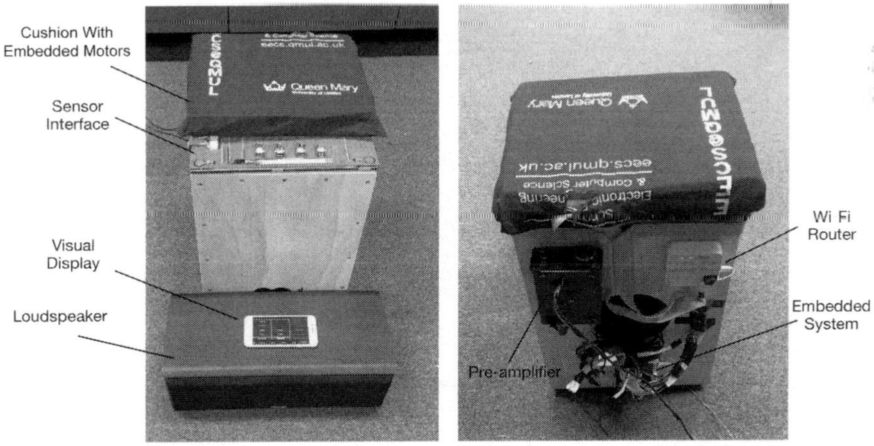


**Figure 11.3** Smart Cajón Prototype Reported in (Turchet et al. 2018a) and (Turchet et al. 2018b) with Components for Smartification.

augmented with sensors, motors for vibro-tactile feedback embedded in a cushion and wi-fi connectivity. The computational unit is the Bela board for low-latency audio and sensor processing[7] (McPherson and Zappi 2015). The audio engine is composed of a sampler and various audio effects (e.g. reverberation, delays, frequency shifter, equalizer). Sensor fusion and semantic audio techniques are utilized to estimate the location and nature of the performer strokes on the front and side panels. The stroke detection engine can be used to map strokes to different sound samples to simulate various percussive instruments, or produce automated score transcription. By mapping musical gestures to tactile stimuli using haptic wearable devices, it is possible to provide audience members with haptic feedback while performers play (Turchet and Barthet 2017a, 2017b). In another example application involving creative audience participation, the Smart Cajón provided notifications to the performer through tactile stimulation responding to audience intent produced via smartphones (e.g. start/stop playing).

## 11.3 A Do-It-Yourself Smart Musical Instrument

This section provides practical guidelines to start building the core technological components of a basic smart musical instrument prototype using affordable hardware (see list below) and open-source software. The software discussed in this tutorial can be freely downloaded from the chapter's website (Internet of Musical Things Routledge Website 2018).[8] We address here the following prototypical functions:

- Wireless synchronization and exchange of messages between the prototype and a connected laptop;
- Downloading of audio content from the cloud—more information on cloud-based technologies for music can be found in the chapter "Leveraging Online Audio Commons Content For Media Production" in this series' Volume 1 (*Foundations in Sound Design for Linear Media)*;
- Classification of two distinct percussive sound timbres using audio content recorded with a microphone and artificial intelligence techniques;
- Using sound detection to trigger samples and automatically generate symbolic information for scores;
- Modulation of sound samples triggered electronically using audio effects controlled by a sensor interface.

### 11.3.1 Hardware Components

Figure 11.4 presents the hardware components used to build the prototype. This tutorial relies on the Bela embedded system described in (McPherson and Zappi 2015). It is based on the Beaglebone Black platform and allows makers to easily create projects involving low-latency audio and sensor processing. The current version features eight analog inputs, eight analog outputs, and 16 digital general-purpose inputs/outputs, as well as one audio stereo input and one audio stereo output. Bela comes with an internal memory of 4 GB and the memory can be extended with an additional SD card. To provide wireless connectivity to Bela (local and remote networks), a small wireless router can be connected to the board's Ethernet port. The router used in this tutorial is the TL-WR902AC model by TP-Link, which supports the IEEE 802.11ac wi-fi standard and features a USB port for 4G dongles enabling direct Internet connectivity.

To capture an input audio signal, a contact microphone can be used (e.g. HotSpot by K&K Sound) and be fixed to a rigid surface such as a tabletop.

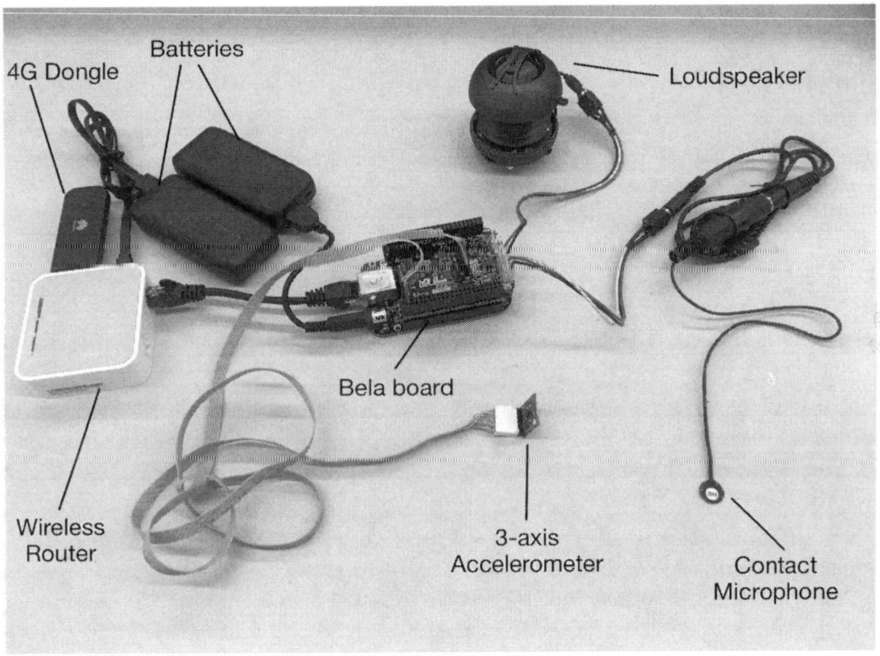


**Figure 11.4** Hardware Components of the Smart Musical Instrument Prototype Discussed in the DIY Tutorial.

A portable speaker (e.g. Xmini Uno by XMI) can be employed for audio reproduction. The sensor interface used in the tutorial is a three-axis accelerometer (ADXL 337 by Analog Devices). Power supply to the overall system is provided using two power banks (5V/2A).

### 11.3.2 Network Setup

To enable communication between the SMI and a laptop, a network needs to be set up. The router should first be turned on. The name and password of the default router network can be found in its technical documentation. Once the laptop is connected to the router's network, the router configuration panel can be accessed (by typing in the address http://tplinkwifi.net/ in a web browser for the TP-Link router used in the tutorial). The name of the network and range of IP addresses supported by the router can be configured. Following the recommendations reported in (Mitchell et al. 2014) to optimize the components of a wi-fi system for live music performance scenarios, in order to reduce latency and increase throughput, the router can be configured in access point mode and security disabled if this does not impart risks.[9] Depending on laptop compatibility, we advise to use the 5 GHz frequency band, which is generally less frequently used, and to limit the support to the IEEE 802.11ac standard only.

### 11.3.3 Software

In our companion website (Internet of Musical Things Routledge Website 2018), we provide software for the Bela embedded system and a laptop connected to the SMI wirelessly. Bela is based on the Linux operating system and open-source programs. We use the default settings coming with the Linux distribution provided with the platform. The cross-platform software Pure Data (Pd)[10] for real-time audio is used both on the Bela and laptop sides. On Bela, we will use the Pd external object abl_link, [11] [12] and objects belonging to the timbreID library.[13] These objects are provided in the software downloaded from the companion website (on Bela, the objects have to be placed in a folder that Pd can access, for instance, ~/Bela/projects/pd-externals on Apple Mac OSs). A screenshot of the main Pd patch running on the laptop is shown in Figure 11.5.

To exchange information between Bela and the laptop connected over a wi-fi local network, we use OSC messages over the User Datagram Protocol. This can be done using the netsend and netreceive Pd objects. Internet connectivity can be obtained by configuring the Bela Linux system and the router connected to the 4G dongle. To illustrate the interaction between a smart musical instrument and the cloud, we will start by downloading two

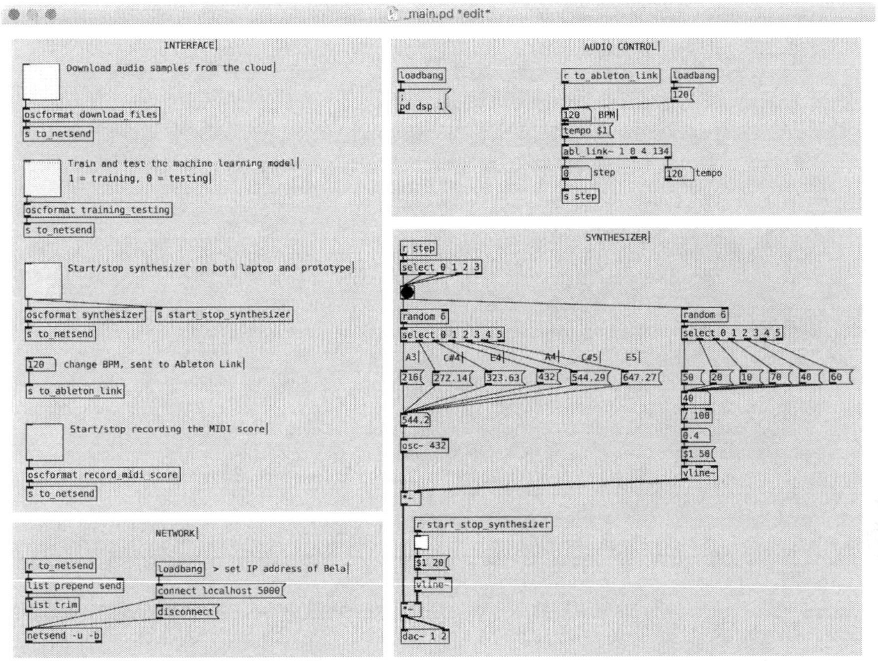


**Figure 11.5** Screenshot of the Main Pure Data Patch Running on the Laptop.

audio samples stored on our server that will be used later. To do so, send the OSC message /download_files 1 from the laptop running the accompanying Pd patch. When this message is received by the prototype, two. wav audio files of a kick and snare drums will be automatically downloaded from a public folder on the www.iomut.eu server hosting them. Specifically, the files are downloaded thanks to a bash script using the Linux command wget, which is called from the Pd patch running on the prototype.

To extract audio features and classify the timbre of an input sound, we used the tools available in the TimbreID Pd library (Brent 2010). Figure 11.6 displays a Pd patch that generates audio feature vectors from signals captured by the microphone, which are then used for training and testing timbre recognition machine learning models. All the Pd objects are configured with a window size of 1024 samples. To detect the timing of a percussive sound, we rely on the bark~ onset detector. To characterize the timbre of the acquired sound, we use the first ten coefficients of the Bark frequency spectrum using the BarkSpec~ object. The peakamp~ object enables to compute the amplitude of the initial portion of the sound. The feature extraction algorithms from the BarkSpec~ and peakamp~ objects

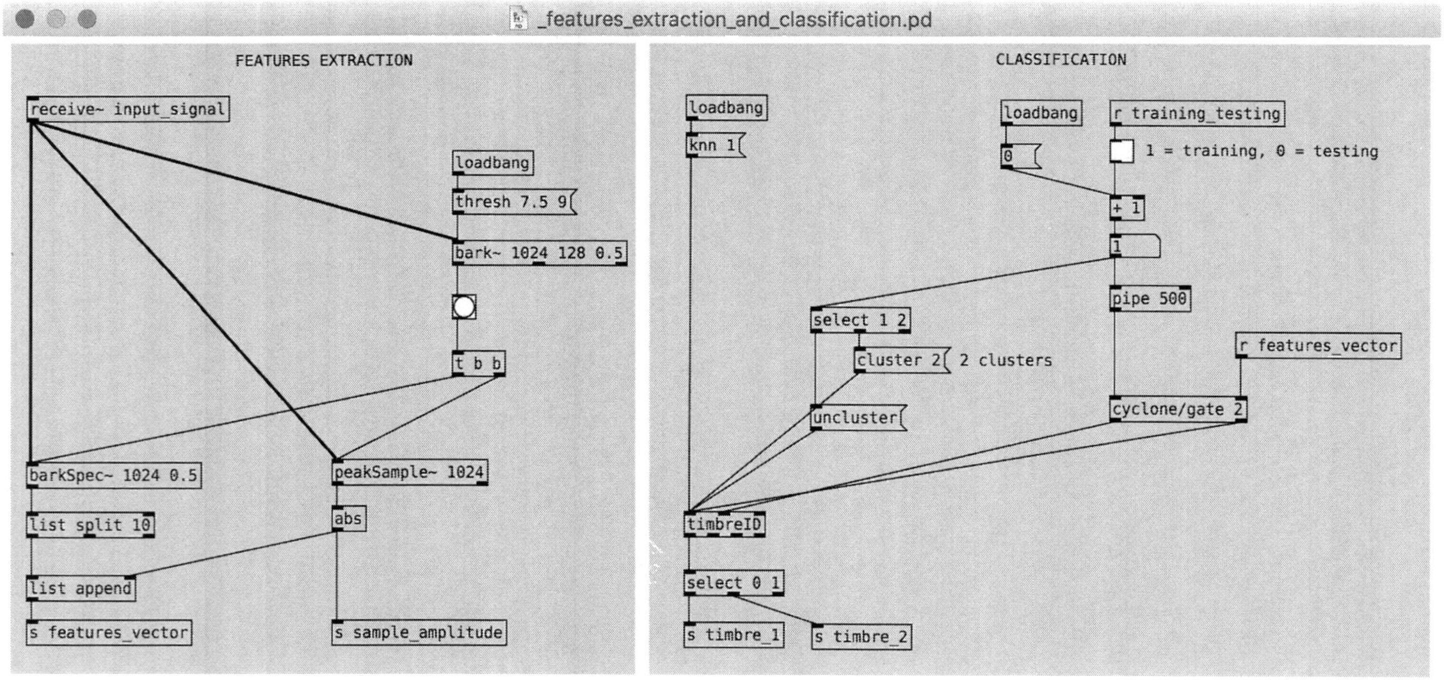


**Figure 11.6** Screenshot of the Pd Patch Running on Bela, Which is Responsible for Extracting Audio Features From the Microphone Signal and for Training and Testing the Machine Learning Model.

retroactively analyze the previous 1024 samples as soon as they receive a bang in input. Such a bang is produced immediately after the onset has been reported by the bark~ onset detector. In this way, the initial portion of the sound is considered during the feature extraction computations.

To classify the timbre of a detected percussive sound, the timbreID object is then used. The object implements the k-nearest neighbors (k-NN) pattern recognition technique, which searches the class most common among k nearest neighbors (k is a positive integer, typically small). In this tutorial, we adopt an unsupervised k-NN method, which means that the timbreID object automatically clusters the incoming vectors to create a model without manual labeling. To train the model, the timbreID object must be set in training mode. This is achieved through the OSC message / training_testing 1, which can be sent using the laptop computer. To train the model, some series of two distinct percussive sounds can be produced on the surface onto which the contact microphone is attached (e.g. one type of sound can be produced by tapping the surface with the fingers, and the other by hitting the surface with a pen). A series of 20 hits for each type of sound can be used to train the model for each type of sound. The OSC message /training_testing 0 can be sent to stop the training mode and start the testing mode (this automatically instantiates two clusters in the audio feature space).

Now that the model has been trained, new instances of the hits that have been learned should trigger the two sound samples previously downloaded, a kick drum in response to bass tones and a snare drum in response to brighter tones. These samples are triggered in real time and their amplitude follows the envelope of the percussive sound produced on the surface being hit. While hitting the surface with one hand to produce sounds, these can be modulated with audio effects by moving the accelerometer with the other hand along its three axes: acceleration along axes x, y and z apply a delay with feedback, a frequency shifter and a reverberation effect, respectively. The orientation of the three axes is reported on the accelerometer.

In order to synchronize the devices over the local network, we use the abl_link~ external object, which implements the Ableton Link protocol.[14] This object is also used in the Pd patch running on the laptop that synchronizes the SMI prototype and laptop according to a common tempo. This tempo can be set up from each of the device. Both of the Pd patches for the Bela platform and the laptop implement a basic synthesizer that randomly triggers notes of short durations on each beat matching a preselected tempo expressed in beats-per-minute (BPM). By sending the OSC message /synthesizer 1 from the laptop, the synthesizers on both the laptop and the prototype start. A change of BPM in the patch will produce slower or faster notes according to the selected tempo.

To accompany the melodic part produced by the synthesizers, a rhythmic part can be produced by hitting the playing surface. When playing, a distinct MIDI note will be generated for each of the two types of percussive sounds. The OSC message /record_midi_score 1, which can be sent from the laptop, starts the recording of a MIDI score. Once a rhythm has been performed, the MIDI recording can be stopped by sending the OSC message /record_midi_score 0. This will generate the MIDI file score.mid, which can be transferred from Bela to the laptop (for instance via the linux command scp).

## 11.4 Conclusions

In this chapter, we presented key concepts related to smart musical instruments encompassing sensory technology, network connectivity and artificial intelligence techniques framed within the emerging Internet of Musical Things field. We discussed properties characterizing smart musical instruments including self-containedeness, personalization, context awareness and proactivity. These properties were illustrated with examples of existing SMIs such as smart guitars and percussive instruments, as well as through speculative ideas providing inspiration for musical instruments of the future. We hope that the DIY part of the chapter will help students, researchers and developers interested in new interfaces for musical expression develop practical skills and start building their own SMIs, which present a promising potential for sound designers and musicians who would like to engage in novel forms of media interaction and audience experience.

## Notes

1 www.mindmusiclabs.com/sensus/
2 https://hyvibe.audio/
3 www.ableton.com/en/link/
4 www.mindmusiclabs.com/
5 www.mindmusiclabs.com/ELK
6 https://hyvibe.audio/
7 Bela is based on the Beaglebone Black board and open-source software (e.g. Linux, Pure Data, SuperCollider).
8 www.iomut.eu/routledge/
9 By disabling security, the wireless network will be unsecured and open for connection by any wireless device.
10 https://puredata.info/

11 https://github.com/Ableton/link
12 https://forum.bela.io/d/244-ableton-link-abl-link-testing/
13 https://github.com/wbrent/timbreID
14 www.ableton.com/en/link/

## References


Armbrust, M., Fox, A., Griffith, R., Joseph, A. D., Katz, R., Konwinski, A., Lee, G., Patterson, D., Rabkin, A., Stoica, I., and Zaharia, M., 2010. A view of cloud computing. *Communications of the ACM* 53, no. 4, pp. 50–58.

Barbosa, A., 2003. Displaced soundscapes: A survey of network systems for music and sonic art creation. *Leonardo Music Journal* 13, pp. 53–59.

Barthet, M., Anglade, A., Fazekas, G., Kolozali, S., and Macrae, R., 2011. Music recommendation for music learning: Hotttabs, a multimedia guitar tutor. In: *Workshop on Music Recommendation and Discovery*. Chicago: ACM, pp. 7–13.

Barthet, M., Depalle, P., Kronland-Martinet, R., and Ystad, S., 2010. Acoustical correlates of timbre and expressiveness in clarinet performance. *Music Perception: An Interdisciplinary Journal* 28, no. 2, pp. 135–154.

Barthet, M., Depalle, P., Kronland-Martinet, R., and Ystad, S., 2011. Analysis-by-synthesis of timbre, timing, and dynamics in expressive clarinet performance. *Music Perception: An Interdisciplinary Journal* 28, no. 3, pp. 265–278.

Barthet, M., Fazekas, G., Allik, A., Thalmann, F., and Sandler, M. B., 2016. From interactive to adaptive mood-based music listening experiences in social or personal contexts. *Journal of the Audio Engineering Society* 64, no. 9, pp. 673–682.

Benacchio, S., Chomette, B., Mamou-Mani, A., and Ollivier, F., 2016. Modal proportional and derivative state active control applied to a simplified string instrument. *Journal of Vibration and Control* 22, no. 18, pp. 3877–3888.

Berdahl, E., 2014. How to make embedded acoustic instruments. *Proceedings of the International Conference on New Interfaces for Musical Expression*. London, pp. 140–143.

Berdahl, E., and Ju, W., 2011. Satellite CCRMA: A musical interaction and sound synthesis platform. *Proceedings of the International Conference on New Interfaces for Musical Expression*. Oslo, pp. 173–178.

Berdahl, E., Salazar, S., and Borins, M., 2013. Embedded networking and hardware-accelerated graphics with satellite CCRMA. *Proceedings of the International Conference on New Interfaces for Musical Expression*. Daejeon, pp. 325–330.

Bevilacqua, F., Rasamimanana, N., Fléty, E., Lemouton, S., and Baschet, F., 2006. The augmented violin project: Research, composition and performance report. *Proceedings of the International Conference on New Interfaces for Musical Expression*. Paris, pp. 402–406.

Brent, W., 2010. A timbre analysis and classification toolkit for pure data. *Proceedings of the International Computer Music Conference*. New York: International Computer Music Association.

Clément, A., Ribeiro, F., Rodrigues, R., and Penha, R., 2016. Bridging the gap between performers and the audience using networked smartphones: The a.bel system.

*Proceedings of International Conference on Live Interfaces*. Brighton: University of Sussex.

Essl, G., and Lee, S. W., 2017. Mobile devices as musical instruments—state of the art and future prospects. *Proceedings of the International Symposium on Computer Music Multidisciplinary Research*. Porto, pp. 364–375.

Fazekas, G., Barthet, M., and Sandler, M. B., 2013. Novel methods in facilitating audience and performer interaction using the mood conductor framework. In: *International Symposium on Computer Music Modeling and Retrieval, Lecture Notes in Computer Science*, Vol. 8905. Cham: Springer, pp. 122–147.

Fazekas, G., Raimond, Y., Jakobson, K., and Sandler, M., 2011. An overview of semantic web activities in the OMRAS2 project. *Journal of New Music Research special issue on Music Informatics and the OMRAS2 Project* 39, no. 4, pp. 295–311.

Fiebrink, R., and Caramiaux, B., 2018. The machine learning algorithm as creative musical tool. In: *The Oxford Handbook of Algorithmic Music*. Oxford: Oxford University Press.

Gabrielli, L., and Squartini, S., 2016. *Wireless Networked Music Performance*. Cham: Springer.

Gregorio, J., and Kim, Y., 2018. Augmentation of acoustic drums using electromagnetic actuation and wireless control. *Journal of the Audio Engineering Society* 66, no. 4, pp. 202–210.

Gubbi, J., Buyya, R., Marusic, S., and Palaniswami, M., 2013. Internet of things (IoT): A vision, architectural elements, and future directions. *Future Generation Computer Systems* 29, no. 7, pp. 1645–1660.

Hunt, A., Wanderley, M. M., and Paradis, M., 2003. The importance of parameter mapping in electronic instrument design. *Journal of New Music Research* 32, no. 4, pp. 429–440.

Internet of Musical Things Routledge Website, 2018. Viewed September 8, 2018 <www.iomut.eu/routledge/>.

Jordà, S., Geiger, G., Alonso, M., and Kaltenbrunner, M., 2007. The reactable: Exploring the synergy between live music performance and tabletop tangible interfaces. *Proceedings of the International Conference on Tangible and Embedded Interaction*, New York: ACM, pp. 139–146.

Leeuw, H., 2009. The electrumpet, a hybrid electro-acoustic instrument. *Proceedings of the International Conference on New Interfaces for Musical Expression*. Pittsburgh, pp. 193–198.

Machover, T., and Chung, J., 1989. Hyperinstruments: Musically intelligent and interactive performance and creativity systems. *Proceedings of the International Computer Music Conference*, San Francisco: International Computer Music Association, pp. 186–190.

McPherson, A., 2010. The magnetic resonator piano: Electronic augmentation of an acoustic grand piano. *Journal of New Music Research* 39, no. 3, pp. 189–202.

McPherson, A., and Zappi, V., 2015. An environment for Submillisecond-Latency audio and sensor processing on BeagleBone black. *Audio Engineering Society Convention* 138, Warsaw: Audio Engineering Society.

Migicovsky, A., Scheinerman, J., and Essl, G., 2014. MoveOSC—smart watches in mobile music performance. *Joint Proceedings of the International Computer Music Conference and the Sound and Music Computing Conference*. Athens, pp. 692–696.

Miranda, E. R., and Wanderley, M. M., 2006. *New Digital Musical Instruments: Control And Interaction Beyond the Keyboard* (Computer Music and Digital Audio Series), Book 21. Middleton, WI: AR Editions, Inc.

Mitchell, T., Madgwick, S., Rankine, S., Hilton, G. S., Freed, A., and Nix, A. R., 2014. Making the most of wi-fi: Optimisations for robust wireless live music performance. *Proceedings of the Conference on New Interfaces for Musical Expression*. London, pp. 251–256.

Overholt, D., 2011. The overtone fiddle: An actuated acoustic instrument. *Proceedings of the International Conference on New Interfaces for Musical Expression*. Oslo, pp. 4–7.

Overholt, D., Berdahl, E., and Hamilton, R., 2011. Advancements in actuated musical instruments. *Organised Sound* 16, no. 2, pp. 154–165.

Pachet, F., Roy, P., Moreira, J., and d'Inverno, M., 2013. Reflexive loopers for solo musical improvisation. *Proceedings of the SIGCHI Conference on Human Factors in Computing Systems*. Paris: ACM, pp. 2205–2208.

Palacio-Quintin, C., 2003. The hyper-flute. *Proceedings of the International Conference on New Interfaces for Musical Expression*. Montreal: McGill University, pp. 206–207.

Rottondi, C., Chafe, C., Allocchio, C., and Sarti, A., 2016. An overview on networked music performance technologies. *IEEE Access* 4, pp. 8823–8843.

Satyanarayanan, M., 2001. Pervasive computing: Vision and challenges. *IEEE Personal Communications* 8, no. 4, pp. 10–17.

Schmeder, A., Freed, A., and Wessel, D., 2010. Best practices for open sound control. *Proceedings of the Linux Audio Conference*, Utrecht, *10*.

Shi, W., Cao, J., Zhang, Q., Li, Y., and Xu, L., 2016. Edge computing: Vision and challenges. *IEEE Internet of Things Journal* 3, no. 5, pp. 637–646.

Slaney, M., 2012. Semantic-audio retrieval. *IEEE International Conference on Acoustics, Speech, and Signal Processing*, no. 4, pp. IV-4108–IV-4111). IEEE.

Turchet, L., 2017. The hyper-mandolin. *Proceedings of the International Audio Mostly Conference on Augmented and Participatory Sound and Music Experiences*. London: ACM, pp. 1:1–1:8.

Turchet, L., and Barthet, M., 2017a. Envisioning smart musical haptic wearables to enhance performers' creative communication. *Proceedings of International Symposium on Computer Music Multidisciplinary Research*. Porto, pp. 538–549.

Turchet, L., and Barthet, M., 2017b. An internet of musical things architecture for performers-audience tactile interactions. *Proceedings of the Digital Music Research Network Workshop*. London: Queen Mary University of London.

Turchet, L., and Barthet, M., 2018. Ubiquitous musical activities with smart musical instruments. *Proceedings of the Eighth Workshop on Ubiquitous Music*, São João del Rei.

Turchet, L., Benincaso, M., and Fischione, C., 2017. Examples of use cases with smart instruments. *Proceedings of Audio Mostly Conference*. London: ACM, pp. 47:1–47:5.

Turchet, L., McPherson, A., and Barthet, M., 2018a. Co-design of a smart Cajón. *Journal of the Audio Engineering Society* 66, no. 4, pp. 220–230.

Turchet, L., McPherson, A., and Barthet, M., 2018b. Real-time hit classification in a smart Cajón. *Frontiers in ICT* 5, no. 16, pp. 1–14.

Turchet, L., Fischione, C., Essl, G., Keller, D., and Barthet, M., 2018c. *Internet of musical things: Vision and challenges*. IEEE Access, no. 6, pp. 61994–62017.

L. Turchet. Smart Musical Instruments: vision, design principles, and future directions. IEEE Access, 7, 8944–8963, 2019.

Weinberg, G., 2005. Interconnected musical networks: Toward a theoretical framework. *Computer Music Journal* 29, no. 2, pp. 23–39.

Wright, M., Freed, A., and Momeni, A., 2003, May. Open sound control: State of the art 2003. *Proceedings of the International Conference on New Interfaces for Musical Expression*. Singapore, pp. 153–160.

Wu, Y., Zhang, L., Bryan-Kinns, N., and Barthet, M., 2017. Open symphony: Creative participation for audiences of live music performances. *IEEE MultiMedia* 24, no. 1, pp. 48–62.

Zhang, L., Wu, Y., and Barthet, M., 2016. A web application for audience participation in live music performance: The open symphony use case. *Proceedings of the International Conference on New Interfaces for Musical Expression*. Brisbane: Griffith College, pp. 170–175.

12

# Text-to-Speech Synthesis

**Esther Klabbers**

## 12.1 Introduction

Text-to-speech (TTS) synthesis has been an area of research for several decades now. As the name suggests, text serves as the input and a speech waveform is produced as output. Numerous books and book chapters have been written about the TTS process (Klatt 1982; Dutoit 1997; Jurafsky and Martin 2014; Sproat 1998; Taylor 2009; van Santen et al. 2013). An excellent overview of slides and video tutorials can be found at Speech Zone,[1] a website maintained by Prof. Simon King, director of the Center for Speech Technology Research (CSTR) in Edinburgh, Scotland.

Present-day TTS research is a multidisciplinary field requiring knowledge of (computational) linguistics, phonetics, acoustics and computer science. It is thus impossible to fill a book chapter, let alone a whole book, with all the information required to fully understand and implement the TTS process. This chapter describes the state of the art in TTS synthesis and the steps involved in going from input text to generated speech and offers various links to relevant open-source tools that address different aspects of the TTS system.

In the last few years the use of TTS systems in real-world applications has become ubiquitous. You just have to mention "Siri" or "Alexa" and people know exactly what you are talking about. Virtual agents such as Siri or robots can interact with people about various topics such as the weather, traffic, or how best to prepare a turkey. Other TTS applications involve GPS navigation in cars, or web/document reading for accessibility, in education or second language learning. But what does it take to build a TTS system?

A TTS system typically consists of two major modules: (1) a linguistic front-end, and (2) an acoustic back-end. The linguistic front-end converts

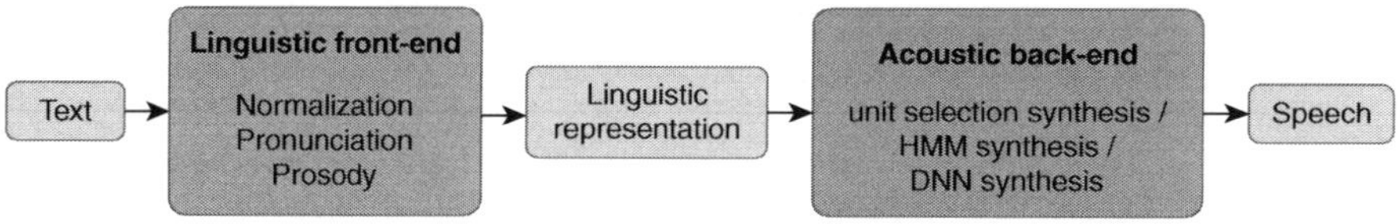


**Figure 12.1** TTS Architecture.

text into a linguistic specification, by analyzing the input text and breaking it up into tokens/words and deriving linguistic features that are relevant to pronouncing the text. The acoustic back-end converts the linguistic specification of the input text into a speech waveform. There are various methods to do so. This chapter will describe the current prevailing methods including unit selection synthesis, statistical parametric speech synthesis and waveform synthesis. The linguistic front-end is language-specific. The acoustic back-end can be largely language-independent as the speech waveforms and the relevant acoustic features are the same across languages. Of course, the actual speech used for building a synthetic voice is language-dependent.

## 12.2 Linguistic Front-End

The linguistic front-end or text analysis module takes in an input text and converts it to a linguistic representation that can be used to generate speech. Most TTS systems synthesize speech one sentence at a time. While paragraph structure plays a role in how humans read texts, the TTS process works faster when feeding it only one sentence at a time and is also simpler. Furthermore, most acoustic databases used for synthesizing speech are recorded as isolated sentences without any paragraph structure in it.

The tasks performed by the linguistic front-end are language-dependent and when designing or training the tasks, input is required from a linguist with knowledge of that language to make sure the resulting linguistic representation is accurate and errors that are made are not too outrageous. There are a few open-source programs available for building TTS systems that include the linguistic front-end. The most well-known program is called Festival (Black et al. 2001), which will be described in more detail in section 12.4.1. There is also a small footprint spin-off from Festival called Flite which is written in C (Black and Lenzo 2001).

### *12.2.1 Tokenization and Text Normalization*

The first step in the linguistic module is to segment the input text into sentences. For English and other Indo-European languages, sentences are separated by punctuation markers such as periods, question marks and exclamation marks. But periods are also used in abbreviations, so the linguistic front-end needs to decide whether a period in the input text marks the end of a sentence or an abbreviation. Usually, a list of known abbreviations for a given language is defined, and pattern matching is performed on the input text to find abbreviations in the text and expand them to their full form before splitting sentences.

When the text has been split into sentences, the TTS will process them one at a time. *Tokenization* demarcates the tokens in a sentence. In the case of Indo-European languages, these tokens are words and they are separated by spaces. East-Asian languages such as Japanese, Mandarin Chinese and Korean need different strategies as the spaces between tokens do not correspond to word boundaries. Word segmentation algorithms can be trained for these languages using pre-segmented corpora as training data. They use sequences of n characters (also called *n-grams*) to decide whether a specific token is the start of a new word or a continuation of the previous word.

Once abbreviations are detected, they are expanded into the full words to be spoken. Sometimes there is ambiguity and an abbreviation such as St. can be expanded to "Saint" or "Street." The linguistic front-end uses information about surrounding words to pick the correct expansion. Besides expanding abbreviations, text normalization also involves normalization of other nonstandard words including numbers, currency amounts, dates and acronyms (Sproat et al. 2001). For example, in English whenever a number is preceded by a currency symbol, the number is pronounced first followed by the currency. If the number is "one," the currency word (pound, dollar) is singular, otherwise it is plural (pounds, dollars). And if there is a decimal in the currency amount, one has to expand "$3.25" to "three dollars and twenty-five cents." As you can see in the dollar amount, the decimal is indicated by a period, but in continental Europe the comma is used as the decimal marker and the period as the thousands marker. When converting a date to words, the language also matters. For US English 06/07/2008 would be expanded to "June seventh two thousand eight" whereas in British English it would be expanded to "July sixth two thousand and eight."

## 12.2.2 Disambiguation

Most languages have words that are spelled the same but can be pronounced differently depending on their role in a sentence. These words are called *homographs*. In order to pronounce them the correct way, they need to be disambiguated (Yarowsky 1995). Most homographs differ in their part of speech (e.g. whether they function as nouns, verbs or adjectives in the sentence), such as *lives* (noun: black lives matter, verb: he lives in a big house). Others have the same part of speech but differ semantically, such as *bass* (bass fisherman versus bass player). Using the context of surrounding words can help to disambiguate them.

### 12.2.3 Pronunciation Prediction

Most TTS systems have a very large pronunciation lexicon to look up the pronunciation of words. A pronunciation consists of the sounds needed to pronounce the word (also called *phonemes*), the lexical stress (DEsert vs deSSERT) and the locations of syllable boundaries. Creating a reliable lexicon for a new language is very costly and tedious. But not all words in a text can be captured in a pronunciation lexicon. New names and words can occur at any time in an input text and even words that are misspelled need to be pronounced. Therefore, the TTS system needs a mechanism to predict pronunciations for unknown words. In the past, these predictions were rule-based but for some languages the letter-to-sound correspondence is so poor that the set of rules quickly becomes too unwieldy and can give unwanted results. With current data-driven methods a letter-to-sound or grapheme-to-phoneme (G2P) model can be trained automatically. Bisani and Ney (2008) give a thorough overview of data-driven methods before deep neural networks (DNNs) became popular (more about DNNs later in this chapter). Novak et al. (2012, 2016) developed a program called *Phonetisaurus* which uses the openFST toolkit developed at Google (Allauzen et al. 2007[2]). This program first uses a many-to-many aligner to align the input letters and output phoneme symbols. Then a joint n-gram model is trained from the training corpus of input:output symbols. The n-grams are a sequence of length n of input and output symbol pairs. In languages such as English where the correspondence between letters and phonemes is not so strict and there are many-to-one mappings, n-grams of four or longer are better. Consider for instance words like “rough,” “through,” and “though,” which all end in the same four letters but have three different pronunciations (sounding like “ruff,” “threw,” and “low”). An n-gram that is five input-output pairs long for the word “rough” would look like this:

r}r o}uh u}_ g}f h}_

where the underscores mark the absence of a corresponding phoneme. The program collects probabilities over all the n-gram sequences found in the training corpus and stores them together in a weighted finite state transducer (WFST). A new pronunciation is predicted by composing a simple left-to-right FST of the input characters with the trained WFST model and finding the shortest path with the highest probabilities to give the output symbols.

More recently, the use of neural networks for solving the G2P problem has become more popular (Yao and Zweig 2015; Rao et al. 2015). The use of sequence-to-sequence models has been shown to perform as accurately as the WFST approach or better. What is interesting to note is that the type of errors it makes can be more random than with the WFST approach. With WFST, one can instruct the many-to-many aligner to only align allowable letter-to-sound combinations, thereby reducing the risk of producing odd or illegal phonetic pronunciations. The neural networks have no such restrictions.

### 12.2.4 Prosody Prediction

Prosody is a suprasegmental phenomenon that helps the listener understand the message better. A speaker uses acoustic cues such as lengthening of certain sounds, inserting pauses and raising the pitch and loudness of the speech in certain places to indicate where the important words are (called emphasis or *accentuation*) and where phrase boundaries occur, which divide sentences into shorter phrases that are easier to process. For the linguistic front-end, it is necessary to predict where these events will occur based on the input text. The type of prosodic information that needs to be predicted can depend on the type of synthesis used in the acoustic back-end. The current prevailing method of unit selection synthesis often only uses a very simplistic prosodic description. Sentences are marked as either declarative sentences or questions. Commas mark the location of phrase boundaries. Some TTS systems also have rules for breaking up long phrases into shorter phrases using part-of-speech tags to find good places to break up the phrase.

Some systems use a very rudimentary distinction to separate function words from content words to determine which words are important and which are not. This is a very coarse distinction and while most of the time function words are not accented, there are cases where they can be accented. There are also words such as "one" that can be used as a function word but also as a number. Additionally, not all content words are accented. Several studies have been performed to predict accentuation based on the input text using information about part-of-speech of the word (e.g. noun or verb) and the words surrounding it, the location relative to the phrase boundary and end of the sentence, the number of words in a sentence, and the distance to a previously accented word (Syrdal et al. 2001; Rosenberg 2009; Fernandez et al. 2014). Named entity recognition and information about frequent accentuation patterns are also important. For instance, in

English one always says "DOWNING street" and "Windsor DRIVE," and even though they are both compound nouns that form a proper name, the accentuation patterns differ.

Finally, predicting prosody is a difficult problem, but predicting prosody for expressive speech synthesis is even more challenging. Analysis of emotional speech has shown that speakers produce higher pitch excursions when producing happy or angry speech and the entire pitch contour is higher when they produce fearful speech (Klabbers et al. 2007). But they can also be more emphatic by producing more accentuated words or inserting more phrase breaks.

## 12.3 Acoustic Back-End

The acoustic back-end or speech generation module takes the predicted linguistic information and transforms it to speech. In the 1970s, speech synthesizers started to generate speech using vocoders. Speech data would be recorded from a single professional speaker and relevant acoustic features would be extracted from the speech signals. A set of rules would be derived from these acoustic features to direct how to generate the feature trajectories in various linguistic contexts. The generated acoustic features would then be transformed to a new speech signal using the vocoder software. The acoustic features represent a source-filter model, describing the excitation part and the spectrum part of the signal. The excitation part models the sound source or the behavior of the vocal cords. When the speech is unvoiced, the excitation consists of noise, and when it is voiced it consists of pulses at pitch intervals corresponding to the speed of vocal cord closures. The spectrum part of the signal describes the effect of the vocal tract and tongue on the excitation. It acts as a filter (hence the name source-filter model) which amplifies certain frequencies in the air flow coming from the lungs through the vocal cords and diminishing others. The spectrum describes these frequency patterns as a function of time and frequency and amount of amplification of the frequencies. There are various methods to discretize the spectrum into a smaller set of parameters.

The first vocoders used a description of the formants and formant bandwidths (regions in the spectrum with high amplitude) as spectral features (Klatt 1980). The problem with formant synthesis is that sometimes the formants are difficult to extract from the acoustic signal reliably making the resulting speech sound less intelligible. In addition, the source model is too simplistic and the description of the spectrum as a set of (usually four) formants and formant bandwidths removes too much relevant information from the spectrum. The resulting vocoded speech can sound very intelligible but is far from natural.

Another popular approach was called articulatory synthesis, describing the articulatory muscles and properties of the tongue, mouth and vocal chords, which are responsible for the changes to the speech signal (Liljencrants 1985). In both cases, rules were handcrafted based on the analysis of a small amount of speech recordings from a single speaker. Similarly to formant synthesis, the models were overly simplistic and thus lacked naturalness. In the 1980s, concatenative synthesis started to emerge, where speech was generated by concatenating prerecorded segments of speech called *diphones*. Diphones represent the transition from one phoneme to the next. For instance, for the English word "cat" you would need the phonemes [k ae t] which would be made up of the diphones silence-k k-ae ae-t t-silence. The cut points in each phoneme occur roughly in the middle.

For a given language, depending on its phoneme set, a diphone inventory of 3000–4000 diphones is needed to cover all possible transitions. The diphones were recorded in a monotone fashion with articulatory neutral phonemes on either side (such as a [t] or [d]) in a carrier sentence to minimize the spectral mismatch at the concatenative joins. Duration and pitch contours were computed by rule and applied to the speech signals via LPC coding or pitch-synchronous overlap-and-add (PSOLA). The prosody rules were handcrafted by analyzing utterances from the same speaker that recorded the diphone inventory. The signal modification and simplicity of the duration and prosody rules had a large effect on the naturalness of the synthesized speech.

### *12.3.1 Unit Selection Synthesis*

With the advances in computing power and speed, the largely rule-based diphone synthesis approach was replaced by a more resource-intensive method called *unit selection synthesis* (Hunt and Black 1996; Conkie 1999; Clark et al 2007). It uses a large acoustic database recorded by a professional voice talent. Most commercial systems nowadays use anywhere from ten to over one hundred hours of speech. The sentences that are read by the professional speaker are selected from a large source of text for the target language, and they are selected to cover a wide variety of phonetic and prosodic contexts. The recordings are segmented into phonemes and additional information is stored in label files to indicate word boundaries, syllable boundaries, half phone boundaries (where the cut point should be in the phoneme) as well as stress.

At synthesis time, the linguistic description is used along with a target cost and a concatenation cost to find appropriate acoustic units in the database to concatenate to produce the most natural sounding speech. The target cost compares the candidate acoustic units in the database (half phones) to the target unit as specified by the linguistic representation.

*Target cost*—The simplest target cost function simply counts for each candidate unit how many linguistic features differ from the candidate unit. If there is a perfect match, the cost will be zero. For the target cost to predict how well the candidate will fit in terms of prosody, the linguistic specification needs to contain information about where in the sentence the candidate unit occurs in the original recording, whether the candidate occurs in a stressed syllable or not and ideally whether the candidate occurs in an accented word or not. The problem with this simple target cost is that candidates with different linguistic specifications can sound very similar whereas candidates with similar linguistic specifications can sound very different. Therefore, often a more complicated target cost is used that predicts the acoustic properties of target units. It will model target pitch, duration, and even energy based on the linguistic specification of the target sentence and compute deviations between the candidates and the target values. A typical unit selection system uses a mix of linguistic and acoustic features to produce a target cost.

*Concatenation cost*—The concatenation cost computes the difference in pitch and spectrum between the top candidate units and the selected target unit that precedes it. The goal is to pick a candidate unit that fits as closely as possible to avoid audible glitches at the concatenation boundary. Some TTS systems perform some type of smoothing at the boundary, to remove some of the glitches, but this requires speech signal modification, which in itself could cause some degradation of the signal.

In order to minimize the chance of glitches at locations where two acoustic units are pasted together, the speech is typically read in a neutral reading style. Unit selection synthesis works very well for a wide variety of popular applications of TTS such as web and document reading, GPS navigation, e-learning, and so forth. However, due to the neutral reading style needed to avoid audible glitches, this method is not as suitable for applications that require more expressive speech such as virtual agents or robots designed to work with children or elderly people. To extend a unit selection system to include more expressive speech, more audio needs to be recorded with the professional speaker. The recordings always need to be performed over a short period of time to ensure that the studio settings and the professional speaker's voice do not change. Creating a new unit selection voice is very time-consuming and costly due to the large amount of recordings that are needed and the post-processing of the database (i.e. segmentation corrections).

### 12.3.2 Statistical Parametric Speech Synthesis

Statistical parametric speech synthesis (SPSS) goes back to the principles used in vocoded speech synthesis. But instead of using handcrafted rules, powerful computer algorithms are now able to perform complex regression tasks to learn the relationship between an input linguistic description of an utterance and the output acoustic features needed to synthesize speech using a vocoder. The model is parametric because it describes the speech using parameters, such as the acoustic features used in vocoders. It is statistical because it describes those parameters using statistics (e.g. means and variances of probability density functions) that capture the distribution of parameter values found in the training data.

The input linguistic description is very similar regardless of the type of SPSS performed (Zen et al. 2009). Each phoneme in a sentence in the training data is described by linguistic features, such as the phoneme identity of the current phoneme and two preceding and following phonemes, the place and manner of articulation of the current phoneme, the position of the phoneme in the syllable as well as higher-level features such as the number of phonemes in the preceding, current, and following syllable, whether the preceding, current or next syllable has lexical stress, where the syllable is located in the current word or phrase, what the guessed part of speech is of the preceding, current, and following word, how many syllables are in the preceding, current, or following word, and so on.

The output acoustic features can be extracted from the speech waveforms using any vocoder, as long as the acoustic features are sufficient to reconstruct the speech signal with high quality. The STRAIGHT vocoder (Kawahara et al. 1999) is a very popular vocoder used in SPSS. It is available for free for academic use. The WORLD vocoder (Morise et al. 2016) is an alternative vocoder which is free to download for academic and commercial use.[3] STRAIGHT extracts acoustic features for every frame (5 ms) and typically uses between 40 and 60 parameters per frame to represent the spectral envelope (Mel-Generalized Cepstral coefficients or MGC), one parameter for the value for F0 (the log fundamental frequency or lf0), and five parameters to describe the spectral envelope of the aperiodic excitation (Band Aperiodicity Parameters or BAP). In order to capture the dynamic quality of the speech signal, they also contain the first order derivative called the *delta* coefficients and the second order derivative called the *delta-delta* coefficients.

The big advantage of SPSS is that the model parameters can be easily adapted to produce voice conversion or different speaking styles. Additionally, the trained statistical models provide a very compact representation.

As such not a lot of space is needed for an SPSS system, and it is very suitable for embedded devices such as GPS navigation in cars or in cell phones.

### *12.3.2.1 HMM Synthesis*

Hidden Markov Models (HMMs) are statistical time-series models, which are used in various fields. In automatic speech recognition (ASR), they have been used for over twenty years to recognize time-series sequences of acoustic features and label them with phonemes or words. Given the success in ASR, the idea to do the opposite, to train HMMs to produce time-series sequences of acoustic features from the linguistic specification was born (Zen et al. 2009; King 2011, Yamagishi 2006).

Figure 12.2 shows an overview of the HMM synthesis process. The HMM training part takes as input the linguistic representation and as output the acoustic features in terms of the excitation and spectral parameters

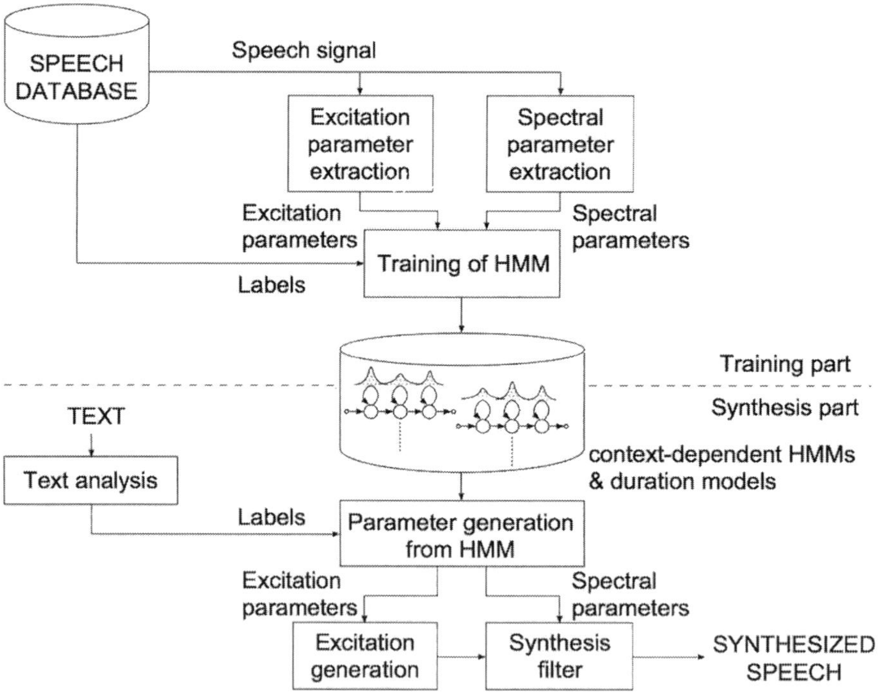


**Figure 12.2** HMM Synthesis.

*Source:* From Zen et al. 2007). Reproduced with permission.

and trains regression models to go from the input to the output. Since each linguistic feature can have many different feature levels, the number of permutations when considering all combinations is enormous. It is very unlikely that any amount of training data can cover all permutations even if only linguistically valid combinations are taken into account (Möbius 2003). Even though the linguistic features operate at different levels of a tree structure (phoneme, syllable, word, phrase), they are flattened into a sequence of features operating at the phoneme level to be used for SPSS. The traditional approach for handling these complex contexts in HMM synthesis is to use a distinct context-dependent HMM for each particular combination of all possible contexts at the phoneme level. But this leads to poorly trained models due to the sparsity of certain context combinations.

To address this problem, top-down decision-tree-based context clustering is used, where states of context-dependent HMMs are grouped into "clusters" and the distribution parameters within each cluster are shared. The decision tree is constructed by sequentially selecting the questions that yield the largest likelihood increase of the training data. The size of the tree is controlled using a predetermined threshold of likelihood increase or by introducing a model complexity penalty, such as minimum description length (MDL, Shinoda and Watanabe 2000). With the use of context questions and parameter sharing, the unseen contexts and data sparsity problems are effectively addressed.

The HMM method trains a context-dependent phone-based HMM model where the spectral features are represented not as feature values but as occurring somewhere on a Gaussian distribution of that spectral feature with a certain mean and variance. When dealing with vector data, a multivariate Gaussian needs to be used with a covariance matrix instead of the variance (Taylor 2009). Because speech is so complex, one Gaussian distribution is not sufficient to describe a feature, so they are usually represented as a mixture of Gaussians with a mean and covariance matrix for each Gaussian in the mixture. The representation is called a Gaussian Mixture Model or GMM. The HMM is different from the models used for ASR in that an explicit duration model is trained to define the durations of each state in the HMM. Therefore, it is often called a Hidden Semi-Markov Model (HSMM).

During synthesis, an utterance HMM is constructed by concatenating context-dependent HSMMs according to the linguistic features of the input text. State durations are determined based on the context-dependent state duration probability density functions. Parameter trajectories are then generated so that their output probabilities are maximized. These parameter trajectories are smoothed using maximum likelihood parameter generation (MLPG) (Tokuda 2000) and are then converted into a synthesized waveform by the vocoder.

The advantages of HMM synthesis are the same as for any SPSS system. The disadvantage of HMM synthesis is that it results in safe and overly smoothed parameters, which can have a negative impact on the synthesis quality. Even with a high-quality vocoder such as STRAIGHT, the resulting speech often sounds quite muffled, buzzy or unnatural compared to the original recordings in the training database. There are several different effects of signal representation and statistical modeling that contribute to the perception of over-smoothing (Merritt and King 2013). The spectral envelope is smoothed by the low-dimensional representation, then again by averaging over consecutive frames and over multiple tokens. The temporal structure of the speech parameters is smoothed because the model represents the trajectory with limited resolution (e.g. five states per phone-sized-unit). The variance of the generated speech parameters is also lower than those from natural speech. This has long been known to significantly reduce the quality of the generated speech and is in some systems mitigated by considering Global Variance (GV). However, GV cannot guarantee perfect restoration of the correct variance of the parameters.

#### *12.3.2.2 ANN Synthesis*

A number of studies have demonstrated that modeling acoustic parameters using artificial neural networks (ANNs) can achieve significantly better performance than decision tree clustered HMMs. While decision trees in HMM synthesis make predictions for each hidden state, and makes them separately for source and excitation parameters, ANN synthesis typically operate directly at the level of the acoustic frame, and trains output source and excitation parameters simultaneously, thus offering an efficient and distributed representation of the complex dependencies between linguistic and acoustic features. A neural network model performs nonlinear regression on the input linguistic features to produce the output acoustic features. A study by Watts et al. (2016) showed that in general listeners preferred the output of ANN-based synthesis over HMM-based synthesis when using the exact same data in both methods.

Figure 12.3 shows a general architecture of an ANN synthesis system. The bottom shows the input linguistic representation. The input to an ANN is the same as used in HMM synthesis, using a flattened list of linguistic features at the phoneme level. These features are represented in a vector for each phoneme by converting categorical features to binary features (asking questions such as, is the current phoneme a plosive yes or no, is the current syllable stressed yes or no). These questions are stored in a question file and the exact questions and values depend on the language. This conversion is called *one-hot encoding*. For each categorical question

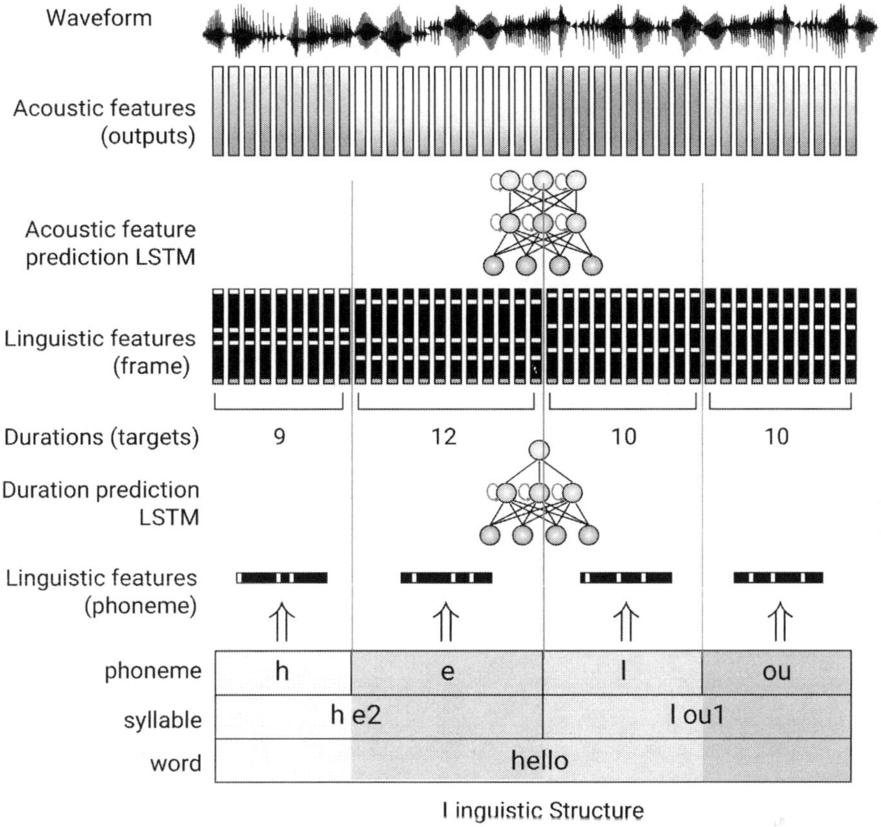


**Figure 12.3** ANN Synthesis

*Source:* From Zen et al. 2016. Reprinted with permission.

there is one 1 value for the yes answer (the "hot" or "true" answer) and several 0 values for the no answers. One-hot encoding does the same thing as creating binary decision trees in HMM synthesis. This input vector is then fed to an ANN that predicts durations for phones or states (just as with HMM synthesis, phonemes are sometimes better represented as a sequence of five states, to better deal with within-phoneme variations). The input linguistic vector and the predicted durations are then fed to another ANN, which predicts the most likely acoustic output features that are then converted to speech using a vocoder. The output waveform is the end result. In this figure, the output sentence is different than the input linguistic description shown. We can deduce from the shape of the waveform that the sentence is much longer and contains many words whereas the input example only shows the single word “hello.”

Different ANN architectures have been used to train both duration and acoustic models. In Figure 12.4 an example is shown of a simple feed-forward neural network (FNN). A FNN connects the units in an input layer to a hidden layer and then to one or more other hidden layers and finally to an output layer. The number of hidden layers and the number of units in each layer can be varied. For the input layer, the number of units depends on the number of linguistic features. The number of units in the output layer depends on the number of acoustic features to be predicted. In the case of the duration model using a five-state phoneme model, the number of output units is five. Since duration prediction is simpler than acoustic feature prediction a simpler architecture with fewer hidden layers and fewer units is sufficient. Each unit has an activation function that relates the output of that node to the input via some function. The activation function has to be nonlinear, otherwise it would just perform linear regression, which is not complex enough for predicting acoustic features. Units in an FNN are connected from left to right from input to hidden layer(s) to output and each connection has a weight associated to it, which is multiplied with the activation function.

The FNN model in Figure 12.4 is a fully connected network. Every unit in the network is connected to every other node in the subsequent layer. The weights are the parameters of the model, which are trained to produce optimal acoustic features. To do this, the ANN model receives input and output acoustic features from a train set and compares the predicted

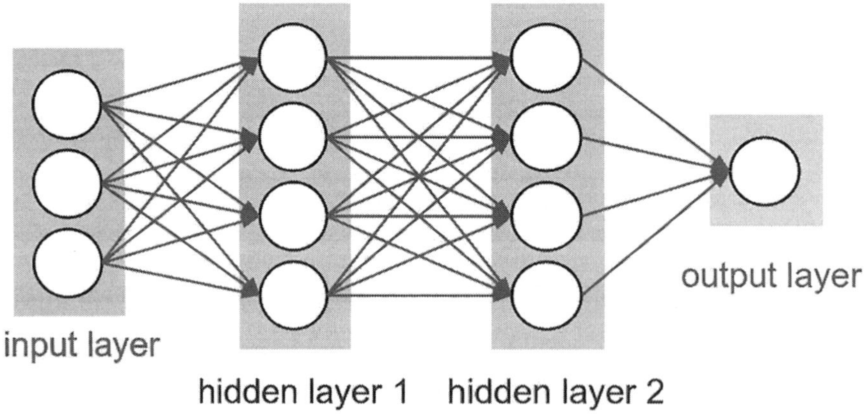


**Figure 12.4** A Simple Feed-Forward Neural Network (FNN).

acoustic features to the actual acoustic features observed in a validation set and a test set. The validation set comparison is done at every epoch in the training cycle to further optimize the weights.

Recurrent neural networks (RNNs), especially long short-term memory (LSTM)-RNNs, provide an elegant way to model speech-like sequential data that embodies short- and long-term correlations (Zen et al. 2016). Typically, RNN models consist of several feed-forward layers and one or two LSTM layers. Some researchers have employed bottleneck features, which are features that are generated by including a compressed hidden layer with fewer units in the network. This compressed representation contains information about what the model has learned. By training a model with the bottleneck layer as output layer and feeding the output features into a new training model as input features, the prediction can be improved (Wu et al. 2015).

While ANN synthesis generally sounds less buzzy and muffled than HMM synthesis, the pitch contours generated by ANN synthesis are still more monotonous than in unit selection synthesis because prediction and annotation of accentuation lacks accuracy and the word-level accentuation feature is the same for each frame in the word, thereby diluting the effect of that feature. There is also no easy way to boost the importance of a single linguistic input feature such as accentuation to the output pitch contour. Some solutions have been proposed to model the pitch separately using syllable-based pitch templates (Ronanki 2016) or using a hierarchical model (Ronanki 2017). Other approaches involve decomposing the pitch contour into pitch and accent contours to model long-term phrase curves separately from accent curves (Langarani et al. 2014, 2015; Suni et al. 2017).

#### *12.3.2.3 WaveNet*

The latest breakthrough in statistical speech synthesis is called WaveNet, developed by Google's DeepMind team (van Oord et al. 2016). Instead of extracting acoustic features from the speech signals, WaveNet generates speech samples directly from the raw audio waveform. It uses causal convolutions to model the conditional probabilities of generating audio samples based on previous frames of audio samples. The resulting synthesis sounds much more natural than comparable unit selection and ANN synthesis using the same voice (van Oord et al. 2016). In order to ensure that the audio has the required language-specific characteristics, the model also needs to be conditioned on the linguistic and prosodic features. The main drawback of WaveNet is that it currently requires a lot of computational power to train models. However, ongoing efforts aim to make the algorithm faster and less resource-intensive (Shen et al. 2018).

Another popular trend spearheaded by Google is end-to-end synthesis in their Tacotron framework (Wang et al. 2017). The idea of end-to-end synthesis is that the input is just the characters in the input text and a convolutional neural network learns the mapping to the appropriate acoustic features. Feature engineering is not considered necessary. This also means that any disambiguation (whether the "ea" in "read" goes to an [ee] or [eh] phoneme) is implicitly learned by the ANN model by giving it enough examples. If you have a trove of training material these types of phenomena should be learned from the data. This is very helpful for low-resource languages where one does not have a lot of linguistic knowledge available. But for languages such as English, it remains to be seen whether this method yields results that are as accurate as other methods that use the linguistic and prosodic information available.

## 12.4 TTS Software and Tools

### 12.4.1 *Festival*

Festival is an open-source toolkit for TTS developed by CSTR in Edinburgh. The code is available on their website.[4] Tools to build new voices are available via the Festvox project from CMU.[5] Festival is written in C++ and has a scheme-based interface for controlling which modules are loaded and for performing several tasks in the TTS chain. One can listen to an online demo on the Festival website.[6] The linguistic front-end from Festival can be used to create linguistic representations to feed into HMM or ANN synthesis systems.

### 12.4.2 *Merlin*

Merlin is an open-source toolkit for ANN synthesis developed by CSTR in Edinburgh.[7] The Github download comes with a demo voice using audio data from the ARCTIC data set.

### 12.4.3 *Other Tools*

- HTS: The open-source toolkit HTS (Zen et al. 2007) has become the standard toolkit for building HMM TTS voices.[8] HTS is built on top of HTK, the Hidden Markov ToolKit (Young et al. 2002, p. 175).
- Sparrowhawk: This was developed as an open-source version of Google's Kestrel text normalization for TTS module (Ebden and Sproat 2015).[9] It segments text into sentences and then tokenizes them. A set

of grammars converts tokens to words for several classes such as currency amounts, dates, and so on. It uses openFST to represent grammars. Example grammars are available for several small languages.
- SpaCy: This NLP toolkit provides python code for a range of NLP tasks such as tokenization, part-of-speech tagging, named entity recognition, and so forth for over 26 languages.[10]

## 12.5 Conclusion

Research into SPSS and all aspects related to linguistic representations and acoustic features is flourishing at the moment. At Interspeech, one of the largest language and speech technology conferences, the number of papers related to these topics has been growing over the last decade. The International Speech Communication Association has made papers from previous conferences and workshops publicly available.[11] For the past few years, ANN synthesis has dominated the annual so-called Blizzard challenge, where researchers working on TTS can use a common data set to create the best TTS system possible using their personal knowledge.[12] As the quality of the synthesis improves, interest in generating expressive speech, voice conversion and singing synthesis grows as well. This chapter has provided an overview of what is going on in the field of TTS synthesis and offers some links to provide starting points for people interested in learning more about the multidisciplinary topics in the field.

## List of Websites Relevant to TTS

Speech Zone: www.speech.zone/
Speech Synthesis Special Interest Group: https://synsig.org/index.php/Main_Page
– Software: https://synsig.org/index.php/Software
– Pointers: https://synsig.org/index.php/Pointers

### *Companies not Mentioned in the Speech Synthesis SIG Pointers Page*

ReadSpeaker www.readspeaker.com
NeoSpeech: https://neospeech.com
Acapela: www.acapela-group.com
IVONA text-to-speech (providing TTS for Amazon): www.ivona.com
Amazon Polly: https://aws.amazon.com/polly/
Cereproc: www.cereproc.com
Google: https://ai.googleblog.com/search/label/TTS

## Notes

1 Speech zone: www.speech.zone/
2 openFST website: www.openfst.org
3 WORLD vocoder: https://github.com/mmorise/World
4 Festival: www.cstr.ed.ac.uk/projects/festival/
5 Festvox: http://festvox.org
6 Festival demo: www.cstr.ed.ac.uk/projects/festival/onlinedemo.html
7 Merlin code: https://github.com/CSTR-Edinburgh/merlin
8 HTS: http://hts.sp.nitech.ac.jp
9 Sparrowhawk: https://github.com/google/sparrowhawk
10 Spacy: https://spacy.io/
11 www.isca-speech.org/iscaweb/index.php/archive/online-archive
12 Blizzard challenge: www.festvox.org/blizzard/

## References

Allauzen, C., Riley, M., Schalkwyk, J., Skut, W., and Mohri, M., 2007. OpenFst: A general and efficient weighted finite-state transducer library. *Implementation and Application of Automata*, pp. 11–23.

Bisani, M., and Ney, H., 2008. Joint-sequence models for grapheme-to-phoneme conversion. *Speech Communication* 50, pp. 5, pp. 434–451.

Black, A. W., and Lenzo, K. A., 2001. Flite: A small fast run-time synthesis engine. *4th ISCA Tutorial and Research Workshop (ITRW) on Speech Synthesis*, Pitlochry, Scotland. https://isca-speech.org/archive_open/ssw4/index.html.

Black, A. W., Taylor, P., Caley, R., Clark, R., Richmond, K., King, S., Strom, V., and Zen, H., 2001. The festival speech synthesis system, version 1.4. 2. Unpublished document. Viewed August 16, 2018 <www. cstr. ed. ac. uk/projects/festival.html>.

Clark, R. A., Richmond, K., and King, S., 2007. Multisyn: Open-domain unit selection for the Festival speech synthesis system. *Speech Communication* 49, no. 4, pp. 317–330.

Conkie, A., 1999, March. Robust unit selection system for speech synthesis. *137th Meeting of the Acoustical Society of America*, Berlin, Germany, p. 978.

Dutoit, T., 1997. *An Introduction to Text-to-speech Synthesis*. Vol. 3. Dordrecht, Boston, London: Kluwer Academic Publishers.

Ebden, P., and Sproat, R., 2015. The Kestrel TTS text normalization system. *Natural Language Engineering* 21, no. 3, pp. 333–353.

Fernandez, R., Rendel, A., Ramabhadran, B., and Hoory, R., 2014, September. Prosody contour prediction with long short-term memory, bi-directional, deep recurrent neural networks. *Interspeech 2014*, pp. 2268–2272. Viewed August 16, 2018 <http://mazsola.iit.uni-miskolc.hu/~czap/letoltes/IS14/IS2014/PDF/AUTHOR/IS141021.PDF>.

Hunt, A. J., and Black, A. W., 1996, May. Unit selection in a concatenative speech synthesis system using a large speech database. *Acoustics, Speech, and Signal Processing, 1996. ICASSP-96. Conference Proceedings, 1996 IEEE International Conference*. Atlanta, GA: IEEE, Vol. 1, pp. 373–376.

Jurafsky, D., and Martin, J. H., 2014. *Speech and Language Processing*. Vol. 3. London: Pearson.

Kawahara, H., Masuda-Katsuse, I., and De Cheveigne, A., 1999. Restructuring speech representations using a pitch-adaptive time—frequency smoothing and an instantaneous- frequency-based f0 extraction: Possible role of a repetitive structure in sounds. *Speech Communication* 27, no. 3, pp. 187–207.

King, S., 2011. An introduction to statistical parametric speech synthesis. *Sadhana* 36, no. 5, pp. 837–852.

Klabbers, E., Mishra, T., and van Santen, J. P., 2007. Analysis of affective speech recordings using the superpositional intonation model. *6th ISCA Workshop on Speech Synthesis, Bonn, Germany, August 22–24, 2007. In SSW6*, pp. 339–344. Viewed August 16, 2018 <www.isca-speech.org/archive_open/archive_papers/ssw6/ssw6_339.pdf>.

Klatt, D. H., 1980. Software for a cascade/parallel formant synthesizer. *Journal of the Acoustical Society of America* 67, no. 3, pp. 971–995.

Klatt, D. H., 1982. The Klattalk text-to-speech conversion system. *Acoustics, Speech, and Signal Processing, 1982. ICASSP-82. Conference Proceedings, 1982 IEEE International Conference*. Paris, France: IEEE, pp. 1589–1592.

Langarani, M. S. E., Klabbers, E., and Van Santen, J., 2014, May. A novel pitch decomposition method for the generalized linear alignment model. *Acoustics, Speech, and Signal Processing, 2014. ICASSP-14. Conference Proceedings, 2014 IEEE International Conference*. Florence: IEEE, pp. 2584–2588.

Langarani, M. S. E., Santen, J. V., Mohammadi, S. H., and Kain, A., 2015. Data-driven foot-based intonation generator for text-to-speech synthesis. *Proceedings of the 16th Annual Conference of the International Speech Communication Association*. Dresden, Germany: INTERSPEECH 2015. September 6, 2015—September 10, 2015.

Liljencrants, J., 1985. *Speech Synthesis with a Reflection-type Line Analog*. Doctoral Dissertation, Department of Speech Communication and Music Acoustics, Stockholm: Royal Institute of Technology.

Merritt, T., and King, S., 2013. Investigating the shortcomings of HMM synthesis *Eighth ISCA Workshop on Speech Synthesis. 8th ISCA Speech Synthesis Workshop. August 31—September 2, 2013*. Barcelona, Spain. Viewed August 16, 2018 <www.cstr.ed.ac.uk/downloads/publications/2013/ssw8_PS2-4_Merritt.pdf>.

Morise, M., Yokomori, F., and Ozawa, K., 2016. World: A vocoder-based high-quality speech synthesis system for real-time applications. *IEICE TRANSACTIONS on Information and Systems* 99, no. 7, pp. 1877–1884.

Möbius, B., 2003. Rare events and closed domains: Two delicate concepts in speech synthesis. *International Journal of Speech Technology* 6, no. 1, pp. 57–71.

Novak, J. R., Minematsu, N., and Hirose, K., 2012, July. WFST-based grapheme-to-phoneme conversion: Open source tools for alignment, model-building and decoding. *Proceedings of the 10th International Workshop on Finite State Methods and Natural Language Processing*. Donostia/San Sebastian: Association for Computational Linguistics, July 23–25, 2012, pp. 45–49. Viewed August 16, 2018 <www.aclweb.org/anthology/W12-6208>.

Novak, J. R., Minematsu, N., and Hirose, K., 2016. Phonetisaurus: Exploring grapheme-to-phoneme conversion with joint n-gram models in the WFST framework. *Natural Language Engineering* 22, no. 6, pp. 907–938.

Oord, A. V. D., Dieleman, S., Zen, H., Simonyan, K., Vinyals, O., Graves, A., Kalchbrenner, N., Senior, A., and Kavukcuoglu, K., 2016. Wavenet: A generative model for raw audio. arXiv, viewed August 16, 2018 <https://arxiv.org/abs/1609.03499>.

Rao, K., Peng, F., Sak, H., and Beaufays, F., 2015, April. Grapheme-to-phoneme conversion using long short-term memory recurrent neural networks. *Acoustics, Speech, and Signal Processing, 2015. ICASSP-15. Conference Proceedings, 2015 IEEE International Conference*. Brisbane: IEEE, pp. 4225–4229.

Ronanki, S., Henter, G. E., Wu, Z., and King, S., 2016. A template-based approach for speech synthesis intonation generation using LSTMs. In: *INTERSPEECH*, pp. 2463–2467.

Ronanki, S., Watts, O., and King, S., 2017. A hierarchical encoder-decoder model for statistical parametric speech synthesis. *Proc. Interspeech 2017*, pp. 1133–1137.

Rosenberg, A., 2009. *Automatic Detection and Classification of Prosodic Events*. PhD thesis, Columbia University Press, New York, NY. Viewed April 11, 2019 < http://www.cs.columbia.edu/nlp/theses/andrew_rosenberg.pdf>.

Shen, J., Pang, R., Weiss, R. J., Schuster, M., Jaitly, N., Yang, Z., Chen, Z., Zhang, Y., Wang, Y., Skerry-Ryan, R. J., and Saurous, R. A., 2018. Natural TTS synthesis by conditioning WaveNet on mel spectrogram predictions. *Acoustics, Speech, and Signal Processing, 2018. ICASSP-18. Conference Proceedings, 2018 IEEE International Conference*. IEEE, Calgary, pp. 4779–4783.

Shinoda, K., and Watanabe, T., 2000. MDL- based context-dependent subword modeling for speech recognition. *Journal of the Acoustical Society of Japan* 21, no. 2, pp. 79–86.

Sproat, R. (Ed.), 1998. *Multilingual Text-to-speech Synthesis: The Bell Labs Approach*. Norwell, MA: Kluwer Academic Publisher.

Sproat, R., Black, A. W., Chen, S., Kumar, S., Ostendorf, M., and Richards, C., 2001. Normalization of non-standard words. *Computer Speech & Language* 15, no. 3, pp. 287–333.

Suni, A., Šimko, J., Aalto, D., and Vainio, M., 2017. Hierarchical representation and estimation of prosody using continuous wavelet transform. *Computer Speech & Language* 45, pp. 123–136.

Syrdal, A. K., Hirschberg, J., McGory, J., and Beckman, M., 2001. Automatic ToBI prediction and alignment to speed manual labeling of prosody. *Speech Communication* 33, no. 1, pp. 135–151.

Taylor, P., 2009. *Text-to-speech Synthesis*. New York: Cambridge University Press.

Tokuda, K., Yoshimura, T., Masuko, T., Kobayashi, T., and Kitamura, T., 2000. Speech parameter generation algorithms for HMM-based speech synthesis. *Acoustics, Speech, and Signal Processing, 2000. ICASSP-00. Conference Proceedings, 2000 IEEE International Conference*, Istanbul, Turkey: IEEE, pp. 1315–1318.

Van Santen, J. P., Sproat, R., Olive, J., and Hirschberg, J. (Eds.), 2013. *Progress in Speech Synthesis*. Cham: Springer.

Wang, Y., Skerry-Ryan, R. J., Stanton, D., Wu, Y., Weiss, R. J., Jaitly, N., Yang, Z., Xiao, Y., Chen, Z., Bengio, S., and Le, Q., 2017. Tacotron: Towards end-to-end speech synthesis. *arXiv preprint arXiv:1703.10135*.

Watts, O., Henter, G. E., Merritt, T., Wu, Z., and King, S., 2016, March. From HMMs to DNNs: Where do the improvements come from?. *IEEE International Conference on Acoustics, Speech and Signal Processing* (ICASSP), pp. 5505–5509.

Wu, Z., Valentini-Botinhao, C., Watts, O., and King, S., 2015, April. Deep neural networks employing multi-task learning and stacked bottleneck features for speech synthesis. *Acoustics, Speech, and Signal Processing, 2015. ICASSP-15. Conference Proceedings, 2015 IEEE International Conference*. Brisbane, Australia: IEEE, pp. 4460–4464.

Yamagishi, J., 2006. *An Introduction to Hmm-based Speech Synthesis*. Technical Report. Viewed August 16, 2018 <https://wiki.inf.ed.ac.uk/pub/CSTR/TrajectoryModelling/HTS-Introduction.pdf>.

Yao, K., and Zweig, G., 2015. Sequence-to-sequence neural net models for grapheme-to-phoneme conversion. *arXiv preprint arXiv:1506.00196*.

Yarowsky, D., 1995, June. Unsupervised word sense disambiguation rivaling supervised methods. *Proceedings of the 33rd annual meeting on Association for Computational Linguistics*. Cambridge, MA: Association for Computational Linguistics, pp. 189–196.

Young, S., Evermann, G., Gales, M., Hain, T., Kershaw, D., Liu, X., Moore, G., Odell, J., Ollason, D., Povey, D., and Valtchev, V., 2002. The HTK book. *Cambridge University Engineering Department* 3, p. 175. Viewed August 16, 2018 <www.dsic.upv.es/docs/posgrado/20/RES/materialesDocentes/alejandroViewgraphs/htkbook.pdf>.

Zen, H., Agiomyrgiannakis, Y., Egberts, N., Henderson, F., and Szczepaniak, P., 2016. Fast, compact, and high quality LSTM-RNN based statistical parametric speech synthesizers for mobile devices. *arXiv preprint arXiv:1606.06061*.

Zen, H., Nose, T., Yamagishi, J., Sako, S., Masuko, T., Black, A. W., and Tokuda, K., 2007, August. The HMM-based speech synthesis system (HTS) version 2.0. In: *SSW, pp. 294–299. 6th ISCA Workshop on Speech Synthesis, Bonn, Germany, August 22–24, 2007*. Viewed August 16, 2018 <www.cs.cmu.edu/~awb/papers/ssw6/ssw6_294.pdf>.

Zen, H., Tokuda, K., and Black, A. W., 2009. Statistical parametric speech synthesis. *Speech Communication* 51, no. 11, pp. 1039–1064.

13

# Sound and UX Design

Michael L. Austin

## 13.1 Introduction: User Experiences and Sonic Experiences

All sound design is user experience design. At its core, one of the primary purposes of sound design is to create and curate the sound that accompanies an experience in order to enhance that experience. For example, the sound designer for a film or television show can use sound design to enhance, or create from scratch, the sonic elements of a film in an effort to enhance the experience of the audience. Companies use jingles and other iconic sounds in their advertising to connect whatever they are selling with our memories in hopes that we have future experiences with their product and remember their particular product or service the next time we visit the market. Sound designers and acoustical engineers work with car companies to design the custom-made sounds of their revving engines or closing doors, both of which are important parts of the car-driving experience. Sound designers create a set of sonic alerts for dating apps on smartphones that are used to notify you when you when someone interacts with your profile, and these sounds are so closely and uniquely associated with your experience with this app that you are able to distinguish between a love connection and a professional connection in your email app; while on the surface, they may seem like elements of interaction design, these sounds also elicit emotional responses that influence your overall experience of both apps—excitement that someone has shown romantic interest in your dating profile or dread that your boss may be sending more work to your inbox.

In this chapter, I discuss the various ways we perceive the designed experiences in our lives and the ways in which sound designers can use audio material to shape, control and define these experiences we have with various products and services on a day-to-day basis. As case studies to explore the ways we perceive various experiences, I will also highlight a selection of universal principles of design that are usually only recognized as visual concepts, but that have enormous influence on user experience design. I will connect these principles to sound design practices and possibilities that illustrate the use of sonic material for UX design. But first, what exactly is user-experience design?

## 13.2 UX Design: Definitions and Disciplinary Intersections

User-experience design (abbreviated as UX, XD, UXD or UED, hereafter UX in this chapter) is somewhat eponymous in that its definition is spelled out in its name: the design of a user's experiences of a particular product, device or application, system, website, kiosk, service, brand, or other object, often through a user interface. This experience extends beyond the use of the products or services themselves to their advertising, packaging and any other point of engagement a user might have with them. While some scholars and practitioners recognize UX design as a discipline in its own right, others view UX design as a subfield within human-computer interaction (HCI), or comparable to—and sometime interchangeable with—interaction design (IxD), user-centered design (UCD), value-based design, and human factors and ergonomics. And due to its lack of a clear-cut and authoritative definition, UX design has become an industry buzzword that at once means everything and nothing.

Donald Norman, usability engineer, cognitive scientist, design scholar and critic, is credited with coining the term "user experience design." In his book, *The Psychology of Everyday Things* (first published in 1988 and renamed the *Design of Everyday Things* in subsequent editions), Norman distinguishes experience design from other related fields, noting that industrial designers place emphasis on form and material and interactive designers focus on understandability and usability, while experience designers emphasize the emotional impact of a design. He further explains the differences among the three design fields, pointing out that none of them is particularly well defined:

*Industrial design:* The professional service of creating and developing concepts and specifications that optimize the function, value and appearance of products and systems for the mutual benefit of both user and manufacture.

*Interaction design:* The focus is upon how people interact with technology. The goal is to enhance people's understanding of what can be done, what is happening, and what has just occurred. Interaction design draws upon principles of psychology, design, art and emotion to ensure a positive, enjoyable experience.

*Experience design:* The practice of designing products, processes, services, events and environments with a focus placed on the quality and enjoyment of the total experience (Norman 2013, p. 7).

Norman was hired by Apple Computer, Inc. in 1993 to serve as "User Interface Architect," but he insisted that his title be changed to "User Experience Architect." In a 1998 email to Peter Merholz, Norman wrote, "I invented the term because I thought human interface and usability were too narrow. I wanted to cover all aspects of the person's experience with the system including industrial design graphics, the interface, the physical interaction and the manual. Since then the term has spread widely, so much so that it is starting to lose it's (sic) meaning" (Merholz 2007).

User experience design is relatively new term that describes some concepts that have been circulating within various fields of design for years. Industrial designer Henry Dreyfuss published his 1955 book, *Designing for People*, a seminal text in the field of human-centered design and a precursor to user experience design. In it, he highlights his ethos for focusing on the experience of the user of a product by including it prominently on the title page of his book:

> We bear in mind that the object being worked on is going to be ridden in, sat upon, looked at, talked into, activated, operated, or in some other way used by people individually or en masse. When the point of contact between the product and the people becomes a point of friction, then the industrial designer has failed. On the other hand if people are made safer, more comfortable, more eager to purchase, more efficient—or just plain happier—by contact with the product, then the designer has succeeded
>
> (Dreyfuss, title page)

In the last few decades, attempts to formally define UX design have centered on the emotional or perceptual experience with a product, but it almost always appears within the context of web design or app development, and seldom in discussions of analog product design. Furthermore, in reading definitions from various practitioners, design firms, blogs, and so forth, it becomes unclear if UX is a facet of design, customer service, consumer psychology, or engineering, or whether or not it is an interdisciplinary field that intersects with these fields or lies in the interstitial space between them. For example:

UX as interaction design optimization for users:

> User experience design as a discipline is concerned with all the elements that together makeup that interface, including layout, visual design, text, brand, sound, and interaction. UE works to

> coordinate these elements to allow for the best possible interaction by users.
>
> (User Experience Professional Association 2005)

UX as form of customer service/satisfaction:

> The goal of user experience design in industry is to improve customer satisfaction and loyalty through the utility, ease of use, and pleasure provided in the interaction with a product.
>
> (Kujala et al. 2011, p. 473)

UX as a dynamic, time-based journey that forms and shapes the users' experience:

> When first encountering a product, a user forms a momentary impression—which evolves over time, typically as the product is used throughout a period. In this process, the user's perception, action, motivation and cognition integrate to form a memorable and coherent story: called "the user experience.
>
> (Interactive Design Foundation, n.d.)

UX is less about usefulness and more about emotional, subjective interactions:

> UX highlights non-utilitarian aspects of such interactions, shifting the focus to user affect, sensation, and the meaning as well as value of such interactions in everyday life. Hence, UX is seen as something desirable, though what exactly something means remains open and debatable.
>
> (Law et al. 2009, p. 719)

> [UX design is] . . . dynamic, context-dependent and subjective.
>
> (Law et al. 2009, p. 727)

In an attempt to solidify professional standards for the ergonomics of HCI, the International Organization for Standardization (ISO)[1] defines user experience as a "person's perceptions and responses resulting from the use and/or anticipated use of a product, system or service" (ISO 9241–210 2010). In a series of notes, the ISO provides more details to further elucidate the qualities of user experience:

Note 1: User experience includes all the users' emotions, beliefs, preferences, perceptions, physical and psychological responses, behaviors and accomplishments that occur before, during and after use.

Note 2: User experience is a consequence of brand image, presentation, functionality, system performance, interactive behavior and assistive capabilities of the interactive system, the user's internal and physical state resulting from prior experiences, attitudes, skills and personality, and the context of use.

Note 3: Usability, when interpreted from the perspective of the users' personal goals, can include the kind of perceptual and emotional aspects typically associated with user experience. Usability criteria can be used to assess aspects of user experience. (ISO 9241–210 2010).

This definition and notes clarify and contextualize a number of things, especially that user experience is derived from the user's subjective lived experience. It is also based on context, some of which is controllable by a company (through advertising and contact with the user prior to use of the product or service and during interaction with the product or service), while other experiences with the product lie completely outside of a company's control (with a user's prior experience with a similar product, for example). Usability is obviously a key part of user experience, and this usability "exists in the experience of the person. If the person experiences a system as usable, it is. A commitment to designing for people means that, at base, we must accept their judgement as the final criterion for usability. . . . The starting point for usability engineering must be the uncovering of user experience" (Whiteside and Wixon 1987, p. 7). How, then, does sound design contribute to good UX design, how can it be employed to make things easier to use, and how does it shape a user's experience into a positive one?

## 13.3 Processing and Designing (Sonic) Experiences

In his book, *Emotional Design: Why We Love (or Hate) Everyday Things*, Donald Norman cites research in cognitive psychology that he conducted with colleagues at Northwestern University in which they outline three levels at which we are able to process affective experiences: "the automatic, prewired layer, called the *visceral level*; the part that contains the brain processes that control everyday behavior, known as the *behavioral level*; and the contemplative part of the brain, or the *reflective level*" (Norman 2004, p. 21). In the remaining pages of this chapter, I will outline ways in which users employ these three levels of cognitive processing to make sense of life experiences in general, and how sound design specifically can shape these experiences to make products and services more inviting, memorable and useful.

### 13.3.1 Visceral Experiences

According to Norman, the visceral level is "pre-consciousness, pre-though," where "appearance matters and first impressions are formed" (Norman 2004, p. 37). Visceral experiences are those that engage the various modalities of human sensation, namely, audition (hearing), somatosensation (touch), olfaction (smell), gustation (taste) and vision (sight). Visceral design relies on our physical senses to interpret information from our environment in order to provide immediate emotional feedback about the experience. Our experiences are built using various combinations of these sensory modalities and our brain's ability to make sense of the observed stimuli to create a unified perception of the event at hand. For example, we experience a rainstorm by seeing lightning and rain, hearing thunder, smelling *petrichor*, feeling raindrops on our skin, and so on.[2] Effective user experience design relies heavily on most of these senses, especially sight and sound, and a user's psychological and physiological response to them, in order to create the desired experience for the user. Beyond this set of five senses that most of us were taught as children, we also experience the world through other modes of perception which receive much less attention, despite the important roles they play in our visceral experience of the world around us—and inside us—such as proprioception (the awareness of the location of one's own body in space and includes the sense of movement, i.e. kinesthetic perception, and a sense of balance), interoception (internal senses, such as feelings of pain, hunger, nausea, etc.), and exteroception (the external sense of the environment outside of our bodies, such as changes in temperature, or thermoception).

Sometimes, our perceptions are deceiving, and we misperceive the realities present in the physical world. Optical illusions are an example of the ways in which our brain can be fooled, resulting in visual perception that does not match the realities of the physical world. Psychoacoustics is the study of the ways in which our brain processes sonic information, transforming it from physical mechanical sound waves in the outer and middle ear, into neurological impulses by the inner ear's organ of Corti (the sensory organ responsible for hearing, located in the cochlea), and transmitting these neurons to the brain in order to be perceived as sound. And as with optical illusions, there are instances in which our perception does not match physical reality. For example, we do not perceive the amplitude, or volume, of all pitches equally. Humans can perceive sounds within the range of 20 to 20,000 Hz, and as we age, we naturally tend to lose the ability to hear frequencies at the top of this range. When played at the same volume, we perceive the sounds in the middle of this range to be loudest, especially those in the voice frequency band (2,000–4,000 Hz).

The scope and limits of our auditory perception should be kept in mind when designing experiences using sonic material.

#### *13.3.1.1 Masking*

Beyond the limits of our hearing system described above, the psychoacoustic effect of auditory masking prevents our brain from perceiving all sounds present within our physical space. There are several forms of auditory masking. For example, *simultaneous masking* occurs when a loud sound covers up another softer sound when sounding together. Masking also occurs when a loud sound happens too close in time before or after a softer sound (called forward masking and backward masking respectively); in these instances, our visceral perception is unable to cope with sudden changes in volume, and the softer sound will not be perceived. Practicing sound designers work to avoid auditory masking so as to prevent the occlusion of sonic information that users would consider to be valuable in a particular experience. For instance, in the case of a smartphone application, user experience designers would work with sound designers to make sure that audio clips are played one after another, rather than simultaneously, and when they do, that one is not drastically louder than the other.

#### *13.3.1.2 Noise*

David Novak describes noise as something "typically separated from music on the grounds of aesthetic value. Music is constituted by beautiful, desirable sounds, and noise is composed of sounds that are unintentional and unwanted" (Novak and Sakakeeny 2015, p. 126). Although noise is often unavoidable and perceived on a visceral level, not all noise is unwanted, and sound design often relies on a great deal of noise in order to create a realistic experience for the intended audience of a designed experience. Extraneous noise is often associated with cities and other built environments, often heard in the form of cacophonous automobile traffic, blaring car horns, police and ambulance sirens, grating industrial machines, ear-splitting jackhammers and other construction noise, and so forth. Taking these sounds as raw material, sound designers for films or video games will often create soundscapes to serve as more realistic ambience, or atmosphere, for the location or setting of a scene. Not only does the noise communicate to the audience that the scene is set in the city (even if the trappings of the city cannot be seen on screen), it also creates a sense of sonic space. In this case, noise serves as a stable background, generally unaffected by the action of the scene, that helps foreground dialog and sound effects that are caused by diegetic actions within the narrative. Further interpretation of this noise can also influence reflective experiences positively or negatively: designed noise can be interpreted as positive if the soundscape causes the listener to fondly remember a trip they took to a particular city, but it could also destroy the sense of immersion within a scene if, for example, a sound designer carelessly includes the sound of sea gulls and European ambulances in a scene set in Kansas City.

##### *13.3.1.3 Sound in the Uncanny Valley (and Other Biases)*

In 1970, roboticist Masahiro Mori proposed a hypothesis about the visceral, uneasy feeling humans experience around objects that are humanoid but not *quite* human enough. Using a graph, Mori explains that an observer's comfort level with robots (and can be extended to mannequins, three-dimensional animations and video game characters, life-like dolls, clowns, etc.) increases as their appearance increases in verisimilitude. When a robot reaches the point of looking *almost* human, but not exactly human, we feel a sense of unease, repulsion and general uneasiness about the humanoid figure. This dip in affinity is what Mori calls *Bukimi no Tani Genshō*, or the Uncanny Valley (Mori 1970/2012).

Imbuing robots with more human-like sonic characteristics enhances user experience in some instances. When robots interact with humans in healthcare situations, people prefer a more human like voice over a robotic-sounding voice (Tamagawa et al. 2011). Other research indicates that robots using non-linguistic utterances, which are sounds such as "chirps, beeps, squeaks and whirrs," can provide social and emotional cues to humans during HCI that are less off-putting and anxiety-inducing (Read and Belpaeme 2016, p. 31), proving that thoughtful sound design can be an effective tool in bridging the Uncanny Valley.

Video game audio and music scholar Mark Grimshaw writes about uncanny sound design in video games, especially games in the horror genre. Grimshaw provides sound designers who want to include or exclude the uncanny affect in or from their projects with a rough context-based "rule of thumb," based on the small body of available research on the Uncanny Valley:

- Certain amplitude envelopes applied to sound affect perceptions of urgency.
- Frequency might have an effect on the unpleasantness of sound and this might lead to negative affect.
- Familiar or iconic sounds can be defamiliarized and this can lead to perceptions of uncanniness.

- Uncertainty about the location of a sound source, its cause or its meaning in the virtual world increases the fear emotion.
- An aural resolution that is lower than a high quality, human-like visual resolution might lead to the uncanny.
- A lack of synchronization between lips and voice for photo-realistic virtual characters leads to a perception of the uncanny (Grimshaw 2009, p. 5).

To illustrate the last of Grimshaw's axioms, sound is sometimes used in designed contexts in order to intentionally trick users into viscerally perceiving something that is not actually happening in the physical world. For example, the *spatial ventriloquism effect* leads an audience to believe that as they watch a film, they hear dialog coming from the mouths of the actors on screen, despite the fact that the loudspeakers projecting the dialog may be scattered throughout the movie theater and are rarely ever located near the actors' mouths. From a sound design perspective, this effect is only effective if the recorded dialogue syncs perfectly with the movement of actors' lips. Very small deviations completely ruin the illusion, and thus, the audience's experience of the film.

In interactive applications, synchronicity of designed sound is particularly important when sonic feedback accompanies particular interactions with an interface. If a user makes a selection on a touch screen and the earcon designed to accompany the action occurs much later, the utility of the sonic feedback is lost, likely causing a negative visceral reaction from the user, such as confusion or frustration. Likewise, sonic feedback that operates on the visceral level works much better when less interpretation is required, especially when the feedback provides vital information that should not be misinterpreted. Imagine a touch screen interface that controls the warning systems of a nuclear power plant. If the reactor core reached a dangerous temperature, putting everyone within miles in mortal danger, and someone in the plant noticed and needed to alert authorities, the sonic feedback from the touch screen of the warning system would need to be calming, yet serious and assertive, and easy to understand as second nature, rather than requiring more than a split second to interpret its meaning. Sonic feedback that included a panicked voice shouting "We're all going to die!" or a series of xylophone tones would be much less helpful and would evoke a less desirable visceral response in the people involved in this hypothetical situation who heard it than would a conspicuously sounding alarm or a clear, authoritative voice that loudly repeated the word "Danger!" every few seconds.

In addition to causing unintentional bias against a product for being too near-human (rather than unhuman or perfectly human), sound design can also viscerally trigger unconscious biases that people may have without their realization of it. People can attribute gender stereotypes to computers based on the types of voices used in their interaction design, where computers with female voices are rated as being more trustworthy with information about love and relationships and male computers are seen to be trustworthy with technical information about computers, even though these computers have access to the same information (Nass et al. 1997). The feeling of discomfort or unease can cause dissatisfaction with a product or service and can become a usability obstacle (Tinwell 2009). Therefore, good user experience design should endeavor to meet customer's expectations, and sound design can assist in meeting these expectations on a visceral level. However, sound designers should think seriously about ways to avoid reinforcing gendered (or other types of unhealthy) biases sonically and should work to create effective ways of challenging them or eliminating them outright.

### *13.3.2 Behavioral Experiences*

Although UX design shifts the focus from usability of a product or service and towards the ways in which the user feels about, remembers and relates to it, a good designer still takes into consideration the ways in which a user will engage with their design, and take into considerations ways in which the design will affect the behaviors of the user. These behaviors, in turn, have a dramatic impact on the overall designed experience. According to Norman, in behavioral design, "function comes first and foremost; what does a product do, what function does it perform? If the item doesn't do anything of interest, then who cares how well it works?" (Norman 2004, p. 74). UX design helps a user understand how a product works and what they are supposed to do with it, and it prevents them from negative experiences born from frustration or confusion.

#### *13.3.2.1 Affordances*

Good UX design facilitates a relationship through which users learn about how a product or service is utilized. These relationships are called affordances. According to psychologist James J. Gibson, affordances are elements within our environment that:

> . . . are what it *offers* the animal, what it *provides* or *furnishes*, either for good or ill. The verb to afford is found in the dictionary, the noun *affordance* is not. I have made it up. I mean by it something that refers to both the environment and the animal in a way

> that no existing term does. It implies the complementarity of the animal and the environment.
>
> (Gibson 1979, p. 127, emphasis in the original)

In our designed environment, affordances demonstrate how something is properly used, and elements of the design facilitate, or at least suggest, that use. For example, the roundness of a doorknob affords holding or grabbing, rather than pushing with a flat hand. Its height affords grasping easily by the general population, most of whom that are average height find the doorknob slightly higher their waist—just the height their hand naturally reaches when their arm is slightly extended. Some doors demonstrate poor UX because they communicate ways to be used that contradict their actual proper use. If you ever find yourself pulling on a door that should be pushed open, this may be due to the fact that instead of a push plate that affords pushing, the door has a handle that suggests the door should be pulled open.[3]

Sounds can also be designed to communicate how an object should be used or that it is being properly used, and they can also promote positive experiences and feelings with the object. Luxury car companies, such as Mercedes and BMW, hire sound designers who work with engineers to design the sounds that cars make in order to craft a luxury experience to accompany the car's luxury price tag. These sounds can also communicate a sense of the car's safety and reliability. According to Emar Vegt, aural designer for BMW, "the sound of the door closing is a remarkable aspect of the buying decision," he says. "It gives people reassurance if the door feels solid and safe" (Baker 2013). Some sounds are designed specifically to encourage positive behaviors when interacting with various objects. For instance, people are prone to exhibit more empathy towards female-sounding robots and were more likely to donate money to a robot research lab when asked by a robot with a female voice over a male-sounding robot (Siegel et al. 2009).[4]

#### *13.3.2.2 Defensive (Sound) Design*

Most, if not all, of the experiences described above are desired experiences and were designed to maximize pleasure, usefulness and other positive qualities. There are times, though, when a UX designer needs to warn a user against a particular action or provide negative feedback whenever the user makes a poor or less optimal choice. Defensive design, also called "hostile design" or "unpleasant design" is a trend in design and architecture in which designers create objects for public use that prevent people from using them the "wrong" way or for an unintended purpose. Referred to by some as "disaffordances," designs in this category are often intentionally made to be unwelcoming, uncomfortable and hard to use to the point of discouraging use. For example, chairs in airports usually include arm rests to discourage passengers from napping across several of them and taking up space where others need to sit. Benches at bus stops and public parks are also sometimes designed with no seat back at all or are cast in odd geometric shapes to discourage sleeping and skateboarding.

Sonic examples of defensive design include stores that play classical music or high frequency noise over their loudspeaker system to discourage youth from loitering on the premises. Germany's national train operator, Deutsche Bahn, has recently resorted to broadcasting atonal music into a railway station in Berlin in an attempt to make drug users feel unwelcome and uncomfortable (Marshall 2018). Sound designers incorporate various audio cues into physical products and software that alert a user when he or she has made a mistake or tries to execute an action that is beyond the permissible scope of use. For example, when using my computer and I try to search for a misspelled file name, or if I try to ignore an important alert window that has just popped up on my screen and try to continue to type, I receive auditory feedback in the form of a "thunk" error sound, discouraging me from continuing to make that mistake. Not only is this an audio cue designed to discourage errors, it is also one of several bespoke sonic icons (or earcons) that comprises the computer manufacturer sonic brand; so while it discourages mistakes and unwanted behaviors, this sound encourages familiarity with the company.

#### *13.3.2.3 Sonic Skeuomorphism*

Relying on past experiences with sound can inform our present behavior when we hear those sounds in other places or situations. Skeuomorphism refers to "a design element, usually ornamental, that mimics design qualities of other versions; digital skeuomorphs mimic design features of physical/analog objects" (Austin 2016, p. 29). Skeuomorphism connects our memories of using an object in one context and uses that connection to direct behavior and to create a meaningful experience in new context. In much the same way as we use metaphors connected to analog objects and actions to help us understand how our computers work, such as "windows," "files" and "cutting and pasting," skeuomorphs borrow analog design elements to more easily connect us to digital spaces. For example, some smartphone calculator apps are designed to look and function much like old-fashioned adding machines or an abacus. Note-taking apps sometimes resemble notebook paper or stationary, and calendars with flippable pages will be decorated with leather trim.

Sounds will often accompany these skeuomorphic elements. If a smartphone's e-reader app is designed with a function that allows users to turn the pages of e-books by swiping across the touch screen, the skeuomorphism can be extended into the sonic experience of the app by including the sound of turning pages with each swipe. In this way, the user's book reading experience is further enhanced beyond the strictly visual or tactile, and if this sensory experience is valued by the user, it will likely encourage further interaction and engagement with the app. Skeuomorphic earcons also communicate in an easily understandable way that a digital task has been accomplished, as when some computers provide audio feedback that sonically mimics the act of crumpling up a piece of paper and the sound of that paper ball landing in a recycling bin, alerting the user that a particular file has been deleted. This sonic feedback also makes the computer-using experience more enjoyable because the user can quickly and easily receive aural acknowledgment that a task has been done, rather than needing to expend time and energy to visually confirm that the file has actually been removed from its former location.

### *13.3.3 Reflective Experiences*

Thus far, I have discussed types of experiences that result from user experience design that engages with users on a physical and emotional level; visceral experiences are those that deal with immediate perceptions and first impressions, while behavioral experiences are related to the ways design guides our actions and uses of a particular design. Other experiences are more mental and emotional, and our understanding of them results from conscious interpretation, rather than natural reactions to them or their utility. Working on a variety of interpretive levels, these reflective experiences are "all about message, about culture, and about the meaning of a product or its use . . . it is about the meaning of things, the personal remembrances something evokes . . . it is about self-image and the message a product sends to others" (Norman 2004, p. 84). Sometimes users choose various products or services not necessarily because of how they look or feel, *per se*, but because of how the look or feel taps into nostalgic feelings about the product itself, the brand that supplies it, or perhaps even about something entirely unrelated that the user associates with the product. The value of a product can often come from what the experience of it communicates to the user or what it communicates about the user or owner, so even when products that work better or cost less are available, people will still purchase and use inferior, more expensive products that serve as status symbols that facilitate a desired experience that has less to do with the product's functionality, durability or price and more to do with conspicuous consumption.

### 13.3.3.1 Sonic Status Symbols and Sonic Branding

As I previously discussed, luxury automobile design includes the creation sonic elements that communicate the safety and security of a particular car to its driver. These sounds are also unique to each car or brand, functioning both as feedback that signals safety, high-performance and trustworthiness and as a mark of luxury, quality, exclusivity and cultural cachet. Engineers at BMW work closely with sound designers to craft each car's sonic signature: "A Mini, for example, is playful and joyful and the sound of the car has to reflect that, so we modulate the exhaust to give a sporty, impulsive sound. By contrast, a 7 Series has to be very quiet. The driver wants to be in his own zone, so there is lots of damping and insulation" (Baker 2013). The sound that the engine of Harley-Davidson motorcycles makes is so iconic and so closely associated with the brand and the overall experience of these vehicles and the motorcycle subculture that surrounds them, the company once attempted to trademark the sound (O'Dell 2000). Smartphones and their associated sounds, such as bespoke, proprietary ringtones and other earcons, can at once notify the user that someone is calling while simultaneously broadcasting to those within earshot what brand and model it is. In these reflexive experiences, various interpretations of the sound have the potential to result in many different experiences with the products. Some users derive great value from sharing their taste in fine motorcycles and smartphones through the sounds they emit. Based on their own reflective experiences in the past, others could find these same sonic experiences to be off-putting or annoying, or they could connect the sounds to products (or people that use them) with less desirable qualities.

## 13.4 Conclusion: Designing Inclusive Experiences With Sound

While most products and services are designed with a particular audience in mind, UX designers work to make sure that as many users as possible derive positive experiences from interactions with their design. In their work on reflective design, researchers of the Culturally Embedded Computing Group at Cornell University developed design strategies that facilitated HCIs that were more democratic, participatory and conscious of the various differences in cultures, abilities and values among potential users (Sengers et al. 2005). These strategies include providing for interpretive flexibility (for example, what seems "optimal" to one user may not be the best for other users), creating designs that allow users to participate and provide feedback, and inverting metaphors and crossing disciplinary boundaries in various approaches to design. An obvious example of the use of sound for this purpose would be crosswalks that signal users to "walk" with an illuminated walking person symbol, and sound to indicate to those who cannot see the signal that it is safe for them to cross. In the context of sound design for embedded media, this could possibly include the creation of tools or functions of a design that allow users to record or upload their own audio clips to be used as assets that provide sonic feedback, choosing for themselves what sounds provide them with the optimum amount of feedback. This approach to sound design could even be as simple as thinking more critically about sonic material and about the ways that various groups and cultures could interpret them differently. For example, in American English, the sound a cat makes is "meow," but in Japanese, cats say "nyan," so the sound designers of cat-themed smartphone apps or video games should carefully consider the impact of using one or the other of these linguistic forms or eschewing both options to use an audio recording of a real-life cat instead. Inclusivity, accessibility and cultural awareness should undergird all UX design, and thoughtful sound design is a valuable mechanism through which designers can further support and promote these values.

## Notes

1 The ISO is the premiere global organization dedicated to establishing voluntary industrial and commercial standards for manufacturing and various other trades.
2 *Petrichor* is the scientific name for the scent that is created when dry soil becomes wet; it is caused when plant oil found in dry soil is released into the air.
3 These poorly designed doors are called "Norman Doors," named for Don Norman, who describes them in the first few pages of his book, *The Design of Everyday Things* (2013, pp. 1–3).
4 For more on the ways in which UX design is used to promote feelings of safety and comfort in designed objects and experiences, see Austin 2016.

## References

Austin, M. L., 2016. Safe and sound: Using audio to communicate comfort, safety, and familiarity in digital media. In: Tettegah, S., and Noble, S. (Eds.) *Design, Technology and Emotions*. London: Elsevier, pp. 19–36.
Baker, D., 2013. Did you know BMW's door click had a composer? It's Emar Vegt, an aural designer. *Wired Magazine*. Viewed August 16, 2018 <www.wired.co.uk/article/music-to-drive-to>.
Dreyfuss, H., 1955. *Designing for People*. New York: Simon and Schuster.
Gibson, J. J., 1979. *The Ecological Approach to Visual Perception*. Boston: Houghton Mifflin Harcourt.

Grimshaw, M., 2009. The audio uncanny valley: Sound, fear and the horror game. *Games Computing and Creative Technologies: Conference Papers (Peer-Reviewed)*. Paper 9. Viewed August 4, 2018 <http://digitalcommons.bolton.ac.uk/gcct_conferencepr/9>.

Interactive Design Foundation, n.d. *User Experience (UX) Design*. Viewed August 2, 2018 <www.interaction-design.org/literature/topics/ux-design>.

ISO 9241-210, 2010. *2.15 User Experience*. Viewed August 2, 2018 <www.iso.org/obp/ui/#iso:std:iso:9241:-210:ed-1:v1:en>.

Kujala, S., Roto, V., Väänänen-Vainio-Mattila, K., Karapanos, E., and Sinnela, A., 2011. UX curve: A method for evaluating long-term user experience. *Interacting with Computers* 23, pp. 473–483.

Law, E., Roto, V., Hassenzahl, M., Vermeeren, A., and Kort, J., 2009. Understanding, scoping and defining user experience: A survey approach. *Proceedings of the CHI 2009 Conference on Human Factors in Computing Systems*. New York: ACM, pp. 719–728.

Marshall, A., 2018. Will jarring music drive drug users from a German train station? *New York Times*. August 22, 2018. Viewed August 26, 2018 <www.nytimes.com/2018/08/22/arts/music/atonal-music-deutsche-bahn-drugs-trains.html>.

Merholz, P., 2007. *Peter in Conversation with Don Norman About UX & Innovation*. Viewed June 24, 2018 <http://adaptivepath.org/ideas/e000862/>.

Mori, M., 1970/2012. The uncanny valley. *IEEE Robotics and Automation* (trans: MacDorman K. F., and Kageki, N.) 19, no. 2, pp. 98–100. Viewed July 14, 2018 <https://ieeexplore.ieee.org/stamp/stamp.jsp?tp=&arnumber=6213238>.

Nass, C., Moon, Y., and Green, N., 1997. Are machines gender neutral? Gender-stereotypic responses to computers with voices. *Journal of Applied Social Psychology* 27, pp. 864–876.

Norman, D. A., 2004. *Emotional Design: Why We Love (or Hate) Everyday Things*. New York: Basic Books.

Norman, D. A., 2013. *The Design of Everyday Things (Revised and Expanded Edition)*. New York: Basic Books.

Novak, D., and Sakakeeny, M., 2015. *Keywords in Sound*. Durham: Duke University Press.

O'Dell, J., 2000. Harley-Davidson quits trying to hog sound. *Los Angeles Times*, June 21, 2000. Viewed August 18, 2018 <http://articles.latimes.com/2000/jun/21/business/fi-43145>.

Read, R., and Belpaeme, T., 2016. People interpret robotic non-linguistic utterances categorically. *International Journal of Social Robotics* 8, no. 31. Viewed July 14, 2018 <https://link.springer.com/article/10.1007/s12369-015-0304-0>.

Sengers, P., Boehner, K., David, S., and Kaye, J., 2005. Reflective design. *Proceedings of the 4th Decennial Conference on Critical Computing: Between Sense and Sensibility*. New York: ACM, pp. 49–58.

Siegel, M., Breazeal, C., and Norton, M. I., 2009. Persuasive robotics: The influence of robot gender on human behavior. *Proceedings of the International Conferences Intelligent Robotic Systems, St. Louis, MO, Oct. 10–15*. Piscataway, NJ: IEEE, pp. 2563–2568.

Tamagawa, R., Watson, C., Kuo, I. H., MacDonald, B. A., and Broadbent, E., 2011. The effects of synthesized voice accents on user perceptions of robots. *International Journal of Social Robotics* 3, pp. 253–262.

Tinwell, A., 2009. Uncanny as usability obstacle. In: Oxok, A. A., and Zaphiris, P. (Eds.) *Online Communities*. Berlin: Springer, pp. 662–631.

User Experience Professionals' Association, 2005. Glossary. *Usability Body of Knowledge*. Viewed August 5, 2018 <www.usabilitybok.org/glossary>.

Whiteside, J., and Wixon, D., 1987. The dialectic of usability engineering. *Proceedings of the Second IFIP Conference on Human-Computer Interaction—INTERACT'87. Stuttgart, Germany*: Elsevier B.V., pp. 17–20.

14

# Toys and Playful Devices

## A Case Study

**Kristin Carlson, Greg Corness and Prophecy Sun**

### 14.1 Introduction

Toys are playful objects, devices and artifacts that children engage with as they are exploring their new-to-them environment. While adults tend to view toys as a special subset of objects designed primarily for children to use in an artificial world of play, imagination and entertainment, Sutton-Smith observes that babies don't play but, rather explore and learn in the real world (1986). From this perspective we may say that children do not perceive toys as separate from their own world, but, instead, toys make up a large portion of the accessible objects in a child's world. Toys are the part of the world with which they are allowed to engage, explore, test, make mistakes and learn (Ayers and Robbins 2005).

Childhood is about developing multiple layers of understanding the physical, social and personal world through exploring and testing new sensory experiences (Ayers and Robbins 2005). For example, the sounds produced by toys such as rattles, crinkly plastic or banging on objects become tools for expanding a child's reach and attention in the world by adding to the sound environment around them. As a child develops, their engagement with toys reflects their developing sense of self and the world (Sutton-Smith 1986). Sound is an important design feature because it contributes to a child's understanding of their environment in many ways such as feedback from a direct action, an ambient background or a playful exploration of interaction and affordance (Johnson 1990). In this respect, toys are not a subset of objects in the world, but rather a regular part of a child's experience of the world. This is what makes toys special and unique as objects to design. As toys are part of a child's everyday experience, and part of their everyday surroundings, they should reflect, support

**Figure 14.1** O Playing With Assorted Toys, 2016.

**Figure 14.2** O Playing With Assorted Toys, 2016.

and encourage their place in their world. This ideal extends to the experience of sound.

Sound is an ever-present part of our world. Our ears never close off sound, even when we sleep. We often hear objects before we see them, and sound timbre and reverb provide us with a sense of space. In this way, sound connects us, inextricably linking and helping us understand and differentiate the nuances of our immediate, physical and cultural environments. As authors David Novak and Matt Sakakeeny assert, sound provides us with important survival skills such as learning to acknowledge changes in our environment that could be consequential to us (2015). Playtime becomes a game of understanding the sonic cues provided by the physical world. Through play we learn to locate sounds, associate sounds with other physical attributes, size weight and material and recognize the association between changes in parameters such as distance with changes in the sound. Just as play is part of our physical development, balance and coordination, sound play is an important part of our sensorial and cultural development (Ayers 2005).

So while we learn to account for and comprehend sound in our environment, we are also learning how to hear in a nuanced manner. We are learning important skills, such as separating out and identifying specific sounds to develop and improve our listening abilities. As a child learns about sound they can break music down into identifiable components such as "high and low, fast and slow, stop (rests) and go (notes) [to] foster in toddlers the awareness of musical dynamics" (Turner 2004). In this way the sounds that a toy can make not only increase the entertainment and immersive quality, but they are also an essential part of how a child learns social and abstract structures. Some selected music activities for young children in the preoperational stage are noted in the book "Musical Growth and Development" (McDonald and Simons 1988):

1. Engage in musical action songs and games, with appropriate actions associated with the words of the song or game. (Combining music and motor control).
2. Develop a repertoire of sounds with associated actions, which are sung often to the children; allow them to initiate the action when they can. (Combining music and motor control).
3. Play sound-discovery experiences in which the sources of sounds are investigated and discovered (both environmental and musical sounds). (Exploring affordances and surprise).

When children develop deeper listening skills such as hearing different instruments, tones and rhythms simultaneously, a child is able to follow multiple woven patterns and comprehend the complexity of what is being played, further understanding the value of a single entity experience. Each creative instance, whether it is in the act of play, or making and or watching others making sounds, is when a child feels sounds rhythmically and with all their senses. Sound toys are the sandbox for developing auditory coordination, like a ball for learning hand-eye coordination (Fromberg and Bergen 2006). For example, a rattle provides a child with the experience of sonic agency through physical action, and electronic toys with buttons and knobs can trigger songs to give children a sense of agency to hear a song on command.

However, sound design for toys is often restrained by many factors, including the focus of the designer, the quality of available parts, affordances of the object and sounds emitted and production costs (Light et al. 2004). Due to these restraints, designing engaging sound interaction for toys can be challenging. Light et al., writing about technology design for special needs children, note that "very limited attention has been given to applications of sound and voices that might enhance the appeal and facilitate learning of Augmented and Alternative Communication technologies by young children". Young children produce and respond to a wide range of sounds in addition to speech in the course of their daily lives. In a study of children's toy preferences, Heppel (1999) reported that ". . . all the children we worked with found, and reported, the sounds made by . . . toys as a key feature. Sounds offer confirmation of action, entertainment and much more but above all else sounds were a key contributor to engagement for these children". We believe that these observations for sound specifically in special needs toys can be extended to commercial toy design as well.

This chapter considers toy sound design and surveys the existing literature on how children learn to hear, perceive and make sense of sound in their environment. We then deconstruct how and why children playfully interact with sound through various toys and objects in their world. We present a case study of three young children in their exploration of sound and toys and suggest strategies to use when designing for playful sound experiences.

## 14.2 Background

### 14.2.1 *Children's Physical Development*

There is a vast amount of research on vision, motor control and cognitive development in young children, yet information on how children engage with sound in their environment is more limited (Berk 2012; Boyd 2017).

Sound is often researched in terms of how it supports the connections between an infant and their parents, such as the investigation of maternal utterances and how infants learn to recognize their parents voices (Franchak et al. 2010; Reissland et al. 2002; Woodward and Hoyne 1999). It has been found that infants and young children learn to use words and sounds as labels for objects with which to communicate (Woodward and Hoyne 1999), and approval and disapproval can be understood by infants through the pitch of their parent's voices (Quam and Swingley 2012). Piaget discusses sound as a mode of imitation to establish a base knowledge for infant's communication, with the discovery of new sounds being assimilated into their current worldview (1952). Yet this approach of enacting as imitation is not mirrored in research on other actions, such as motor development, which is sometimes attributed to imitation but also enables a child's own discovery.

The consideration of prominent developmental milestones can help orient and provide perspective in the discussion of sound in interactive toys. Research shows that a newborn baby's vision begins first in black and white and then color develops over time (Boyd 2017). Newborns can see objects best when they are approximately eight to ten inches away from their face. By five months, a child is developing their perception of depth and color. Mobiles are designed to capture visual attention and may have some sound element to them and are often used by parents. In contrast, sound perception is actually more advanced than vision at birth. A newborn can accurately identify sound on their right or left side of their body even if they cannot see what is creating that sound (Gravel 2005). And though they do not tend to seek out the sound source until they are older, this is believed to be due to an inability to shift, roll or flip over until approximately four to six months old, along with their developing perception of depth (Gravel 2005). Furthermore, unlike the development of vision, which starts out black and white and later develops into a perception of color (light frequencies), newborns can hear the full range of frequencies for humans (20 Hz–20 kHz)(Werner and Boike 2001). This research supports the notion that an infant's hearing is fully developed and suggests that their hearing is very susceptible to background noise. For designers, such research can inform our decisions in a number of ways. For example, background music can be both a beneficial and problematic element because a child is making associations early on between sounds and objects in their world. What this means is that they are learning to differentiate sounds in order to tell who is around them and where they are in relation to those sounds. Overall, what the research on child development suggests is that sound is a significant part of a child's experience and should be given full consideration when designing toys.

### 14.2.2 Interaction Through Sound

Research comparing parent-child interactions with traditional toys versus electronics have found some varying results. Sosa found that the quantity of language-based utterances diminished with the use of electronic toys and concluded that, thus, electronic toy play should be discouraged (Sosa 2016). It has also been noted that "the incentive behind the use of classical music or folk tunes stems from commercial considerations and global marketing interests rather than developmental issues" (Sulkin and Brodsky 2015), even though some research has seen benefits of the "Mozart Effect" (Sulkin and Brodsky 2015). Current trends suggest that mothers will more often play a soundtrack or use a mobile with audio to help their infants fall asleep than personally singing to them (Baker and Mackinlay 2006; Young 2008). Young has noted the shifts in our current culture to reach to more electronic and digital options for children's engagements than in previous eras, which has changed interactions between parents, children and their toys (2008). While many new toys do explore the implementation of educational aspects that are not possible to include in traditional toys, they are often implemented in forms for visual and motor skills, not sound skills (Yilmaz and Goktas 2017).

Miller et al. explore additional angles to electronic toys and children's communication through parent-child interaction, and rather than only focusing on vocalizations as in Sosa's work (2016), they look at nonverbal communication and time spent playing and the amount of focus on the toys. This approach found that parents responded with more direct feedback when playing with traditional toys than with feedback toys, as found with Sosa (2016). However, it was also found that children had more sustained attention on a feedback toy than with a traditional toy. Parents also provided less redirective behavior with feedback toys, though it is unclear if this affected the children's sustained attention with feedback toys or not. Radesky and Christakis also state that the tendency is for a parent to hand a digital toy to a child for solo entertainment, not as an interactive platform between the child and parent, noting that "the technical or interactive aspects of toys can engage the child but disengage the parent" (2016, p. 113). It was noted that "the flashing lights and sounds of the feedback toys clearly capture and hold attention" and, as noted above, "fill the auditory space" which decreases infant communicative behaviors (Miller et al. 2017, p. 643). In another study on interactive play with a technology-augmented toy, the social gameplay with the toy's features elicited both infant laughter and parent language (Bergen et al. 2010). The reviewed authors' discussion of these findings focus primarily on toys as creating additional stimulation that is more attention-grabbing, and not providing any educational or social development benefits. We are interested in how new concepts for toy design could support a developmental understanding of sound in a child's world as a form of agency, expression and social interaction, including interaction with caregivers and peers.

### 14.2.3 *Agency and Affordances: Shifting Through Perspectives*

One aspect of a child's engagement with sound made by toys is the impact (or agency) the sound affords them to have on their environment. For the child, playing with a loud toy may attract attention or just change the environment. Similarly, a sound, a cry, or a scream, can provide a primary means to affect change and get attention. We use the term "agency" to describe the child actively making choices, expressing self and acting as a controlling agent in the situation (Doll et al. 1996). In this way a child begins to explore their agency in the world through their engagement with toys and sound. As Doll et al. state: "exercising choices such as this at an early age is an important foundation to the development of self-determination" (Doll et al. 1996). Of course, as a child grows, they develop their sense and skill in applying agency (Fromberg and Bergen 2006). A significant aspect of their development is their perception of affordances in their world.

The term *affordance* is defined as the methods of interaction that are offered by an object and was originally coined by Gibson in reference to how our senses are perceptual systems attuned to visible and invisible information that extends beyond our bodies and limbs, onto objects within our environments (1962). Gibson notes that there are many forms of information in our environments that we first perceive and then interact with and use based on how we perceive it (Greeno 1994). Don Norman writes extensively about the way in which affordances are designed into our environment, such as door handles (Norman 2013). He states that a door handle suggests, in the way it is designed, how the user should interact with it: whether the door needs to be pushed or pulled or twisted or leveraged or slid to open.

Door handle design is the classic example of a design leveraging an affordance perceived through visuals that reflects what we know throughout experience in the world. Unfortunately, Gibson does not address sound directly very often (Gibson 1962). However, we may extrapolate a sound equivalent in the case of toys to discuss three levels of sonic affordance. First, a button on the side of a truck suggests being pushed and from the child's experience they may expect a sound to be triggered. This is one example of how affordances built into a toy's design through shape, color, texture and other qualities may suggest a use including the triggering, recording and manipulation of sound (Johnson 1990). At this level the toy affords the engagement with sound as the handle affords the engagement with the door. A second level in considering sonic affordance is what the making of the sound affords. Just as the door handle affords the opening of the door, the pushing of a noisy toy affords the gaining of attention as does the repeated triggering or nuanced playing of an instrument. Any object that can produce noise, or amplify a voice can be used as a toy and can help the affordance of getting attention. These affordances may not be purposefully designed into a device such as a fork or cup but are discovered by a child through their use and the response of those around them.

A third level of affordance when considering sound may also lie in considering the sound itself. Gibson discusses the notion of active and passive touch. For Gibson this is the difference between touching an object as a way to feel it and gain information or to tough as a byproduct of holding (Gibson 1962). We may infer a sonic equivalent in the experience of actively attending/ listening to a sound versus passively letting sounds fade into the background. Through their experience and play the child learns that some sounds, when listened to, provide information about their environment, while others provide a pleasant background and still others can be ignored (Berk 2012). The sound of a pet running in another room may entice the child to go and see what is happening, while a car passing outside is ignored. In the context of toys, a sound that is part of the toy mechanism may attract attention at first but latter be ignored as a passive noise of little importance. This parsing of sounds in the environment is part of the child's development and experience (Ayers and Robbins 2005).

A child may approach object affordances differently from adults for various reasons including their age, stage and developmental perspective, as well as their individual personality and how they operate in the world (Fromberg and Bergen 2006). Children of various ages have different experiences to draw from and so rely upon differing degrees of discoverable affordances. This means a child may engage with objects in a variety of ways, utilizing discoverable affordances more than the designed or expected actions for that particular toy. For example, a child may sit on a ball instead of throw it; stand on a piano rather than push buttons or keys with their fingers; roll a toy lawn mower across a room to chase and scare dogs rather than to perform a domestic task that mimics a caregiver. As noted earlier, babies just explore and learn (Sutton-Smith), and there is no such thing as playing that is separate from exploring the world (1986). The playfulness of affordances is important to children's toy design because it is the core feature that supports exploration and learning.

To support implementation of agency and affordances in toy and interaction design, Fromberg and Bergen illustrate a variety of ages and what

**Figure 14.3** T and C Playing With Buckets.

types of actions engage children of that age through play (2006). Children aged one to two years explore what they can do with objects, which leads into their years of experimentation (Vondra and Belsky 1989, p. 176). Children aged 12–18 months delight in play that responds to their actions, such as "jacks-in-the-box that pop out when a string is pulled, books that emit words or music when a button is pressed, or an app on a phone or other electronic device" (p. 13). Meanwhile, three-year-olds focus more on manipulative materials (such as clay), and their sense of mastery increases as they can control the play activities (p. 13). These categories of exploring affordances of objects, enjoying surprise and investigating feedback to their actions include mechanical, electronic and digital toys.

### 14.2.4 *The Role of Feedback and Sound in Learning*

As children develop, they make associations between sounds and events in their environment by perceiving a causal connection (Berk 2012). They develop spatialized hearing to be able to locate their parent in a room, and eventually learn cause and effect. For example, if a child steps on a twig and breaks it, their action creates feedback of a cracking sound. Through cause and effect children quickly learn agency, how they can create sound, control how it is created and make an effect on their environment (Fromberg and Bergen 2006). For feedback to be effective, it needs to be instantaneous for a child to associate it with the proposed action. For example, some electronic sound toys have an extended delay between the button press and the sound played, which is enough time for the child to be distracted by something else, and then surprised when the sound appears. The VTech Touch and Learn Activity Desk uses touch on the desk to trigger different sound files, though there is sometimes a delay, which provides incorrect feedback to what the child pressed, particularly if they pressed it multiple items. We have observed how latency in a digital toy can quickly cause the child to lose interest, and often disrupts the expected feedback cycle for a child's understanding of interaction. Similar experiences exist in art toys for children, such as the Crayola Mess Free drawing sets that use clear markers on special paper to reveal color. Because there is a delay in the chemical reaction on the paper, younger children do not see their own agency when drawing and can lose interest quickly.

Alongside causation, repetition can play a strong role in influencing how a child builds associations. As a child begins to isolate and recognize recurring sounds they are also perceiving patterns in their environment, similar to Piaget's concept of assimilating information into their environment and then accommodating new information to build up the prior knowledge (McDonald and Simons 1988). They begin to develop associations to sounds. For example, white noise is used early on to help newborns sleep because there is an association with noise heard from the mother's body when in utero (Berk 2012). The sound of a parent's voice can be quickly recognized and associated with the feeling of safety, warmth and food (Berk 2012). As a child grows older they are able to identify sounds and associate them in more complex and nuanced ways (McDonald and Simons 1988). In some instances, a child with limited language skills is able to anticipate the starting of a favorite show, based on a repeated TV station announcement, before any visual or auditory material of the specific show is visually shown.

Similarly, children learn cultural associations by observing those around them, in play and at work. Children learn such cultural assumptions as

HAPPY IS UP, SAD IS DOWN and will apply these in their explorative play with new toys and environments (Antle et al. 2009). Lakoff and Johnson refer to these associations as orientational metaphors (2003). Orientational metaphors are based on spatial orientations, such as up-down, in-out, front-back, on-off, deep-shallow, near-far, and then applied to more abstract concepts such as emotion or parameters of sound (e.g. LOUD is HIGH). In a related vein, ontological metaphors are based on our interaction with physical objects and our own bodies. For example, BASS MUSIC is HEAVY, or LOUD SOUNDS GET IN THE WAY (Bakker et al. 2009). In these ways, children learn language metaphors that support and guide their understanding and experiences with sound.

### 14.2.5 Designing Digital Toys for Children

Most literature that explores designing technologies for children are very focused on the visual interface design. In Chaisson and Gutwin's article "Designing Children's Technologies," the authors focus on HCI design for children and how interaction can support children's learning by addressing topics of literacy, mental development, motivation and social interaction. Their evidence relies on how well children can see objects and text on a screen, the size of buttons and quality of feedback for their actions. However, the authors do include many additional important features such as the importance of children feeling empowered and in control of their interaction, how metaphors can support comprehension and that the use of extrinsic rewards are often more successful with children than intrinsic rewards. One more sound-based project, *Our Little Orchestra* by Browall and Lindquist, explores how children collaborate with sound in technology in ways that are not screen-, panel-, or mouse-based (2000). They created a sound mixer in the shape of a birthday cake using the metaphor of round shapes to support togetherness. A similar work that is more recent is the *Blipblox Audio Exploration Module*, a kid's synthesizer for encouraging play with electronic music. Features include a low pass filter, eight oscillator modulation schemes, two envelope generators, MIDI and two low frequency oscillators, to name a few (https://blipblox.com/). *MusicPets* is another project that designed plush toys to store audio in a DJ-like scenario (Tomitsch et al. 2006). Plush toys have a slider to change pitch, and three buttons for recording, creating pauses and inserting chords. The duration of the sound created is visualized through an LED strip. An interesting feature of this work is that the designers allowed the children to change the design of the interface on the toys, which surprised them when selections were made that went against the author's assumptions of interface metaphors: "For example, our first prototype featured a round red

**Figure 14.4** H Playing a Piano, 2017.

record button, which was changed into a yellow rectangular button by the children" (pg.2). The authors emphasized the importance of consulting children in their design process to explore opportunities for interfaces that would resonate with the children's way of engaging.

While the above projects focused on tangible interactions with sound, the *SoundMaker* project by Antle et al. looked at the use of metaphor in interacting with sound. The *SoundMaker* system enables children to manipulate musical sound parameters of volume, tempo and pitch through body movement (Antle et al. 2008; Bakker et al. 2009). Two different mappings were created between the movement and the sound: an embodied metaphor context and a non-metaphor context. The embodied version uses knowledge built into a children's intuitive understanding of the world (e.g. running fast mapped to tempo, standing tall [high] or low mapped to pitch change), where the other version could be learned but was not built on existing intuitive knowledge. Findings showed that children were more accurate when using the interface with the embodied metaphor mapping compared to children using the non-metaphor mapping.

## 14.3 Toy Categorization for Design, Agency and Interaction

To better understand design considerations for toys, we have identified three axes based on the experience of a child playing with them. Our first axis focuses on whether the primary use of the toy emphasizes sound or other forms of play. For example, an instrument such as a toy piano or a microphone can be categorized as a primary use, as it is designed for making noise. In contrast, a toy truck that can make some sounds is primarily designed as a truck, to be pushed. Yet a truck can have a secondary role of making sound. A second axis, especially significant to secondary toys, concerns if sound is an intentional part of the interaction or a byproduct. By intentionality we refer to whether a toy's sound is intentionally designed for interaction, such as a button that is designed to be pressed and then plays a sound. Unintentionally designed sound refers to sound that is not specifically designed into the toy, but can be discovered through affordances. For example, a ball making a dull whacking noise when thrown against a wall can be explored specifically for the sound it makes, though it was not intentionally designed for that interaction. The division of toys into primary and secondary categories reflects the initial approach the child has to the toy and their continued engagement. Primary sound toys are engaged with the intent to make noise. The child plays with the toy as long as the toy's noise and affordances are of interest. Secondary sound toys, in contrast, were more often observed to have two modes of play: with and without noise. We found children more likely to use the sound of a secondary toy in a manner that was outside the initial design (i.e. using a siren on a car to annoy a sibling). This alternative use of secondary toys also points to why intentionality is a bigger factor in these toys, since the dual mode of play encourages discovery.

The final axis is whether a toy is primarily digital or analog. For this separation, we focus on the sound mechanism of a toy. For example, a pull toy with a ratchet on the wheels is considered analog, whereas a truck with a button to trigger an engine sound is considered digital. These categories should not be seen as the only method for grouping sound toys, but they do allow us to focus on the experience of the child based on observations of how the child used the toy. Based on these three axes we discuss the intersections of these categories and describe toys that fit within them.

### 14.3.1 Analog Sound in Toys

Analog sounds are widely noted in toy history, from banging pots and pans, to baby rattles, to rubber band boxes, to wind chimes, as well as purposeful instruments such as kazoos and slide whistles. These toys, whether intentionally designed or not, have affordances towards sound that are commonly experienced by children.

Analog unintentional toys are objects that often played with by children where sound is not part of the object's intentional design. While these affordances are sometimes discovered by children themselves, parents may also guide their focus towards unintentional toys from their own

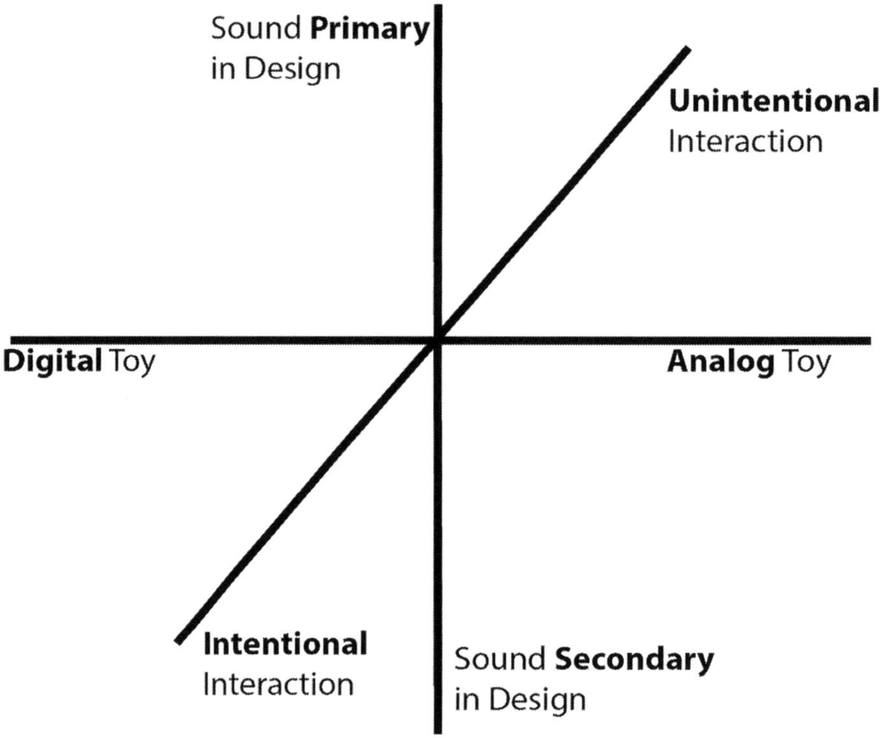


**Figure 14.5** Toy Axis for Design, Agency and Interaction.

experience in the world. This would include banging on pots and pans, popping bubble wrap, or the whirl of the toilet paper holder as a child spins it off. These objects have a natural resonance that can produce loud or strange noises. Children have clear agency and impact on these objects. They tend to quickly recognize the connection between their actions and the sound made when they engage with these toys. Often their play includes discovering what sounds are possible, lending an emergent quality to the interaction based on discoverability. Rocking horses can be unintentional analog toys, in that they are meant to be movement toys, but the springs creak and whine when ridden. Children generally figure out that a spoon is to be held and used to bang, and that pots resonate when they are banged on. They may explore banging pots or bowls on pots as well. We have observed that pots tend to be heavy and harder to pick up and control than a wooden spoon.

**Figure 14.6** T Riding a Rocking Horse.

### *14.3.2 Analog Intentionally Interactive Sound Toys*

Analog intentional toys refer to toys that are specifically designed to trigger sound. These include the bubble popper, a long toy that can be pushed and has gears attached to the wheels to *pop* small balls inside, which makes noises. Musical instruments also fall into this category. For example, a tamborine could be a toy that works both as a drum and a tambourine, and can be shaken as well as banged on with a mallet. However, a tamborine was specifically designed to create sounds and be handled by a child or an adult to make noise.

### 14.3.3 Primary Versus Secondary Sound in Analog Toys

Primary sounds in analog toys are toys designed with sound as the focus. This would include a wooden caterpillar or duck on a string that makes a click-clack noise as it is pulled. Wind chimes, though not usually considered a toy, have a primary focus on sound as their function. Analog toys that make sound as a secondary focus are harder to pinpoint, because sound is still intentionally designed, but it is not the focus. An example of secondary sound in an analog toy would include the wooden fruit and birthday cake toys that are held together by velcro. The focus of these toys is to cut them with a toy knife, but the velcro used to connect the pieces makes a ripping noise when cut. In this case sound is an intentional part of the design, but it is not the primary focus of play.

### 14.3.4 Digital Intentionally Interactive Sound Toys

The use of intentional sound in digital toys is frequent, as toys would not be digital without some sort of light or sound trigger. Many digital toys are designed to be intentionally interactive with sound, including buttons to press, strings to strum, surfaces to hit, or plush toys to squeeze. For example, Tickle Me Elmo is intentionally designed to giggle when squeezed, with this interaction as the primary focus of the toy.

### 14.3.5 Digital Unintentionally Interactive Sound Toys

Unintentional interaction in digital toys is quite challenging to find, because digital toys tend to be heavily designed (intentionally). Due to the frequent use of sound samples in digital toys, there are few examples of this phenomena digitally. An example of this category would be the noise made by a toy's motors (such as the Code-a-pillar) or other electronic noise associated with the toy but that is not intended to be noticed. Such toys were difficult to find and hence were not a significant part of our study.

### 14.3.6 Primary Sound in Digital Toy Design

There are a variety of consumer-grade children's toys intentionally made to engage a child through triggered sound as the focus of the toy. For example, electronic instruments such as a guitar, drum set or tiny piano are durable and enable children to bang on them multiple times while trying to produce triggered sounds much like the adult version of the instrument that they resemble. These devices may also expand on the functionality of their adult counterparts. For example, a cat piano like the B. Meowsic Keyboard similarly has options to play the keys as typical piano notes, or to trigger a variety of songs, and additionally to trigger a *meow* sound that changes tone depending on the key (http://www.mybtoys.com/toys/lively/meowsic/). The piano also has a microphone option to capture the child's own vocalizations and can play them back while playing the instrument.

Other current interactive toys include the Code-a-pillar, a centipede looking toy that allows children to organize different parts of the body to help it move in certain directions. The Code-a-pillar plays songs depending on the procedural advancement through the pieces as the whole toy moves. Kaltenbrunner documents toys on his website that includes interactive games with tangible pieces that allow children to explore sound (2018). For example, Zoundz gives users a variety of tangible objects that they place on a hot spot on the board to trigger sounds. Users can play with different combinations to create their own music, and manipulate the tempo, volume, and echo. Thus, children can manipulate a variety of different sounds and how they layer together.

### *14.3.7 Secondary Sound in Digital Toys*

Digital secondary sound toys use sound as a background texture or affordance that is not central to the design of the toy. This could include computer games that have background music while the child's focus is on a different interactive component.

We use these categories to consider the engagement of children with sound toys and how the design encourages specific modes of play. We observed that children do not always follow the guidelines set out for them. Further criteria such as the quality of sound, discoverability of sounds, possible emergence of interaction, style of interaction (small/large motion), latency of feedback, number of sounds, number of sounds and amount of agency all contribute to the child's engagement but are more at an individual toy level.

## 14.4 Case Study

We present an ethnographic case study in which we observed children playing with a variety of toys and analyzed their interactions based on a response to and agency with sound. We regularly observed three children from the age of one to four years old and their interaction with sound toys. We also observed children in other age groups less regularly; 5–8 years old playing at a birthday party, 8–11 years old playing at a family gathering, and 12–14 years old playing video games at a school function. These observations enabled us to see the development of patterns that start in young children and grow over time. We focus the scope of this discussion on designing for young children ages one to four, as it is a highly imaginative time for exploring the affordances of toys that then develop into more realistic and pragmatic interactions as a child ages. This section categorizes and describes a variety of existing toys and how sound is incorporated. We articulate design principles that support a child's sound experience using toys and provide case study evidence.

Although we observed a number of children, we focus on three in particular whom we will refer to as H, T and O. There was one male and two female children in total. These children range in ages from one to four years old and were observed over the course of one year. By looking at these younger children we could see their foundations developing and emergent patterns.

### 14.4.1 *Observed Child Behaviours*

We first considered the physical design of toys. For example, toys designed for children need to be robust and durable, resilient to rough play and made of materials that are safe to touch, bite and lick (see Figure 14.7). They should not contain sharp edges or components that would poke, cut or hurt

**Figure 14.7** O and H Playing With Assorted Toys, 2016.

a child. Electronics should be durable and all batteries or hardware should have no way of making contact with the child. Above all, the toys need to be fun, engaging and have affordances that encourage play and agency from the child. There are also practical manufacturing concerns such as the cost of materials. For example, we have found that many sound-based toys rely on poorly made speakers with high-pitched and tinny sounds. While to an adult this may be difficult to listen to for long periods of time, we found that all the children we observed were quite intrigued with these sounds. We also noticed that we tend to speak to young children and animals with high-pitched voices, which may explain better than a manufacturing flaw or cost issue why toy designers choose these speakers.

### *14.4.2 Experiential Use Versus Designed Use*

The strongest behavior we observed concerns the connection between the designed intent of a toy and the engagement of a child with that toy. We observed that younger children especially focused their engagement on what they *could* do with a toy more than what the children was supposed to do with that toy. This is not surprising to any parent who has stopped a toddler from chasing a family pet with a toy lawn mower or bubble popper. Or for example, when a child bangs on a pot with a spoon very enthusiastically but presses the play button on a music toy only to walk away and play with something else instead. Older children tend to explore the designed use of a toy, such as swinging a foam sword at their sibling in the way that a sword is intended to be used, or playing a video game in the way it was intended whereas younger children tend to explore the affordances of a toy by investigating the sensory experiences and opportunities that it can provide.

Let's consider the examples of child H and child T. In one session, H was playing with an electronic keyboard. H tried repeatedly to turn on and off a series of prerecorded sounds but seemed to get only one tone to play. He appeared to lose interest after a minute of trying and instead turned the keyboard over and used it to create a physical bridge between two chairs.

In another session, T was playing with an older child. The coveted toy was a fire engine with a small button on the side that triggered the sound of a siren. By the size and location of the button, we may assume that the main intent of the toy was to push it as a truck, occasionally pressing the button to make the siren noise. It should be noted that the size of the truck and the location of the button would require the child to do this with a second hand. The toy had little special interest until one of the children discovered the siren button. Now both wanted the toy. Through parent intervention, T gained possession and held the toy in her lap repeatingly pushing the siren. In this case the designed use of the toy was quickly abandoned for a new experiential use of the toy.

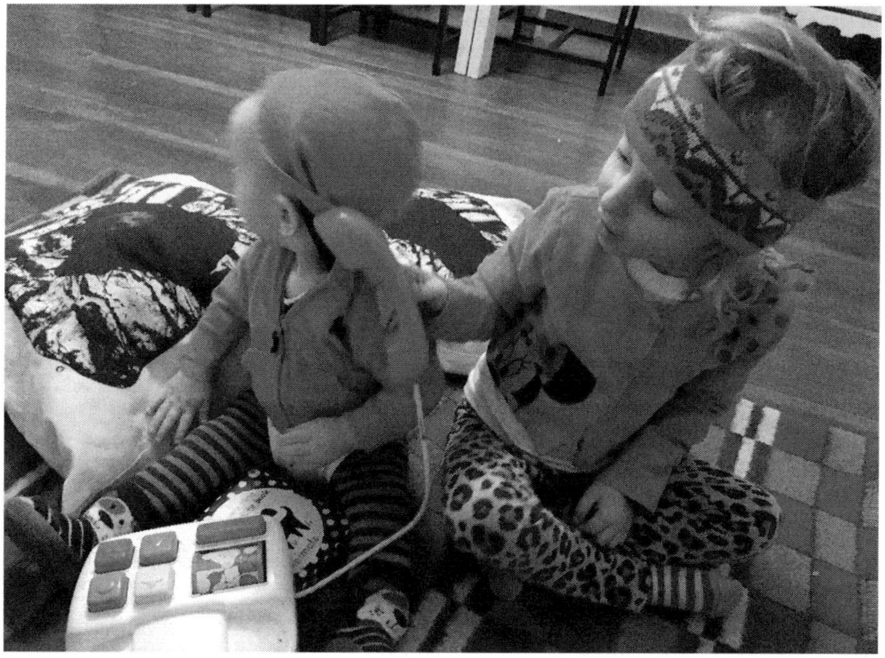

**Figure 14.8** O and H Playing With a Phone, 2017.

The next sections discuss the shift between *look what I can do*, as a child discovering through their experience that they have agency in the world, and the change to *I'm an artist*, as a child's understanding increases about the designed intent of a toy and they develop more complex interactions with it.

### *14.4.3 Look What I Can Do*

Understanding the new use T had for the fire truck is somewhat speculative. But a suggestion may come from observing how she shows her parent her new ability to trigger sounds in other toys. This behavior reflects both a pride in her work and a desire for agency in her environment. Often T will trigger a song on a music toy and walk away to play with something else. If her parent stops the song or changes it she will go back and retrigger the song, suggesting that controlling the background music in her environment is part of her developing agency. H shows a similar tendency with an electric piano toy and a Speak and Spell. In multiple sessions H would trigger a sound, dance and then leave an electronic keyboard. H would return and retrigger a sound once the object had stopped playing or if his sister changed the tune.

Even when the toy's song is intended as a secondary part of the play the child's use may be more about their sonic agency. For example the Code-a-pillar toy plays a song while running through its movement sequence. T at first was frightened by the sound and movement but quickly grew to enjoy the music. She would then trigger the music and carry the toy with her, not allowing it to move as intended. Instead T focused her engagement on the music she enjoyed. This agency in their environment was seen to be preceded by a time of having parents watch them trigger the sound, indicating a pride in their new process of discovery.

### *14.4.4 I'm [More of] an Artist*

The triggering of background music/sounds/noise is an affordance of digital toys. We observed that a child emanated pride while in the act of triggering sounds from an acoustic toy. Further, this pride was also observed to develop over time. While we observed O and H playing with a digital piano, H triggered the song for his background. He was content to let it play. However, O who was older, used other controllers on the piano to modulate the sound. O's interest extended to a more nuanced interest in the sound. This developing interest extended with O's pride and concentration. This development is also observable when comparing the two playing with an acoustic guitar (see Figure 14.9). Though H enjoys the

**Figure 14.9** O and H Playing With an Acoustic Guitar, 2017.

sensation of making a noise, O's engagement is more consistent, generative and prolonged. O was focused on the process of producing a composition of sounds and not just pushing buttons to create a noise. We observed this development of a more nuanced and complex interest in instruments even in the older children, especially a 7- and 13-year-old who took regular violin lessons. The 7-year-old was able to learn technique and follow formal instruction, but was still playing and exploring through the instrument whereas the 13-year-old, who had been playing for eight years, no longer showed an interest in exploring the sensory-based affordances of the instrument because that child was situated in a rigorous learning environment to develop performance technique. However the 13-year-old was beginning to explore new instruments, such as drums and guitars, and was playfully exploring their affordances.

## 14.5 Discoverable Affordances

### 14.5.1 *To Have Agency and Power*

We observed T with a bubble popper, a toy that when pushed makes a popping noise and bounces small colored balls in a bubble. The intended use of this toy, to push and make noise, was initially fun because of the surprise of the noises to the child. However, once the trick was *figured out* it was quickly discarded. Later on the child discovered that the family dogs were dramatically affected by the noise, which led the child to see the toy as a hilarious extension of their self and the power they could have on the dog's behavior. This extension of self can be seen in all the observed children's engagement with various toys, both digital and analog. While exploring affordances and opportunities for interaction, children will develop their use of toys to extend and empower their position. Part of this extending is discovering the effect of the sound in a new situation and using it to their advantage to have impact on their world. This is true for all categories of toys.

### 14.5.2 *To Make New Noises*

Digital toys are often designed for direct interaction with a limited number of sounds, which can mean that these toys provide little opportunity for the child to discover new contexts for the sounds. This is not as true for acoustic sounds, where sound can happen unintentionally more often. The classic example is the child banging their spoon on a table. When the spoon accidentally hits a plate, a new sound is created. The child shows recognition of the new sound, exploring and practicing it. This same behavior was observed in T's play with new percussion instruments. When playing with percussion spoons she was at first surprised that she could hit her leg or the table and get new sounds. She then explored different materials and different ways of hitting, including scraping. During her exploration she was excited to show her new discoveries to her parents. This was also observed in O and T's explorations with the surfaces and textures of various sized metal bowls and a cupcake tray (see Figure 14.10). Both children engaged with the objects in percussive ways such as hitting the top and sides and by flipping the bowls upside down. They tried sequences and solos and sometimes played together. These explorations often reflect the

**Figure 14.10** O and H bowl Drum Session, 2017.

child's exploration of agency in their environment, extending themselves into new tools and devices and investigating new sensory experiences and intentionalities.

### 14.5.3 To Get Attention

Part of the appeal with sound toys is the obvious impact a child can have on their surroundings. T playing with the bubble popper or a vacuum can rile up her dogs, giving her power over their behavior and interaction with her environment. However, sounds can also be used to intentionally get attention from others by ensuring children are making an impact on their environments. For example, the Music Blocks toy is a digital toy that plays Mozart's "Night Music" with an inserted card (like a video game console). There are six blocks with different shapes on each side, that trigger different instrument parts of the music, and different portions of the music, so that a child can *play* the song and have agency on it, but can also trigger the different instruments individually to refine their ability to hear complex music. While educationally this is a helpful toy for deconstructing complex musical structures and teaching them to children in a fun and playful way, we have yet to observe it being used in this manner. The Music Blocks toy has been observed more as a tool for getting attention, through the continual playing of a mish-mash of music samples to be loud, noisy and obnoxious. While the intention of the toy may be better utilized when children are older, aside from getting attention, it has not been of interest to the observed child.

### 14.5.4 To Play in New Ways

T's behavior of showing off her discoveries speaks not only to her interest in new noises and her agency, but also to her interest in discovering something new about the physical world she lives in. Different objects make different noises. Her exploration was similarly observed in her engagement, and frustration, of playing with a kazoo and a slide whistle. She picked up quickly that both objects required being blown into. She even, likely from the quality of the sound, picked up the idea of humming into the kazoo. However, she was observed humming into the slide whistle and blowing into the kazoo especially when moving between the two. Though she seemed to have some confusion about which process went with which object, she was picking up that different objects could be made to sound in different ways. As an extension of this, she continues to try to figure out how a transverse flute works. She continues to closely watch adults play and attempts to experiment, but is still stymied at how to make a sound (much less one that is like an adult's).

A child's persistence to learn how physical objects work is not confined to analog instruments. Observations of O show that this developed understanding of objects also includes digital objects. O was observed over several sessions exploring a new and an old toy and developing new expertises with both. As mentioned earlier, O's engagement with a digital piano is not limited to just triggering or replaying sounds, as with her younger brother, H. O has discovered that she can change the sounds by using other buttons and sliders on the object to create more complex compositions. The behaviors of O and T demonstrate the process of learning how to engage with an object, guiding them to adhere more to the designed use of the toy. In older children this expertise develops from just producing sonic agency in their environment into the nuanced expertise of creating specific sounds. And yet this expertise is still a form of agency of self-expression that recognizes the affordance of the object within a nuanced cultural setting.

## 14.6 Conclusion

In this chapter we surveyed the relevant literature on how children learn in their everyday environment. We also examined both scholarly explorations of toy design and commercially produced toys. We discussed how toys are a regular part of a child's experience of the world, part of their everyday experience and part of their everyday surroundings. Children are constantly learning new skills, but they are learning just by experiencing and exploring their world. Toys help them reflect on, support and establish their place in their world.

We define three options as design considerations: whether the toy is analog or digital, whether the interaction with sound in the toy is intentionally or unintentionally designed, and whether the sound is designed into the toy in a way that it is the primary focus of the toy, or a secondary focus. These design considerations rely heavily on both the rapidly changing child's behavior in interaction with their toys and the discoverable affordances of the toys.

We found that children's behaviors with their toys are more about experiential use, as with babies, who engage and explore with toys as just another part of their world. However as they get older they begin to show a bit of *look what I can do*! to their parents and begin to exert more agency and control on their environment. Over time children become even more nuanced in their interactions as they realize they can control objects with even more complexity, and can say more of *look, I'm an artist*! This increasing interest in their own agency and the control that they can have is important in a child's development, and it is also important in toy design.

**Figure 14.11** O and H Playing With a Keyboard, 2017.

When observing three young children, we saw cases of pure interest in agency, as well as explorations to create new and unusual noises. Children then figured out that they could use this control over new noises as a way to get attention from others, and then to explore play in new ways.

As sound designers for toys, we benefit from keeping in mind the sonic world of our audience: the child. Though their ears are well developed and they hear sounds better than their adult counterparts, learning how to perceive and separate sound is still a challenge. The modern world is filled with a complex barrage of sonic cues and every day we respond to sounds to understand the physical and social world and the space around us. Through their play with toys, children are constantly learning these skills.

### 14.6.1 Speculative Design and Future Work

Our future work includes developing a variety of prototypes for a toy where the interaction with sound changes over time. With this concept the toy can *grow* with a child and change its form of interaction every three to six months. Ideas for this work include sound as background noise for newborns, which slowly separates into a variety of spatialized sounds as they get older, into tangible devices with playful affordances for the toddler years.

For newborns, a weighted plush toy with controlled white noise playing would help some babies sleep better, or a sleep sack with sewn in

**Figure 14.12** Sound-Based Toy Prototypes.

speakers. For older babies and toddlers, exploring fun objects that can capture and manipulate sounds would support the exploration of affordances (see Figure 14.12). For example, a box that can be talked into and then shaken up, mixing up the sounds. Or a large syringe or tube to *suck up* sounds and then spit a manipulated version out again. Also a bubble wand or butterfly net could be used to *capture* many sounds and colors, and dumped out again in different orders and with different layerings. The authors are currently implementing these designs into prototypes and will continue this work. The intent of such designs is to acknowledge the child's exploration and development. We encourage designers to focus the design on "emergence" and "possibilities" rather than "scripted" or "intended" play. Toys should be designed to correlate with the development of the child, to be engaged with through discovered affordances and yet still be learned and encourage pride of use and development of skill.

## References


Antle, A. N., Corness, G., and Droumeva, M., 2009. What the body knows: Exploring the benefits of embodied metaphors in hybrid physical digital environments. *Interacting with Computers*. 21, no. 1–2, pp. 66–75. https://doi.org/10.1016/j.intcom.2008.10.005.

Ayres, A. J., and Robbins, J., 2005. *Sensory Integration and the Child: Understanding Hidden Sensory Challenges*. Los Angeles, CA: Western Psychological Services.

Baker, F., and Mackinlay, E., 2006. Sing, soothe and sleep: A lullaby education programme for first-time mothers. *British Journal of Music Education* 23, no. 2, pp. 147–160.

Bakker, S., Antle, A. N., and van den Hoven, E., 2009. Identifying embodied metaphors in children's sound-action mappings. *Proceedings of the 8th International Conference on Interaction Design and Children*. New York: ACM, pp. 140–149.

Bergen, D., Hutchinson, K., Nolan, J. T., and Weber, D., 2010. Effects of infant-parent play with a technology-enhanced toy: Affordance-related actions and communicative interactions. *Journal of Research in Childhood Education* 24, no. 1, pp. 1–17.

Berk, L. E., 2012. *Child Development*. 9th ed. Chennai: Pearson India.

Blipblox Audio Exploration Module, n.d. Viewed February 6, 2018 < https://blipblox.com/ >.

Boyd, K., 2017. Infant vision birth to 24 months of age. *The American Academy of Ophthalmology*. Viewed February 2, 2018 <www.aao.org/eye-health/tips-prevention/baby-vision-development-first-year>.

Browall, C., and Lindquist, K., 2000. Our little orchestra: The development of an interactive toy. *Proceedings of DARE 2000 on Designing Augmented Reality Environments*. New York: ACM, pp. 149–150.

Doll, B., Sands, D., Wehmeyer, M., and Palmer, S., 1996. Promoting the development and acquisition of self- determined behavior. In: Sands, D., and Wehmeyer, M. (Eds.) *Self-determination Across the Life Span*. Baltimore, MD: Paul H. Brookes Publishing Co, pp. 65–90.

Franchak, J. M., Kretch, K. S., Soska, K. C., Babcock, J. S., and Adolph, K. E., 2010. Head-mounted eye-tracking of infants' natural interactions: A new method. *Proceedings of the 2010 Symposium on Eye-Tracking Research & Applications*. New York: ACM, pp. 21–27.

Fromberg, D. P., and Bergen, D. (Eds.), 2006. *Play from Birth to Twelve: Contexts, Perspectives, and Meanings*. 2nd ed. New York: Routledge.

Gibson, J. J., 1962. Observations on active touch. *Psychological Review* 69, no. 6, pp. 477–491.

Gravel, J., 2005. Guide to your child's hearing. *The Better Hearing Institute*. Viewed February 6, 2018 <www.betterhearing.org/sites/default/files/hearingpedia-resources/Guide_to_Your_Childs_Hearing.pdf>.

Greeno, J. G., 1994. Gibson's affordances. *Psychological Review* 101, no. 2, pp. 336–342.

Heppel, S., 1999. *eTui Public Report . . . What We Have Learned So Far*. Viewed April 27, 2003 <www.ultralab. ac.uk/projects/etui/documentation/reports/D5_1(i).htm>.

Johnson, M., 1990. *The Body in the Mind: The Bodily Basis of Meaning, Imagination, and Reason*. Chicago, IL: University of Chicago Press.

Kaltenbrunner, M., 2018. *Tangible Musical Interfaces Website*. Viewed February 8, 2018 <https://modin.yuri.at/tangibles/>.

Lakoff, G., and Johnson, M., 2003. *Metaphors We Live By*. 1st ed. Chicago, IL: University of Chicago Press.

Light, J., Drager, K., and Nemser, J., 2004. Enhancing the appeal of AAC technologies for young children: Lessons from the toy manufacturers. *Augmentative and Alternative Communication* 20, no. 3, pp. 137–149.

McDonald, D. T., and Simons, G. M., 1988. *Musical Growth and Development: Birth Through Six*. New York: Schirmer Reference.

Miller, J. L., Lossia, A., Suarez-Rivera, C., and Gros-Louis, J., 2017. Toys that squeak: Toy type impacts quality and quantity of parent—child interactions. *First Language* 37, no. 6, pp. 630–647.

Norman, D., 2013. *The Design of Everyday Things: Revised and Expanded Edition*. New York: Basic Books.

Novak, D., and Sakakeeny, M. (Eds.), 2015. *Keywords in Sound*. Durham and London: Duke University Press.

Piaget, J., 1952. *Play, Dreams and Imitation in Childhood*. New York: W W Norton & Co.

Quam, C., and Swingley, D., 2012. Development in children's interpretation of pitch cues to emotions. *Child Development* 83, no. 1, pp. 236–250.

Radesky, J. S., and Christakis, D. A., 2016. Keeping children's attention: The problem with bells and whistles. *JAMA Pediatrics* 170, no. 2, pp. 112–113.

Reissland, N., Shepherd, J., and Cowie, L., 2002. The melody of surprise: Maternal surprise vocalizations during play with her infant. *Infant and Child Development* 11, no. 3, pp. 271–278.

Sosa, A. V., 2016. Association of the type of toy used during play with the quantity and quality of parent-infant communication. *JAMA Pediatrics* 170, no. 2, pp. 132–137.

Sulkin, I., and Brodsky, W., 2015. Parental preferences to music stimuli of devices and playthings for babies, infants, and toddlers. *Psychology of Music* 43, no. 3, pp. 307–320.

Sutton-Smith, B., 1986. *Toys as Culture*. New York: Gardner Press.

Tomitsch, M., Grechenig, T., Kappel, K., and Költringer, T., 2006. Experiences from designing a tangible musical toy for children. *Proceedings of the 2006 Conference on Interaction Design and Children*. New York: ACM, pp. 169–170.

Turner, J. B., 2004. *Your Musical Child Book*. Milwaukee, WI: String Letter Publishing.

Vondra, J., and Belsky, J., 1989. Exploration and play in social context: Developments from infancy to early childhood. In: Lockman, J. J., and Hazen, N. L. (Eds.) *Action in Social Context: Perspectives on Early Development*. New York: Plenum, pp. 173–203.

Werner, L. A., and Boike, K., 2001. Infants' sensitivity to broadband noise. *The Journal of the Acoustical Society of America*, 109, no. 5, pp. 2103–2111.

Woodward, A., and Hoyne, K., 1999. Infants' learning about words and sounds in relation to objects. *Child Development* 70, no. 1, pp. 65–77.

Yilmaz, R. M., and Goktas, Y., 2017. Using augmented reality technology in storytelling activities: Examining elementary students' narrative skill and creativity. *Virtual Reality* 21, no. 2, pp. 75–89.

Young, S., 2008. Lullaby light shows: Everyday musical experience among under-two-year-olds. *International Journal of Music Education* 26, no. 1, pp. 33–46.

15

# The Sonic Internet of Things

## Sound Design for Smart Objects

Vincent Meelberg

### 15.1 Introduction

In the age of the Internet of Things, things are no longer passive objects that only become activated once a human subject picks them up and uses them. Today, things can be active, make decisions and make changes to themselves or to other things, systems or environments, even without the explicit intervention of humans. Things can interact with each other and with human subjects in order to complete a particular task or set of tasks.

The Internet of Things, or IoT, is a term used to refer to the augmentation of everyday physical objects using information and communication technologies. "Things" here refer to embedded systems that are connected to the internet (Turchet et al. 2017). As a result, in the IoT ". . . all that is real becomes virtual: each person and thing has a locatable, addressable, and readable counterpart in the Internet" (Atzori et al. 2014, p. 97). Communication and interaction is possible because all actants (things and human users) are equipped with unique online identifiers, which allow them to communicate and interact with each other, that is, with the human end user or other nonhuman actants in the network.

The nonhuman actants taking part in the IoT are often referred to as smart objects. Miorandi et al. (2012) define smart objects as entities that have a physical embodiment, possess a minimal set of communication functionalities as a result of possessing a unique identifier and may have the ability to sense physical phenomena such as temperature, light and movement, or to trigger actions that have an effect on physical reality. Smart objects sense, log and interpret what is happening within themselves and the outside world, act independently and communicate with each other as well as with human end users (Kortuem et al. 2010).

Atzori et al. (2014) point out that, because of the active role smart objects play in the network consisting of human and nonhuman actants, smart objects can actually be considered objects with an actual social consciousness. These objects have the ability to converse with human end users and other smart objects, while being aware of the environment they are in. This is why smart objects are more than just sensor nodes, Kortuem et al. (2010) explain: "they are interactive tools designed to help people accomplish tasks in the real world. As such, smart objects' interactive input and output capabilities are key to their success" (p. 50). Interaction thus is crucial for an object to actually become smart and social.

As far as interaction with human end users is concerned, sound is one of the most effective methods to establish a relation between smart objects and human subjects. Sound arrives more quickly in the human brain than do visual stimuli. Moreover, humans are conditioned to respond to sound and, as a result, generally respond more strongly to sonic cues than cues transmitted in other media. Another advantage of using sound as a medium for interaction in smart objects is that users do not need to look at the object in order to monitor its activities (Saffer 2014). Sound thus allows us to receive messages while doing other things at the same time.

Despite these advantages, however, users often switch off the sound of their smart objects, because they are annoying to them. Take the sounds accompanying typing on the virtual keyboard of a smartphone, for instance. Even though these sounds may be helpful in providing confirmation that a particular action by the user (pressing a key) was successful, many users feel that these sounds are redundant or even irritating, and that they can perform the actions effectively without this kind of sonic feedback.

Of course, this does not mean that there are no successful implementations of sonic cues in the digital era. Iain McGregor (2017), for instance, mentions the trash emptying sound of paper being crumpled. This auditory icon can be found in many operating systems, McGregor observes, and is often left on by end users, as it is relatively easy to comprehend and remember. As a result, the success of sonic cues, Karmen Franinovic and Stefania Serafin argue, "rely on ways in which designers may successfully create socially meaningful, physically engaging, and aesthetically pleasing sonic interactions" (2013, p. 14).

It is the lack of social meaningfulness, in particular, that is responsible for the ineffectiveness of many sonic cues of smart devices. More specifically, the context in which these sounds occur often is not taken into account, McGregor (2018) maintains. Returning to the example of the virtual keyboard, one of the reasons users generally switch this sound off is because they often type on their phones in public places, such as trains or buses, and they do not want to bother other people by making unnecessary sounds.

Also, many sonic cues sound similar, and therefore it is sometimes difficult to identify which device is producing a sonic cue at a particular time.

This chapter will discuss sound in the age of the Internet of Things. More specifically, I will address the challenges of designing sound for smart devices in an age where there is a ubiquity of information and signals produced by interconnected devices. Therefore, the question I will address is this: How can sound improve the interaction between people and smart devices? In order to develop strategies for productive sonic interaction design for smart devices I will first focus on the materiality of interaction. As in the IoT materials are used "to manifest the computer as part of the world" (Wiberg 2018, p. 156), a representation-driven interaction design approach is less productive. According to Mikael Wiberg, the so-called material turn in interaction design was "a move away from any separation of 'world' and 'representation' (of that world), and a move toward a complete integration of computing in our everyday world (including the integration of computing in everyday physical objects)" (2018, p. 4). As smart devices in fact have transformed computing from a phenomenon that was separated from everyday life into an activity that is integrated in physical objects that people interact with on an almost daily basis, it makes sense to concentrate on the materiality of interaction in the interaction design of smart objects. Therefore, I will examine the possibilities the materiality of sound offers to function as a facilitator of interaction with smart objects.

In order to account for the possible contexts in which the sonic cues of smart objects will sound, I will also incorporate narrative strategies in my discussion. In particular, I will explore how sound events may contribute to what I call contextual narrative closure. Narrative closure is the experience of coherence and completeness of understanding after having experienced a narrative (Bruni and Baceviciute 2013). Contextual narrative closure thus can be considered the experience of coherence and completeness of understanding after having experienced a series of events, including sonic events, within an environment. A sonic event, such as a sonic cue emitted by a smart object, may be the actuator of a contextual narrative "told" by the environment, and if this sonic event contributes to contextual narrative closure—the experience of coherence and completeness of understanding—it can be said that this sonic event may be socially and contextually meaningful.

## 15.2 Material Interaction With Smart Objects

One of the main challenges when designing smart objects is creating an effective user interface. As a smart object is an entity that has a physical embodiment and that is able to communicate with other smart objects as well as with human end users, the effectiveness of this communication, that is, the kinds of interactions the smart object affords, codetermines the usefulness of that object. This means that the user interfaces that enable interaction with smart objects need to be carefully designed, as it is through the design of those interfaces that users experience the smart object either as an integrated whole, or as more or less loosely coupled parts and pieces, where its physical appearance, the user interface, and the functions it is able to perform appear as separate entities (Wiberg 2018, p. 107).

Kranz et al. (2010) refer to the implementation of user interfaces into physical objects as embedded interaction. More specifically, embedded interaction is the "technological and conceptual phenomena of seamlessly integrating the means for interaction into everyday artifacts" (p. 46). In order to achieve this, physical objects need to be augmented with sensing, actuation, processing and networking capabilities. Furthermore, objects need to be outfitted with input and output facilities that enable communication and interaction. As a result, these objects become part of what Kranz et al. call an "ecology of networked, self-configuring, and discoverable objects [that] constitutes a virtual overlay on the physical world" (2010, p. 47).

Users, for their part, need to become accustomed to interact in a meaningful way within this ecology of smart objects. The manner in which these objects afford users the opportunity for physical interaction needs to make sense to them. Moreover, it is through physical, embodied interaction with objects that meaning is created, manipulated and shared (Dourish 2001, p. 126). Consequently, whether or not a smart object makes sense to users depends on the ways in which this object allows for embodied interaction and the kinds of meaning that is created through this interaction.

Paul Dourish points out that embodiment is the common way in which human subjects encounter physical and social reality in the everyday world. Human subjects "encounter, interpret, and sustain meaning through their embodied interactions with the world and with each other" (2001, p. 127). Embodiment, therefore, is about the relationship between action and meaning, it is the manner in which human subjects make their world meaningful. Action and meaning are inherently inseparable.

Karmen Franinovic and Christopher Salter also acknowledge the importance of embodiment in the creation of meaning. They define meaning as an act in which "the perceiver's interaction with the external world is guided by direct action in local situations" (2013, p. 98). Taking the relation between meaning and interaction as a starting point, Franinovic and Salter characterize interaction as follows (2013, p. 109): it is a spatiotemporal-material process that is instigated by an act of poiesis or making. This process is situated, concrete and embodied, as well as performative (it is an act, a performance, that has actual effects in the real world) and emergent. Finally, interaction is nonrepresentational, which means that it is not reducible to an act of mimesis, imitation or purely symbolic processing.

This characterization of interaction as situated, embodied, performative and nonrepresentational is in line with the move in interaction design from a representation-driven approach towards a focus on the materiality of interaction. Traditionally, interaction with computing devices implied interacting with representations of reality. Wiberg (2018) observes that this representation allowed for easy manipulations, calculations and visualizations, yet these activities dealt with representations only, not with actual reality. This kind of interaction entailed a separation between reality and the computing device, a separation that would prevent the integration of computing in our everyday world, an integration that is crucial in the design of smart objects. The so-called material turn in interaction design was a deliberate move away from any separation of "world" and "representation" of that world (Wiberg 2018, p. 4).

The material turn resulted in a shift from mediating tools and objects towards more direct, embodied forms of interaction, forms that engage directly with the physicality of the smart objects themselves. The focus on materiality suggests a focus on physical actions, as well as on the directness of engaging with materials and the feedback this material can generate. In short, this turn moves towards a more embodied manner of interacting with these objects and has transformed interaction design into the "design of interaction through material configurations, and as such, it is increasingly about the ways we can configure materials to enable interaction with and through materials" (Wiberg 2018, p. 1).

It is not a coincidence that the material turn in interaction design coincided with the emergence of the IoT. After all, the IoT has resulted in the introduction of computing in everyday life. Computing no longer is something separate from everyday reality, but instead an integral part of that reality. In smart objects, computing is manifested in material form, and physical materials are part of their user interfaces. Wiberg maintains that the IoT would be impossible if computing and things were kept apart as distinct entities: "computational devices, such as smartphones, would make no sense if not also considered in relation to the form factor of the device" (2018, p. 8). Smart objects are not only smart because they have computing power, but also, and perhaps even primarily, because they allow for smart, embodied interaction that make sense to their users via direct manipulation of the physicality of the objects themselves. As a result, interaction design for smart objects is concerned with the physical properties of materials and the possibilities these properties offer for meaningful embodied interaction.

Consequently, in order to use sound as a means for embodied interaction with smart objects, a focus is needed on the particular physical properties of sound, considered as a material. Can these properties contribute to a more direct interaction with smart objects, or does sound mainly have the potentiality to represent reality? Put differently, can sound be used as a manifestation of computing in material form?

## 15.3 Sound as Interactive Material

In order to explore the possibilities the materiality of sound offers to function as a facilitator of interaction with smart objects, it first needs to be clear what the material qualities of sounds are. What do we mean by "materiality of sound"? Can sound be considered as a material, even though sound is often characterized as ephemeral and intangible?

Christopher Cox believes it can. He maintains that the materiality of sound consists of "its texture and temporal flow, its palpable effect on, and affection by the materials through and against which it is transmitted" (2011, p. 149). Paul Simpson adds that "the sound itself is precisely sound's materiality, its body, its timbre, and about the resonance these produce" (2009, p. 2559). Sound thus is vibration, which is a temporal phenomenon, and the manner in which these vibrations manifest themselves as resonances constitutes the materiality of sound.

At the same time, resonances are responsible for the communicative and performative qualities of sound. Sounds resonate in listeners' bodies. They do so not only in their ears but also as something that is felt (Gershon 2013; Goodman 2010). Sounds can literally move the bodies of listeners. Not only because bodies resonate as a result of being exposed to sounds, but also because these resonances can induce autonomous bodily reactions at an unconscious level. Take a sudden, loud sound, for instance. As soon as listeners perceive this sound, their bodies shiver. A shock is running through their bodies, a shock that can literally be felt but cannot be controlled by the listeners who inhabit these bodies.

According to Gilles Deleuze (2003), sensations such as being startled by experiencing a sudden loud sound can be understood as an affect. In accordance with results obtained in recent empirical and cognitive research on perception, Deleuze argues that affect is an autonomous reaction of an observer's body when confronted with a particular perception. In this view, perception induces bodily reactions called affects in an observer. At first this affect has no meaning or signification yet, because it is entirely physical (Massumi 2002). Yet, this affect cannot be ignored and instead functions as a motivation to reflect on the sensations they are experiencing. The materiality of sound, considered as resonances that can be felt, thus has affective qualities that, in turn, are responsible for the performative potentiality of sound, that is, the capacity of sound to instigate a change in listeners and to grasp their attention. Sound has this ability not because it represents or signifies something other than itself, but because of what it does, how it operates and what changes it effectuates (Cox 2011, p. 157). And it is precisely because of its performative potentiality that sound has the capacity to incite interaction.

Bruce Walker and Michael Nees (2011) identify four different kinds of interactive functions that sounds can have based on their performative potentiality. Firstly, sound can act as an alarm, alert, emphasis or warning. These sounds generally are simple, overt and clear. Furthermore, they are event-based instead of continuous, and relatively information-poor. Alerts are indicators of what Dan Saffer (2014) calls system-initiated actions, such as a process that has ended, a condition that has changed, or an indication that something is wrong. Apart from indicating an ending, change, warning or caution, these sounds usually do not provide additional information. Sound for emphasis is typically used for reinforcing a user-initiated action, that is, as a sonic confirmation that an action a user performed actually took place. The sounds the virtual keyboard of a smartphone produces, mentioned in the introduction of this chapter, is an example of such a sound.

Secondly, sound can be used to monitor processes. These sounds are dynamic and continuous and take advantage of the fact that small changes in auditory events are easily detected by listeners. Walker and Nees point out that monitoring requires listeners to attend to a sequence of sound over a course of time and to detect events that are represented by sounds and identify the meaning of the event in the context of the system's operation (2011, p. 19). Changes may thus be identified by listeners because of changes in resonances they may experience, but what these changes mean depends on the context in which they appear.

The third interactive function of sound is data exploration. This function is not dissimilar to monitoring processes, but here the goal is exploration and discovery rather than observation and surveillance. Finally, sound can be used for art, entertainment, sports and exercise. Here, alarms, alerts and monitoring may be combined for artistic or entertainment purposes.

Walker and Nees (2011) further distinguish between different ways in which interactions between sounds and listeners generate meaning. Firstly, there may be an analogical relationship between a sound and the action or process it represents. The materiality of the sound resembles this action or process. Such sounds are called auditory icons, and their intended meaning is relatively direct and should require little or no learning. Eoin Brazil and Mikael Fernström add that auditory icons mimic everyday non-speech sounds that listeners might be familiar with from their daily experience of the real world, which means that the meaning of the sounds rarely has to be learned as they metaphorically draw upon the previous experiences of listeners (2011, p. 325).

A second way in which sonic interaction can generate meaning is to use sounds as symbolic representations of actions or processes. These so-called earcons, Walker and Nees (2011) explain, are made by "systematically manipulating the pitch, timbre and rhythmic properties of sounds to create a structured set of non-speech sounds that can be used to represent any object or concept through an arbitrary mapping of sound to meaning" (p. 23). Because the mapping of sound to meaning is arbitrary, earcons can represent virtually anything. The possible downside, however, is that listeners have to consciously learn the meaning of earcons, in contrast to auditory icons, where the relation between sound and meaning is more intuitive.

Spearcons, lastly, are created by speeding up a spoken phrase even to the point where it is no longer recognizable as speech. As such, they can represent anything, similar to earcons, as any spoken phrase can be transformed into a spearcon. They are, however, non-arbitrarily mapped to their concept, just as auditory icons are, because they are tied to the meaning of the original spoken phrase (Walker and Nees 2011, p. 24).

As opposed to earcons and spearcons, which are primarily representation-based, auditory icons foreground the materiality of sound. The manner in which the sonic vibrations of an auditory icon manifest themselves as resonances mainly determines the meaning of this sonic cue. McGregor (2017) gives two examples of how the materiality of sound can be productively used in the design of sonic cues. The first example McGregor discusses is the Doppler effect, which suggests movement. A second example is the use of presence frequencies. A rise in pitch indicates an increase in tension and therefore something potentially important. A drop in pitch, on the other hand, elicits a feeling of resolution and suggests that the danger has passed and that it is safe to ignore. In both examples it is the manner in which sound resonates in listeners, and the tensions and resolutions this elicits with them, that is responsible for the meanings generated by the auditory icons. As auditory icons foreground the materiality of sound, a material-centered approach to sonic interaction design of smart objects would thus primarily focus on auditory icons instead of on earcons or spearcons.

Elisabeth Mynatt (1994) distinguishes between four factors that need to be taken into account when designing auditory icons: identifiability, conceptual mapping, physical parameters and user preference. Identifiability depends on the balance between how common the sound is, which contributes to the recognizability of the sound, and the relative uniqueness of the sound, which determines how easy it is to identify the sound among other sounds. Saffer (2014) stresses that identifiability of auditory icons is crucial and warns against using the same or similar sounds for dissimilar events. This is especially relevant for alert sounds that could be triggered independent of user actions. If users are not consciously looking at or paying attention to a smart object, they cannot be sure of which action just happened when the auditory icon sounds.

Conceptual mapping refers to the manner in which the sound is mapped to the aspect of the user interface it is supposed to call attention to. Physical parameters refer to the material qualities of the sound, and thus ultimately to the manner in which an auditory icon manifests itself as resonances. User preference, finally, denotes the emotional responses of an auditory icon by users, responses that are evoked by the affective and performative qualities of the sounds that make up the auditory icon.

Mynatt points out that these factors are not only relevant for the design of individual auditory icons, but also for designing sets of sounds. She observes that "[t]here is little chance that a successful set of icons would result from designing the auditory icons independently of each other" (Mynatt, as quoted in Brazil and Fernström 2011, p. 334). When the context, both sonic and otherwise, in which the auditory icon will sound is not taken into account, the effectiveness of this icon will be diminished. The sounds need to make sense in their context, not just in isolation. This is particularly important when it concerns the design of auditory icons for smart objects, as these objects are supposed to be integrated into our everyday lives. Consequently, these sounds need to be comprehended and understood within the context of our everyday life.

## 15.4 Contextual Narrative Closure

Auditory icons that are designed for smart objects need to be comprehended and understood within the context of our everyday lives in order to be effective. One way to arrive at a degree of understanding of events, sonic or otherwise, is to create stories around them. David Herman observes:

> As accounts of what happens to particular people in particular circumstances and with specific consequences, stories are found in every culture and subculture and can be viewed as a basic human strategy for coming to terms with time, process, and change.
>
> (2003, p. 2)

Kitty Klein adds that "[n]arrative has often been viewed as the product of a universal human need to communicate with others and to make sense of the world" (2003, p. 65). Stories are important both in grasping the world and in communicating this grasp. Narratives function as accounts with which human subjects can make the events they undergo discursive or, in other words, turn them into experiences.

This implies that narrativization, the interpretation of a series of events as a narrative, is not a separate artistic category, but instead is an everyday activity that human subjects perform on a regular basis in order to understand the world around them. Almost everything can be turned into a narrative, including smart objects and their behavior. Helena Barbas argues that such objects, in particular, have narrative potentialities: "If a Thing has the capacity to register information [such as smart objects], it will surely be—as it has been—used to tell a story" (2015, p. 102). Luis Emilio Bruni and Sarune Baceviciute (2013) also point out that all representational media in general, and interactive media in particular, have a narrative capacity, but they explicitly link the narrative potential of objects to the manner in which these objects are designed, and the goals their designers had when designing them. They claim that the narrative potential of an object suggests "that every system embraces a specific goal and thus an intrinsic communication cycle between a system (and its designer) and its users" (2013, p. 13). Even though Herman's and Klein's accounts regarding the narrative potentialities of an object do not presuppose that the person who created that object needs to have had the intention to actually tell a story through that object, it may nevertheless be relevant to incorporate the notion of authorship in the discussion of the design of auditory icons for smart objects.

The primary goal of designing such auditory icons is to create sonic experiences that make sense to users in everyday contexts, and this process of making sense can be conceptualized in narrative terms. When it comes to interactive objects, however, designing a narrative structure offers additional challenges, and may lead up to what Bruni and Braceviciute call "the narrative paradox." In this paradox, "the relationship between authorship and interactivity is seen as being inversely proportional i.e.: the problem of having a free roaming interactive world and an author-controlled narrative at the same time" (Bruni and Baceviciute 2013, p. 14). The more freedom users have to interact with an object, the less control the author has over the development of the narrative that is told via that object.

In the case of smart objects the control of the author/designer is even less, as they cannot fully prescribe or predict the contexts in which smart objects will function. As a result, users may interpret the auditory icons of smart objects in unexpected and surprising ways, Mynatt points out: "Sound seems to lend itself to storytelling. Evaluators of auditory interfaces are often amazed at the stories users will build to explain the sounds that they hear. . . . The metaphorical power of auditory icons is immense, but controlling these mappings is a difficult design task" (1994, pp. 110–111). The imagination of users is difficult to control when it comes to the narrative interpretation of sounds.

More precisely, it is difficult to control what Bruni and Braceviciute call narrative intelligibility. Narrative intelligibility refers to the process in which the audience receives or generates meaning in a way that is close to what is intended, desired or expected by the author. Narrative closure, on the other hand, refers to the process where "the audience may construct its own meaning out of what is being mediated, independent on whether that meaning corresponds or gets close to what is intended by the author" (Bruni and Baceviciute 2013, p. 18). Narrative closure thus can be understood as "[t]he experience of coherence and completeness of understanding after having experienced a narrative, even though the narrative's substance is not understood very closely to the way it was intended by the author or creator" (p. 19). This implies that narrative intelligibility presupposes closure, but narrative closure may be reached without having understood the intentions of the author or creator of the narrative. Consequently, the process of narrativization that I introduced above can be considered a process that leads to narrative closure (although it does not exclude the possibility of narrative intelligibility, either).

Causality, or at least the suggestion of causality, is very important for narrative closure (Gerrig and Egidi 2003). A narrative can be experienced as coherent and complete because the events it consists of can be interpreted as being related in a causal manner, regardless of whether this relation is a reality or a projection of an apprehending subject. Objects that can be interpreted as containing events that are somehow—metaphorically or otherwise—causally related might be more easily grasped in a narrative manner.

The same holds for the narrativization of a sequence of sounds. More specifically, it is the interplay of tension and resolution elicited by the acoustic qualities a sequence of sounds may display, such as dynamics, tempo and rhythm, that gives the suggestion of causality, which in turn may lead to experiencing this sequence as coherent and complete. As a consequence, this sequence can be considered as being narrative, or at least as having narrative potential. It is the materiality of the sounds, the resonances responsible for the tensions and resolutions that listeners may feel, that results in the interpretation of a sequence of sounds as a narrative.

Although some auditory icons that are designed for smart objects may consist of a sequence of sounds, and thus in theory may have the potential to be interpreted as a narrative, often the duration of these icons are too short to be interpreted as narratives themselves. This does not mean, however, that short auditory icons do not have any narrative potential at all. Many sounds may have narrative-evoking qualities, even though these sounds themselves, in isolation, cannot be considered narratives. A single sound can trigger narratives (Back and Des 1996) but generally cannot itself be interpreted as an actual narrative, that is, as a sequence of events that are somehow causally related. A sound of a scream, for instance, can elicit stories about fear or danger, but the scream itself is not necessarily a narrative. Rather, it is the specific way in which the sound may affect listeners that is responsible for its evocative narrative nature. The narratives that are triggered in this manner are called micro-narratives (Back and Des 1996). An auditory icon that consists of a single short sound therefore can have the potential to trigger micro-narratives, not because of what this icon represents, but because of what it does with listeners, the ways in which the materiality of the sound of this icon is performative and affects them.

The shape and contents of a micro-narrative depend on the context in which it is triggered by an auditory icon. The environment in which the icon can be heard contributes to narrative. The story is "told" both by the auditory icon that triggered it as well as the environment in which the icon sounds. The way these narratives emerge is not unlike what Jennifer Stein and Scott Fisher (2013) call ambient storytelling, which refers to the context-specific and location-specific stories that emerge over time through daily interactions between an environment and their inhabitants. And as long as these stories can be experienced as coherent and complete, narrative closure is possible.

To summarize, an auditory icon can function as an actuator of a micro-narrative "told" by the environment in which this icon can be heard. The performative nature of the materiality of this sound, its potential to affect listeners via resonances, can trigger a micro-narrative, one that is co-constituted by the environment in which the icon is sounding. Contextual narrative closure, the experience of this micro-narrative as coherent and complete, thus may emerge as a result of the interaction between a listening subject, the auditory icon and the context/environment in which the icon appears.

McGregor (2017, 2018) provides examples of how auditory icons of smart objects can contribute to contextual narrative closure. He stresses that sounds designed for smart objects need to be context-aware in order to be effective. McGregor suggests that sounds might be introduced at subtler levels, and then turned up if they go unnoticed, rather than the more traditional approach of needing to be turned down or off entirely. Smart objects should actively monitor preexisting auditory icons in order to avoid masking or perceptual conflicts, and sounds could even be programmed to make fewer sounds over time to acknowledge that a user is familiar with the device and operating it correctly.

In these examples smart objects need to exhibit spatial, acoustic and temporal contextual awareness in order to function properly. If they do, they can be considered effective, because these types of contextual awareness contribute to the experience of their auditory icons as triggering micro-narratives, and thus as evoking context-specific and location-specific stories that are coherent and complete. As a result, the act of perceiving an auditory icon is turned into an experience that is understood by the users of the smart object that emitted the sound. The sounds take into account the spatial, acoustic and temporal characteristics of the environment and ensure that the interaction between users/listeners, the sounds and the environment result in micro-narratives that tell the "right" stories, that is, stories that result in contextual narrative closure and thus can be understood within that environment.

## 15.5 Conclusion

The central question of this chapter was how sound can improve the interaction between people and smart objects. As interaction design for smart objects is primarily concerned with the physical properties of materials and the possibilities these properties offer for meaningful embodied interaction, it made sense to focus on the particular physical properties of sound, considered as a material, in order to address this question. The manner in which sonic vibrations manifest themselves as resonances constitutes the materiality of sound, and the ways in which these resonances affect users contribute to the sound's interactive potentiality.

Smart objects use auditory icons to communicate the result of the computations they perform to users. The resonances these icons consist of make these computational results observable. It is in this sense that sound can be considered a manifestation of computing in material form.

Observability does not automatically imply understanding, though. Hearing a sound does not inevitably lead to sonic comprehension. Moreover, many sounds, in particular those that are produced by smart objects, are rarely observed in isolation. These sounds thus need to be understood within the context, the environment, in which they are heard.

One way to arrive at a degree of understanding of phenomena, especially temporal phenomena such as sounds, is to narrativize them. An auditory icon, in particular, can function as an actuator of a micro-narrative created by the interplay of this icon and the environment in which it can be heard. The experience of this micro-narrative as coherent and complete, and thus contextual narrative closure, may emerge as a result of the interaction between a listening subject, the auditory icon and the environment in which the icon appears. In this way the icon makes sense in time to a user, and a level of narrative understanding is achieved.

Despite the risk of running into a narrative paradox created by discrepancies between the intentions of the designers and the narrative interpretation of users, it may nevertheless be productive to adopt a contextual narrative approach when designing auditory icons for smart objects. This approach forces designers to take into account the environment in which the smart object will function, as well as temporal considerations. What other sounds can be heard in this environment? How does the environment change over time? Is there a learning curve for users as far as sonically interacting with the object is concerned? How do all these factors influence the manner in which the auditory icons evoke micro-narratives and, thus, how these icons are understood over time by users?

A material and contextual narrative approach may also be useful for research on sound design in the age of the IoT. This approach offers a conceptual framework within which the performative aspects of sounds designed for smart objects, that is, what these sounds do and how they interact with and affect users, can be analyzed. This kind of research, in turn, may lead to new insights into the ways in which sound, as material, can be used to create new possibilities for embodied interaction between users and objects.

## References

Atzori, L., Iera, A., and Morabito, G., 2014. From 'smart objects' to 'social objects': The next evolutionary step of the internet of things. *IEEE Communications Magazine* 52, no. 1, pp. 97–105.

Back, M., and Des, D., 1996. Micro-narratives in sound design: Context, character, and caricature in waveform manipulation. In: Frysinger, S. P., and Kramer, G. (Eds.) *Proceedings of ICAD 96 International Conference on Auditory Display*. Palo Alto: Xerox Palo Alto Research Centre. Viewed May 2, 2018 <http://icad.org/websiteV2.0/Conferences/ICAD96/proc96/back5.htm>.

Barbas, H., 2015. Cloud computing and (new) mobile storytelling in the internet of things. In: Gonçalves, P. J. S. (Ed.) *Proceedings of Euromedia 2015, 27–29 April 2015, Lisbon, Portugal*. Ghent: EUROSIS, pp. 96–103.

Brazil, E., and Fernström, M., 2011. Auditory icons. In: Hermann, T., Hunt, A., and Neuhoff, J. (Eds.) *The Sonification Handbook*. Berlin: Logos Verlag, pp. 325–338.

Bruni, L. E., and Baceviciute, S., 2013. Narrative intelligibility and closure in interactive systems. In: Koenitz, H., Sezen, T. I., Ferri, G., Haahr, M., Sezen, D., and Çatak,

G. (Eds.) *Interactive Storytelling: Proceedings of the 6th International Conference, ICIDS 2013, 6–9 November 2013, Istanbul, Turkey*. Heidelberg: Springer, pp. 13–24.

Cox, C., 2011. Beyond representation and signification: Toward a sonic materialism. *Journal of Visual Culture* 10, no. 2, pp. 145–161.

Deleuze, G., 2003. *Francis Bacon: The Logic of Sensation* (trans. D. W. Smith). London: Continuum.

Dourish, P., 2001. *Where the Action Is: The Foundations of Embodied Interaction*. Cambridge, MA: MIT Press.

Franinovic, K., and Salter, C., 2013. The experience of sonic interaction. In: Franinovic, K., and Serafin, S. (Eds.) *Sonic Interaction Design*. Cambridge, MA: MIT Press, pp. 39–78.

Franinovic, K., and Serafin, S., 2013. Introduction. In: Franinovic, K., and Serafin, S. (Eds.) *Sonic Interaction Design*. Cambridge, MA: MIT Press, pp. viii–xiv.

Gerrig, R. J., and Giovanna, E., 2003. Cognitive psychological foundations of narrative experiences. In: Herman, D. (Ed.) *Narrative Theory and the Cognitive Sciences*. Stanford: CSLI Publications, pp. 33–55.

Gershon, W. S., 2013. Vibrational affect: Sound theory and practice in qualitative research. *Cultural Studies ↔ Critical Methodologies* 13, no. 4, pp. 257–262.

Goodman, S., 2010. *Sonic Warfare: Sound, Affect, and the Ecology of Fear*. Cambridge, MA: MIT Press.

Herman, D., 2003. Introduction. In: Herman, D. (Ed.) *Narrative Theory and the Cognitive Sciences*. Stanford: CSLI Publications, pp. 1–20.

Klein, K., 2003. Narrative construction, cognitive processing, and health. In: Herman, D. (Ed.) *Narrative Theory and the Cognitive Sciences*. Stanford: CSLI Publications, pp. 56–84.

Kortuem, G., Kawsar, F., Fitton, D., and Sundramoorthy, V., 2010. Smart objects as building blocks for the internet of things. *IEEE Internet Computing* 14, no. 1, pp. 44–51.

Kranz, M., Holleis, P., and Schmidt, A., 2010. Embedded interaction: Interacting with the internet of things. *IEEE Internet Computing* 14, no. 2, pp. 46–53.

Massumi, B., 2002. *Parables for the Virtual: Movement, Affect, Sensation*. Durham: Duke University Press.

McGregor, I., 2017. Sound and the internet of things. *A Sound Effect*, December 7, 2017. Viewed May 2, 2018 <www.asoundeffect.com/internet-of-things-sound-design/>.

McGregor, I., 2018. Internet of things needs to use sound in ways computers and phones never have. *Phys.org*, March 15, 2018. Viewed May 2, 2018 <https://phys.org/news/2018-03-internet-ways.html>.

Miorandi, D., Sicari, S., De Pellegrini, F., and Chlamtac, I., 2012. Internet of things: Vision, applications and research challenges. *Ad Hoc Networks* 10, pp. 1497–1516.

Mynatt, E. D., 1994. Designing with auditory icons. In: Kramer, G., and Smith, S. (Eds.) *Proceedings of the 2nd International Conference on Auditory Display (ICAD1994), 7–9 November 1994, Santa Fe, New Mexico*. Atlanta: International Community for Auditory Display, pp. 109–119.

Saffer, D., 2014. *Microinteractions: Designing with Details*. Cambridge, MA: O'Reilly.

Simpson, P., 2009. 'Falling on deaf ears': A postphenomenology of sonorous presence. *Environment and Planning A* 41, pp. 2556–2575.

Stein, J., and Fisher, S., 2013. Ambient storytelling experiences and applications for interactive architecture. In: Weyn, W. (Ed.) *Proceedings of AMBIENT 2013: The Third International Conference on Ambient Computing, Applications, Services and Technologies, 29 September—3 October 2013, Porto, Portugal*. Antwerp: Artesis University College of Antwerp, pp. 23–28.

Turchet, L., Fischione, C., and Barthet, M., 2017. Towards the internet of musical things. In: Lokki,T., Pätynen, J., and Välimäki, V. (Eds.) *Proceedings of the 14th Sound and Music Computing Conference, 5–8 July 2017, Espoo, Finland*. Aalto: Aalto University, pp. 13–20.

Walker, B. N., and Nees, M. A., 2011. Theory of sonification. In: Hermann, T., Hunt, A., and Neuhoff, J. (Eds.) *The Sonification Handbook*. Berlin: Logos Verlag, pp. 9–39.

Wiberg, M., 2018. *The Materiality of Interaction: Notes on the Materials of Interaction Design*. Cambridge, MA: MIT Press.

16

# On the Importance of Listening

**Antti Ikonen**

## 16.1 Introduction

Before we begin, I would like to tell about myself in order to provide a backdrop to what will follow. When reading, I think it is useful to be aware of both the expertise and the limitations of the author in question. My background is not in academia: I started my career as a sound designer in the early 1980s without any formal education in the field, mainly because at the time there was not any training of that kind available in my home country. I obtained my designer's skills through practice, and I initially ended up a university lecturer on the basis of my hands-on experience as a composer and sound designer in numerous artistic and commercial projects and productions. Academic studies followed later in my life, although I still consider myself an artist, designer and pedagogue rather than a researcher or scientist. Furthermore, the thoughts and ideas of this chapter are to a large extent based on my experiences and findings from the various practical projects I have participated in where I have been responsible for the audible dimension of the outcome. In a broader scope, if I am to mention one great source of inspiration, it is undoubtedly *Acoustic Communication* by Barry Truax (especially the second edition). "Clearly, the old dualisms of science and art, or that of objective and subjective criteria, do not serve us well in attempting to formulate principles of acoustic design" (Truax 2001, p. 110). I am sure that other parts of this book series will provide numerous theoretical references that will hopefully complement the practice-based content I will introduce in this chapter. Finally, I would like to express my deep gratitude to Ava Grayson, whose comments and suggestions have made this text much better than it would have been without her invaluable help.

### 16.1.1 Listening to the Actual Environment

It is self-evident that sound recording and reproduction technologies together with telematic media—from analog broadcasting to contemporary digital, rhizomatic networks—have completely changed our sonic environment and listening habits. The impact of industrialization, electricity and motor-driven vehicles on our soundscape had already inspired and excited the futurists in the early twentieth century. Today, after the invasion of ubiquitous microprocessor technology, mobile devices and wearable audio, the change goes on. Although music as such does not fall within the scope of this chapter (the relationship between music and technology dating back to the dawn of mankind), it is worth keeping in mind how digital technology such as MIDI, sampling and DAWs—as well as the internet-driven new wave of global popular culture—have affected (popular) music in very special ways.

When designing sounds for any purpose or context, we need to be aware of our audience, the people who are hearing, listening, using, consuming—as well as interacting and living with—the designed sounds. It is often quite tempting to approach the design task at hand by trying to figure out the supposed smallest common denominator concerning the taste of the "target audience," or focusing on the surface level of the sounds by using the current audio trends as reference. Often these two approaches go hand in hand, and in commercial productions the client may have rather straightforward opinions about how the work they are paying for should sound. However, instead of learning the latest audio gimmicks and their corresponding software plug-ins, listening is the most important tool in sound design. The importance and value of contemplative listening should be understood in terms of searching for references and inspiration for creative design, as well as in the actual practical audio engineering process, regardless of the equipment used for the work.

Our actual sonic environment provides the richest, the most accurate and most reliable reference material for any kind of sound design. The privilege of being a sound designer is to be able to enjoy the endless resource of sounds that surround us all the time and wherever we go. Sound is free and open source, and any moment of contemplative listening can contain learning. A wide range of experiences is available to an aware listener, from irritating noise (which may force one to leave immediately) to fascinating aural perceptions that may make one stop for a while to just listen, even changing one's route in order to find the source of an interesting sound. In addition to space and place, listening is inextricably connected to time. Soundscapes change over time, typically in periodic cycles of day and night (24 hours) and throughout a year. Furthermore, in inhabited places, the seven-day weekly cycle is usually clearly audible in the soundscape. For perceiving the properties of a certain specific soundscape, whether indoors or outdoors, one needs to be prepared to spend a good amount of time listening to it.

Designing and creating soundscapes for different forms of media (radio, movies, games, spatial installations, etc.) is in most cases more feasible to do by reconstructing rather than by reproducing. In art and fiction, the aim isn't usually to create a replica of an existing sonic environment, but instead to design something resembling it enough and adding elements that support the narration and aesthetics of the whole production. When reconstructing or recreating a soundscape, it is of course essential to gain knowledge of the ingredients and the behavior of the soundscape in question: this cannot be fully achieved without listening for an extended period of time. Furthermore, soundscapes and aural interiors can be listened to in a reduced way, without thinking of what might be causing the sounds or what they might mean but rather treating these sounds as ambient music. Instead of trying to recognize the (origins of the) sounds, the focus is on the physical properties, spatial placement and temporal behavior of the sounds, as well as how the sounds are being "orchestrated" as a whole in terms of dynamics, frequencies, timbre and spatiotemporal structure. When listening to a soundscape, it is important to pay attention to features like density, intensity, pulse and regularity when aiming at understanding the outline of a composition on the macro level. The grasp of a soundscape as a whole, obtained through contemplative listening, can then be applied to designed sound. Understanding human engagement with sound certainly improves the quality of sound design in art, media and entertainment, but it is especially important when designing the audible dimension of an environment that people don't attend just for leisure or cannot help but be exposed to.

An interesting listening experience can be extended by recording the sound(s) in question and listening to the recording later on. Many of us have a portable recording device (i.e. a smartphone) with us all the time, and recording audio snapshots can be a rewarding habit for anyone. Without an external microphone, smartphone recordings are monophonic and all in all often not suitable to be used as production material. On the other hand, snapshot photos are taken without them being considered deficient at all. An equivalent kind of "intrinsic value" should be given to sonic souvenirs, regardless of the motivation behind or the level of ambition of the act of recording. Since the early twenty-first century, interactive soundmaps have been an increasingly significant platform for both hobbyists and professional recordists' materials, making vast amounts of audio snapshots accessible for wider audiences. Being able to listen to geotagged recordings online can certainly be an inspiring and ear-opening experience, providing ideas and inspiration for sound design. However, one must remember that all recorded sound contains decisions made by the recordist—that is, another person—and in the case of larger amounts of material, it is important to consider who the people are who have made the recordings in question. In academic debate, some criticism has been posed to the idealization of soundmaps, asking how democratic, inclusive and non-hierarchic they actually are.

Considering other complementary material for real-time listening on location, there are rather established conventions for documenting soundscapes visually in relation to geography: an isobel map shows areas having certain level of background sound (at certain times) in decibels, similar to a standard map displaying the shape of the landscape in terms of altitude. However, these kinds of visualizations are used more in research and urban planning than in art and design, since they do not directly relate to or reflect human experience. The practice of soundwalks and listening walks dates back to the 1970s, and since then various kinds of (more or less) map-based diagrams have been made in an attempt to document the experience for further reference. Compared to graphical representations, written descriptions—even poetic ones—are often more useful in notating and memorizing sounds and soundscapes, or as an indirect source of what is being heard and listened to. Once again, converting a sonic experience to any verbal form obviously requires in-depth listening in the first place. Different individuals and different listening situations benefit from different methods of complementing documentation.

## 16.2 Listening to Audio Material

Sound recording and reproduction are an art and science of their own, and far beyond the scope of this chapter. However, practically all audio (recorded sound) we hear, with the exception of live broadcasts, is processed and edited in some sort of studio environment. This environment can be anything from a large dedicated building to an app on a smartphone, so it is important to pay attention to how the possibilities of audio technology affect the way sounds are treated. The history of recorded sound is a fascinating story, though way too vast to be handled here. However, it is worth mentioning that the earliest methods for capturing sound have interesting similarities to contemporary digital recording and audio processing techniques. For decades, the official canonized history of audio mentioned Thomas Alva Edison's 1877 tinfoil cylinder recording as the original means of capturing sound. The method of encoding sound with Edison's phonograph was mechanical (i.e. the vibration of sound waves was converted into a linear engraving on a surface made of suitable material). However, as early as 1857, French printer Édouard-Léon Scott de Martinville patented a device called phonautograph, which was able to produce an image of a sound wave on paper. The device was supposed to provide ways to analyze sound, especially the human voice, for scientific purposes. However, there was no way—and no actual intention—to play back the images (phonautograms) as sounds. Over 150 years later in 2008, scientists at Berkeley Lab (US) were able to convert a phonautogram back to its audible form for the first time. Perhaps due to the lack of a "built-in playback" feature, the phonautograph is still missing from some timelines of audio recording history. Nevertheless, the paradigm of observing sound as an image dates back to the mid-nineteenth century.

When digital audio recording and professional editing studios in the consumer market became available (including digital records, i.e. compact discs) and began to replace—although not completely supersede—analog technology and audio formats during the1980s, the hottest debate was regarding the perceived quality of the sound. The difference between analog and digital audio is indeed often audible, and the ranking order of these two categories of sound reproduction might be an eternal question. Much less attention has been paid to how the visual user interfaces of DAWs (digital audio workstations) and audio editing software that display sound as graphical image have changed various phases of recording, editing and processing audio material, and therefore the acousmatic sounds we encounter in so many facets of our everyday life.

In the age of magnetic tape, the only visual interface to audio was the mixing console with its VU-meters, little lights indicating signal clipping, or some functions of the mixing console itself, and, of course, the positions of sliders and knobs that an experienced sound engineer could "read" and interpret, both in terms of details and in a holistic way. An oscilloscope might have been placed in the middle of the control speakers, but mostly for detecting possible anomalies in the signal instead of providing visual information of how the audio actually sounded. During the process of audio production in a recording or a postproduction studio the focus was on the act of listening, and discussions and negotiations on decisions and points of action happened on the basis of what was heard (and what was not), taking people's ability to "store" the listening experience in their memory and recall it when needed as self-evident. There was no way to see the position of a certain event in the sound continuum, although one might have written down the minutes and seconds of the time display or the numbers of the three- or four-digit tape counter. Effects such as reverb and echo were chosen, added and dosed on the basis of listening, even if not starting completely from scratch (such as through utilizing a template from a previous production). Since copy-paste was not an option, the recordist, audio engineer and producer were forced to listen to each individual detail as a unique sound, then judge whether or not it fulfilled its purpose in terms of aesthetics and function.

Although analog audio and analog media formats refuse to fade, very few audio professionals would be willing to return permanently to the times before digital, nonlinear audio editing. Who wouldn't appreciate the possibility to move audio material between devices much faster than in real time? Furthermore, making music from building blocks of various sizes is an integral part of studio work regardless of genre, and repetition is an essential element in many well-established styles of music. However, it is interesting to note whether vertical listening is the chicken or the egg in (pop) music. If the focus in producing music—and by all means any other kind of audio—is based on designing, tweaking and fine-tuning an individual building-block component of the actual piece under construction, is it not also likely that the attention of the listener is steered towards the same direction?

Today, the computer is not only the heart of a recording and postproduction studio, it is also the studio. It is worth spending a few moments touching on our relationship with computers and what we actually expect from them. Regardless of the intended use, the most typical sales arguments for buying a new computer relate to increased efficiency, usually synonymous with speed: the processor speed, the speed of the hard drive and the speed of performing various tasks. In many cases, the increased speed is indeed useful, but with audio software gaining efficiency, it is sometimes used for adding features and functions that do not necessarily help the user to actually listen to the sounds they are working with. Almost all traditional outboard audio gear, like compressors, reverb boxes and graphic equalizers, have their virtual equivalents bundled into audio editing software. Despite their skeuomorphic user interfaces, these virtual versions offer loads of drop-down-menu presets that save the user the trouble of turning virtual knobs and trying to listen how each of them affects the sound. The presets are often labeled with names that may have quite a strong influence on how the virtual effect processors will be used. Sound is, of course, also a fashion, and there is nothing wrong with that as such. Nevertheless, we should be aware to what extent the factory presets and the workflow patterns afforded by the software will actually affect the results of our audio work.

However useful computer-based graphical and visual representations of audio are, staring at them while listening can be quite disorienting and not very helpful when trying to find the right balance, duration and flavor for the audio at hand. Watching the graph of an audio file or multi-track project moving against the timeline as on-screen minutes, seconds and frames is often rather unnerving, as if the computer should in its mighty power somehow speed up the actual time spent for getting things right with the sound. The easiest way to get back on track with listening in studio is simply to close one's eyes and concentrate entirely on the sound, as well as listening for longer than routine would require. In addition, having a decent break and then listening to the material with EQs and other effects off (or bypassed in order to return to the original raw material) is often ear-opening. Listening after turning the volume of the control room monitors completely down and then raising the levels just beyond the threshold of hearing is very useful, too. Again—needless to say—this should be done while keeping one's eyes closed, or at least without looking at the computer screen. After a few intensive hours spent in a studio, listening to something other than the actual project at hand helps to refresh and recalibrate one's hearing. Finally, when the ears begin to feel like they are equipped with built-in limiters, switching off the loudspeakers and taking a rest indeed proves that silence is golden. During the pause, it might be a good moment to take a look at the graphs of the audio files in order to look for unwanted peaks or other visible faults in the material, or perhaps study the whole project file in terms of macro-level structure and consider whether it seems to be in balance. Scott's original concept of investigating phonautograms without being able to listen to them is perhaps not such a bad idea after all.

### *16.2.1 Sound in Interior Design*

Questions considering the impact of aural environments and sonic experiences on our daily well-being are probably as old as our civilization. When the first cities with concentrated populations appeared on the planet, multitudinous sounds of human activities created a soundscape that was new and different from the natural one. Since the late nineteenth century, all other manifestations of culture and society, as well as our audible milieu, have gone through radical changes caused mainly by technology in its various forms. In the post-industrial digital age, the global crescendo—the ever-increasing loudness of urban environments and lifestyle—may already have passed its peak in the so-called first world. However, digital technologies have also caused new acoustic phenomena with their swarms of artificial signals that don't have direct connection to concrete physical actions or materials. These sounds generally do not fall under the category of traditional noise problems since they are usually not very loud or long-lasting. Perhaps it is for this reason that not much attention has been paid to this relatively new layer of sounds around us.

Despite being invisible, sound should not be considered an immaterial element or ingredient of our world. Hearing is strongly connected to our perception of space and time, and sounds could and should be taken into account whenever planning, designing, altering or creating environments, whether interior or exterior. It is interesting to note the ways in which sound is special, and to ponder the consequences of this idiosyncrasy when sound is brought into consideration as an analog to building materials, colors and light. Except for acoustics and soundproofing, interior sound design is still an unstable field when compared to the integration of sound in the design processes of many other domains. When discussing and negotiating the visual appearance of a space, the meanings of the words and expressions are rather exact: the color of a certain corridor wall, certain surfaces being glossy or matt, the coating of the floor, and so on. Although the sensing and experience of colors and textures is of course dependent on the space and its other properties, with visual and tangible elements of an interior it is always possible to have at hand at least some sort of samples and examples of suggestions and options: even the lighting of a space can be demonstrated by comparing images created for this purpose. On the other hand, when trying to have similar kind of discussion about sound in terms of spatial design, demonstrating the ways in which space and time (particularly the latter) affect experience is very difficult, if not impossible. Because of this, only a well experienced person can imagine how certain sounds would possibly feel in certain conditions beyond the imminent. Nevertheless, the challenge of communication remains the same despite the expertise and professionalism of a project's sound designer.

I will now introduce an actual large-scale interior sound design project in order to delineate the characteristics of sound as an element, material and subject of discussion in a bigger planning context. A new hospital for children (from newborns to teenagers) is being designed and built in Helsinki, Finland, and aside from patient rooms and facilities for medical operations, large areas of this hospital are being equipped with loudspeakers. The loudspeakers are used to play an ambient soundscape specially designed and tailored for the hospital. Needless to say, this kind of undertaking is quite multifaceted, containing aspects from audio technology to customer relationship management and from aural aesthetics to labor code. Under my guidance, the project was realized by a group of Sound in New Media major and minor subject students in Aalto University. In the context of this chapter, I will concentrate on the importance of listening within the design process and evaluation of the outcomes rather than going deeply into technical details or administrative issues.

## 16.3 Listening to the Client

Despite the somewhat esoteric nature of sounds compared to other elements of a built environment, the need of a mutually approved commission between the designer and the client is self-evident. Getting acquainted with research regarding auditory sense is highly recommended when designing functional sounds, and more or less essential in cases where the people exposed to the sounds will have little or no ability to control their exposure to the sound environment. In formulating the design brief, a shared understanding of the fundamental guidelines that can be driven by research-based knowledge is often crucial. Furthermore, acoustics and psychoacoustics do provide terms and concepts for discussing sounds, and human experiences of physical properties of sound have had corresponding words and expressions in different languages since time immemorial. For example, the experience of faster and slower frequencies of sound—that is, the pitch differences between tones—can be described with adjectives like high versus low, thin versus thick, or bright versus dark; the experience of the amount of air pressure, the loudness of a sound, can be described as either weak or strong, soft or hard. All in all, ordinary and familiar sounds and sonic experiences can be discussed in a meaningful way without much effort.

When discussing and describing sound perception, music is a special case. Different genres and styles of music have their own specific vocabularies. Music can be described in terms of its internal mathematical relations (melody, harmony and rhythm) or by referring to established compositional forms or categories (e.g. symphony, hymn or jingle). To a large extent, verbalizing music is based on comparing and juxtaposing compositions, recordings and concerts in a way that requires knowledge and understanding of the genre in question. In addition, music is often described from the place of emotions or associations aroused in the listener, thus extending the scope of discussion to that of cultural and societal phenomena and their corresponding connotations projected onto music. Last but not least, in addition to personal, informal and journalistic verbalizations, an immense amount of academic research on music has been written and published over the centuries.

In the case of designing sounds for the New Children's Hospital, the design brief was created over a relatively long period of time through several discussions with the representatives of the hospital, doctors, architects, electricians and several subcontractors in charge of different parts of the infrastructure of the building. Designing the concept and defining the technical specifications were happening simultaneously. In terms of the audio content to be created for this context, there were no existing cases to refer to as an example. An additional challenge was in balancing the desired design affordances of the soundscape by taking the different users of the building into account. Finally, the visual directions for the interior of the hospital, which were already approved by the hospital board, were to be used as the audio guidelines as well. These visual instructions consist of references to natural environments with hints of fantasy elements, with each floor having a different description mapped to geographical altitude "from beneath the sea to space."

## 16.4 Listening to the Concept

The aim of creating a tailored soundscape for the New Children's Hospital was to provide a positive effect for the visitors of the building, taking into account both the children patients and their parents or caretakers. The desired properties of the soundscape were immersion, transparency, friendliness, sympathy, playfulness, augmentation of actual space and combining the natural with the imaginative. The soundscape needed to make the entire hospital less frightening and alienating to the children, hence encouraging their parents to relax at least a bit. Furthermore, all the people working in the building—nurses, doctors, caretakers and employees of the canteens, to mention a few—needed, of course, also to be considered integral members of the hospital's acoustic community.

The original plan for the technical implementation of the soundscape was to use each loudspeaker as a stand-alone sound source with some sort of simple audio player connected to it. Through discussion, the steering group of the project accepted an alternative solution that was rather ambitious: all the loudspeakers would be connected to one computer containing a generative audio engine that would create ever-changing ambiences from the given audio ingredients. The idea is similar to a kaleidoscope: the colorful pieces of glass inside the tube remain the same, but as the kaleidoscope is being turned the abstract images seen from the peephole will never repeat exactly. In the audible domain, the concept of wind chimes is somewhat similar: even though the amount of the metal tubes (sounding at different pitches) of the chimes is limited, an endless and ever-changing melodic line is created as the pitches are generated in random order from the chimes moving in the wind.

The concept was explained to the hospital representatives using nature ambiences as reference: for example, when listening in a forest, the components of the soundscape are rather easy to list, and although the sounds appear in a rather predictable way (following natural time cycles and rules of probability) the result is still almost totally random, never repeating itself.

The design concept and the plan of implementing it were not difficult to explain to the team of students, who were already competently trained in sound design. However, the audio guidelines did not contain straightforward descriptions of the desired aural aspects within the various spaces and areas of the building. Without this guidance, there was no kind of list available indicating which sounds the design team should collect or create. Nevertheless, when the time came to start the actual production phase, the members of the team began to bring all kinds of sounds to the classroom, which were listened to, discussed, evaluated together and combined and tested spontaneously. This kind of "sound tasting" was useful and actually indispensable in achieving shared insight into what should be pursued. Some musical elements such as light and transparent pads, little motifs played with acoustic instruments, and the discreet clinks of bowls and bells with selected pitches were tested and blended with the sounds of water, wind, birds and animals.

## 16.5 Listening to the Work in Progress

Despite the long history and tradition of verbalizing music, it is often difficult to agree on commissioned music solely through writing or discussion. Often the brief from the client includes examples of music to support the verbal communication, containing something "concrete" to listen and refer to. In most cases however, the piece of music is supposed to have a beginning and an end, and the end-user platform or playback device can be deduced from the intended use of the piece. Evaluating a commissioned piece of music—or any linear audio—is not a difficult task, since it can be played from the beginning to the end as many times as necessary, and then discussed. There can also be one or more alternate versions of the piece prepared in advance, and these versions can then be listened to, compared and discussed. Even game music and other game audio, which are mostly nonlinear and adaptive, follow an inherent logic connected to the gameplay and the elements of the game, making it possible to discuss the styles and moods of the music contained therein even if the duration and order of the individual pieces are not predetermined.

In the case of entirely generative audio and music, the situation is quite different. If a generative soundscape is designed to remain forever changing, it is obvious that there is no possibility to "listen through" the entire work. When evaluating generative systems, the focus is on their behavior. In the case of the soundscape of the New Children's Hospital, the idea was to mimic the behavior of natural aural environments, each of them containing a limited amount of distinguishing sounds and following the cycles of both the nychthemeron and year. These features can be programmed to an audio engine rather easily, but they still do not directly define how different locations in the hospital building should actually sound. The general objectives of the sound design can be verbalized rather easily, and with words such as those listed earlier in this chapter, the congenial impressions and atmospheres can be described at least indirectly.

In the working stages of production, when soundscapes are still under construction and being tested outside the actual end-user location, time is the most important factor in listening. Earlier in this chapter, I suggested that the timespan of contemplative listening has shortened, due to both fundamental changes in our acoustic environment as well as our generally visually driven workflow in contemporary recording and postproduction studios. Our team was unanimous about the need to listen to the audio material in an exceptional way: in order to be able to evaluate the behavior of the demo versions of the soundscapes and observe their effects on the listener, people took the project files with them to various places, playing and listening to them from their portable devices or domestic audio appliances as long as possible. The experiences from these listening situations were once again shared within the project group, and many decisions regarding subsequent steps were made on the basis of our findings.

When testing generative audio (music, other sounds or a combination of both), the playback in the testing situation should be as generative as in the final setup—watching the colorful images created on the fly with a kaleidoscope is different from watching video footage recorded from a kaleidoscope. The human ear is quite sensitive to looping: if identical sequences of sound start to appear, the illusion of a natural or "living" ambience disappears easily, but in the case of generativity, unwanted repetition is easy to avoid. However, since the generative audio engine used in these kinds of projects is usually created digitally using object-oriented programming environments like MAX, PureData or SuperCollider, one must once again remember the importance of listening carefully to what is being generated. Although the paradigm of generative audio is fundamentally different from linear DAWs or sequencers, the visual interface of sound might again draw the author's attention away from the audible aspects of the system.

## 16.6 Listening to Place and Space

After the period of trying out audio material in various unconventional listening situations, a six-speaker demo setup was built in the intersection of two corridors in one of the less crowded lobbies on the Aalto University campus. During the course of one week, the members of the team took turns sitting in the lobby and reading, writing or just listening while observing occasional passersby and people who happened to take a seat at one of the available tables. The audio material beautifully accommodated the somewhat secluded corner of a large campus building, becoming almost inaudible when a bigger group of people filled the space with the sound of their chatting and footsteps, and gradually rising again above the threshold of hearing when only the low hum of ventilation was accompanying it. The initial acoustics of the demo environment were quite dry, and the team discussed the potential use of reverb in the final soundscape versions, especially the jungle- and mountain-themed floors.

Another equally important purpose of the demo setup was to play the current versions of the different soundscapes to the representatives of the hospital. An additional listening demo was arranged elsewhere for those who could not attend the first one. The test listeners were remarkably happy with what they heard, saying how they understood that the material would sound different in the actual listening environment (i.e. in the hospital), where the actual fine-tuning and embedding of the audio material would happen. In the end, the deficiency of the two aforementioned test listening situations was the amount of time: it was way too short to fully imagine the reality of actually dwelling in the soundscape. Nonetheless, our team received official permission to proceed with the sound design according to the initial plans.

In the case of site-specific sound design (and sound art), the listening conditions are unique by definition. The size, shape and surface materials of the final listening space, as well as the placement of the loudspeakers, are usually not worth trying to simulate elsewhere unless it is especially within the focus of the project. The listening situation is also affected by all other activity in the space and other sounds being produced there, including the behavior of the listener: do they gravitate towards the loudspeakers or travel farther away, does the soundscape make them relax and concentrate on listening or does it activate them in some unexpected way? The listeners of a soundscape are also a part of it, and the way they act and interact with and within the soundscape can never be precisely predicted. I am especially interested in the reactions of the children.

## 16.7 Listening to the Users

As mentioned earlier, the needs and preferences of the people inhabiting the hospital building are neither necessarily identical nor even similar. This is, of course, a typical situation in almost any acoustic community with a "normal" amount of diversity: in cities, different kinds of people—young and old, pedestrians and motorists, secular and religious—experience the soundscape and express themselves with sounds in myriad ways. Conflicting interests cannot be avoided completely, but an ideal situation within an acoustic community is the ability for it to balance and repair itself.

When designing sounds for a children's hospital, the children should undoubtedly be the first priority when deciding on the design affordances of the aural environment. In terms of amplitude and frequency, children's hearing is more sensitive: they can hear higher pitches and softer sounds. We could assume that in the acoustic realm there would be an additional zone or sphere inaudible to adults, consisting of sounds that only the children can perceive. This could probably be substantiated in laboratory conditions, but the actual listening situations in the hospital have numerous other factors affecting the perception of sound. However, it could have been possible to give the "inaudible dimension" of the soundscape a bigger role in the design after more material had been collected concerning children's experience of the entire body of sounds. All in all, from the earliest demos and sketches of the sounds it felt obvious that low frequencies in general would not be desirable since they tend to create an eerie and unnatural atmosphere when introduced artificially into spaces. The comments collected from the listening tests only strengthened this intuition.

The idea and purpose of the aural augmentation of the hospital interior was to bring joy, playfulness, friendliness and a bit of fantasy or adventure to the child patients. However, it is essential to keep in mind that people working in the building are being exposed to the sounds several hours per day (or night) and tens of hours per week. According to surveys, nurses often experience the sonic environment of their workplaces as tiring, disruptive and even awkward. It is likely that almost anyone who has spent time in a hospital recognizes the sounds of medical equipment, the signals of which punctuate the acoustic environment. Although the noise level of beeping machines is not harmful as such, it is quite obvious that the unavoidable audio prompts—from single strenuous blips to a cacophony of auditory displays—are stressful to listen to without being able to alter or escape them. In the amoeba-like process of creating the design brief and trying to involve all possible stakeholders, the nurses were left with second- or third-hand information about the sound design project. The representatives of the nurses seemed to have the impression that some kind of PA loudspeaker unceasingly playing Muzak would be installed in the hospital premises, which—understandably—sounds more or less like a horror scenario. The representatives of the nurses insisted that each floor must be equipped with a dial or knob that could be used for turning the sounds off. In order to appease this vitally important group within the hospital's the acoustic community, the signal chain was built accordingly, even though it created extra work.

From the designer's perspective the aim was—and will continue to be as the soundscape develops further—to adjust each section of the acoustic environment without compromising the actual goals of the soundscape in a way that no one will desire to turn the sounds off. The irony in the "mute button" is that in each of the floors, this control is located in the nurses' office or similar space where the soundscape is not audible. Due to this, the person who would be adjusting the volume or turning it completely off will not hear what adjustment they will be making. To make a comparison, it would be similar to a dimmer controlling the lights of a space located outside the line of sight of the lights being adjusted. Nevertheless, there is no question of the importance of employees having ownership and influence on their working environment and conditions.

## 16.8 Listening to the Building and the Acoustic Community

The sound design project for the New Children's Hospital in Helsinki, Finland, is planned to have a long lifespan. First of all, the soundscape—or at least the option to have a designed sound environment in every common space within the hospital—is a permanent, built-in feature of the building. The first version of the soundscape is a result of the process described in this chapter, albeit while keeping listening as the focus and leaving technology to its appropriate place as an invisible servant for the goal of the design.

The success of the design and its implementation will be evaluated by carefully listening to the entire building, and more importantly, giving an inclusive voice to the whole acoustic community of the building, using interviews and ethnographic methods in order to understand how the soundscape is experienced and creating strategies to further improve it. Once again, using metaphors from the material world, the computer-based generative system allows its creators to alter both the contents of the kaleidoscope—on the level of individual grain—in addition to the rules determining the probabilities of different shapes that appear. In the audible domain, one could imagine wind chimes that could be altered in terms of the amount of chimes, the positioning and length of each tube. These changes could be made without limitation, and the impact of wind speed and direction that set the tubes in motion could be mapped in any possible way. Having all these options at hand, listening to the imaginary wind chime is crucial before deciding what to do next or diving into trial-and-error improvisation. If you are the only person listening to the wind chimes, it is enough to listen long enough to be able to make the adjustments that satisfy your own ear. If you are designing the wind chimes for someone else, your understanding needs to be adjusted to incorporate others.

## Recommended Reading

Krause, B., 2012. *The Great Animal Orchestra*. London: Profile Books Ltd.

Pallasmaa, J., 2012. *The Eye of the Skin: Architecture and the Senses*. 3rd ed. Hoboken, NJ: John Wiley & Sons Ltd.

Russolo, L., 1987. *The Art of Noises*. Hillsdale & New York: Pendragon Press.

Truax, B., 2001. *Acoustic Communication*. 2nd ed. Westport, CT: Ablex Publishing.

# Index

Note: page numbers in *italics* indicate figures, and page numbers in **bold** indicate tables on the corresponding pages.